Instrumenta

Third Edition

Franklyn W. Kirk

Nicholas R. Rimboi

AMERICAN TECHNICAL PUBLISHERS, INC.
HOMEWOOD, ILLINOIS 60430

Library of Congress Catalog Card Number: 74-75992

ISBN 0-8269-3422-6

3 4 5 6 7 8 9 - 75 - 19 18

Printed in the United States of America

Preface to the Third Edition

Instrumentation has served as a practical introductory text for over ten years. Its three-part program of *introduction, comprehension,* and practical *application* has proven itself as a valuable learning tool.

Part One introduces the student to the basic principles of instrumentation. It includes a discussion of the various instruments employed in industrial applications. *Part Two* deals with operating principles. Instruments are discussed in greater detail to provide fuller understanding. *Part Three* gives the student examples of the actual application of instruments used in process control.

This enlarged 3rd Edition of *Instrumentation* has been recast in a new, easy-to-read format. New illustrations have been generously included to familiarize the student with current equipment and developments. All of the chapters have been expanded to present the advances in instrumentation. *Review Questions* and a *Words to Know* section appear at the end of each chapter to aid the student.

Appendices have been completely revised. *Appendix A,* entitled *Instrumentation Symbols,* provides the latest symbols available from the Instrument Society of America. *Appendix B* shows the most commonly used electrical symbols. *Appendix C* consists of tables which give useful information to the inquiring student. The *Glossary* has also been expanded to include new and changed terminology.

FRANKLYN W. KIRK
NICHOLAS R. RIMBOI

Contents

Page

Part Two

Page

Page

Part Three

Page

part one

PART ONE deals with instruments for measurement, transmission and control. They are described with emphasis on their external appearance, and their operating principles are for the most part only briefly touched upon. The purpose of this section is to indicate the wide variety of instruments that are used in modern industrial control processes, and to introduce the readers to the variables which are measured and controlled by these instruments.

Lift-off of the Saturn IB rocket which carried three astronauts (Charles Conrad, Jr., Dr. Joseph P. Kerwin, and Paul J. Weitz) to the skylab. Skylab is the orbital workshop which was designed to enable scientists to gain new knowledge for improving the quality of life on the earth. Complex and advanced instruments played a key role in the success of this and several succeeding skylab missions. (NASA)

Introduction

Chapter

1

Measuring and process control instruments are essential parts of any industry, from the manufacture of breakfast cereal to the building of huge jet airliners. Improvement in measuring and process control devices has provided essential improvements in the quality and quantity of services and goods necessary in society today.

Instrumentation also provides a means for environmental control, disposing of wastes and by-products of industry to avoid pollution of our cities and to protect precious natural resources. It has also been used to help correct pollution problems stemming from earlier abuses. Thousands of intricate computers and controlling and measuring devices were prime reasons men were able to travel to the moon.

Because of the great strides in science and technology in the modern world, we frequently overlook the instruments which play important roles in our daily lives. The use of instruments in satellites and various installations provides us with weather and other important information. Most homes and apartments are equipped with thermostats which control the temperature and maintain comfort, regardless of the season. The thermostat shown in Fig. 1-1 provides a fuel saving device. This device enables a person to preset the thermostat, which automatically lowers the temperature during sleeping hours and raises the temperature to the original setting at a fixed time. Automotive instrument panels can be equipped to provide a wide range of important information to the motorist. The panel, shown in Fig. 1-2, can measure vehicle speed, motor speed, and provide information on the charging rate of the battery, the level of fuel, and the temperature of the engine. Instruments control the temperature of food. As a safety measure, instruments can be used to control lighting in various parts of a house or other building. Hand calculators are now common in homes and offices. This list, of course, is by no means complete.

This text concentrates on the instruments used in modern industry and process control. These precise, effec-

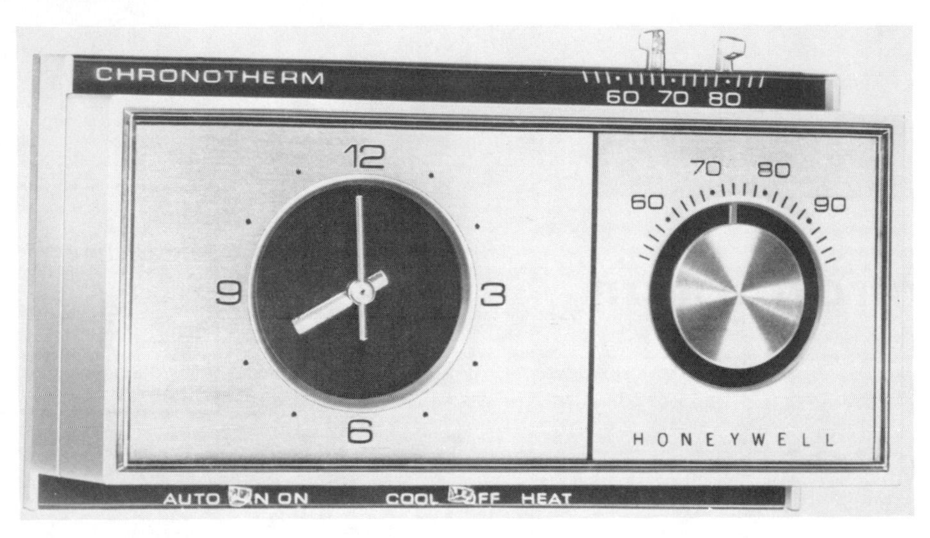

Fig. 1-1. This thermostat has a built-in timer which allows for automatic temperature adjustments at predetermined intervals. (Honeywell Process Control Div./Fort Washington, Pa.)

Fig. 1-2. This automotive instrument panel provides the motorist with vital information. (Ford Motor Co.)

tive, and diversified instruments make possible the massive production feats essential to today's society. It should be noted immediately that the principles of measurement and control do not vary, whether the particular application is in modern industry or a part of a monumental undertaking, such as a moon landing. Improvements in measuring and control aid progress in both industry and science.

Industrial Instrumentation

Instrumentation is the technology of using instruments to measure and control the physical and chemical properties of materials. The term *process instrumentation* refers to instruments used to measure and control manufacturing, conversion, or treating processes. Fig. 1-3 shows an operator monitoring the processing of oily waste water. A *control system* combines measuring and control instruments to provide automatic remote action. The resulting process is termed a *controlled process.*

Measured and Manipulated Variables. Instruments can not always directly measure and control the properties of a process material. Variables, such as temperature, pressure, flow, level, humidity, density, viscosity, etc. frequently affect the process. A variable that must be measured or controlled is called the *measured* or *controlled* variable. A variable which is used to affect the value of the mea-

Fig. 1-3. The control center enables a single operator to monitor the entire process system. (Honeywell Process Control Div./Fort Washington, Pa.)

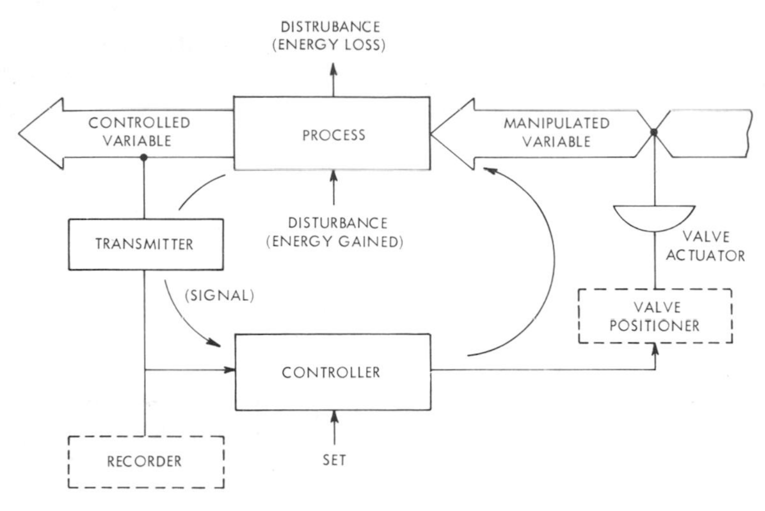

Fig. 1-4. A closed-loop control system. (The Foxboro Co.)

sured variable is called the *manipulated variable*. If water is used to control the temperature of the process material, for example, the temperature of the process material is the controlled variable, and the temperature and quantity of the water are the manipulated variables.

Fig. 1-4 is a diagram of a closed-loop control system. A closed-loop control system provides feedback to the controller about the value of the controlled (measured) variable. An open-loop system, such as an automatic washing machine, does not provide feedback. The transmitter in the closed-loop system provides a signal (feedback) to the controller which takes an appropriate action to return the value of the controlled variable to set point.

A control system can have more than one measured variable. It can be important, for example, to maintain temperature, flow, and level in a given process. Some of the measured variables, however, are not measurements of physical or chemical properties of the process material. Flow measurement can be taken from the example just used.

Often the flow rate of a substance depends on the temperature, density, or viscosity of the substance. Oil, for example, has a higher flow rate at higher temperatures, hence it flows faster in an operating engine than when poured out of a can. Frequently in industrial processes, however, the measurement and control of flow rate establishes the rate at which a substance is subjected to various process changes. Often it is not the flow rate of the process material itself that is of prime importance, but the flow rate of a second substance which is supplying or removing energy. If water is used as the cooling or heating medium in a process, it is the flow rate of the water and not the process material that is of prime importance.

The measurement of level in process control is another example. Level can indicate the size of granular material in a container. But the measure and control of level are most often required to maintain the continuation of the process. Likewise, it can be important to measure and control the level of a second substance rather than the process material itself, like available fuel oil or cooling water.

It can be seen that a control system requires more than measuring and controlling the physical and chemical characteristics of the process material. The various conditions involved in the process itself must be measured and controlled. This information must be continuously available, and the performance of the control units continuously monitored. Measurements can be observed directly on indicating instruments, or recorded continuously on charts, depending on the nature of the process application. Control devices are monitored through the use of visual or audible alarm signals.

Indicating Instruments. Indicating instruments are generally equipped with scales graduated in units of the measured variable. An instrument indicating level would be graduated in feet or inches, or in metric units. The shape and size of the scale vary, depending on the needs of the specific application. In some applications, the scale might be linear, in others, circular. Regardless of these physical characteristics, however, the scale must be capable of correctly indicating the value of the measured variable. The indicating instrument shown in Fig. 1-5 is a typical circular indicator and is graduated in inches of water for the measurement of pressure. Usually, the scale is made so that the value of the measured variable is read by estimating the position of the indicator between two successive divisions.

Fig. 1-5. A typical indicating instrument. (Wallace & Tiernan Div., Pennwalt Corp.)

Digital Meters. Digital panel meters are being used with great frequency in industrial processes because of their accuracy, response, speed, and readability. They can be used to measure or control any of the measured variables. They generally feature a liquid crystal display of the type shown in Fig. 1-6. A chip contains the entire digital circuitry which is hermetically sealed to prevent burnout and hence an inaccurate reading. Other types of displays are available for process environments in which liquid crystal displays might be unsuitable. Digital meters are available in various sizes, ranging from the meter shown in Fig. 1-7A to the large multirange precision indicator shown in Fig. 1-7B. Fig. 1-7A shows a digital thermocouple indicator. The indicator in Fig. 1-7B, designed to readout temperature, is available in dual or triple ranges. Fig. 1-8

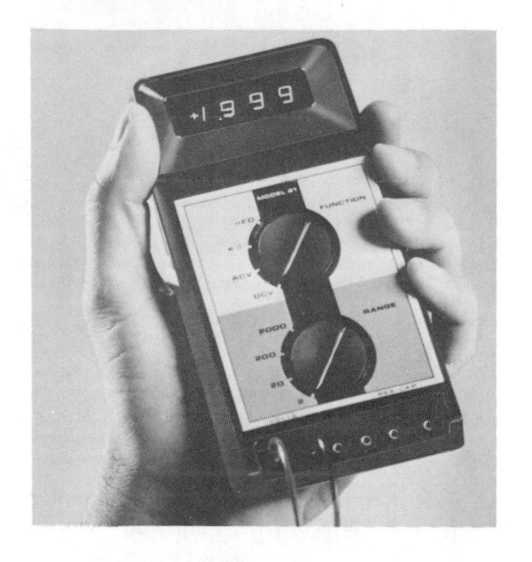

Fig. 1-6. A portable digital multimeter with LED (light emitting diode) readouts. It is used to troubleshoot electronic instruments. (Data Technology Corp. / Santa Ana, Calif.)

Fig. 1-7A. A digital thermocouple indicator for indicating temperature. (Ircon Inc.)

Fig. 1-7B. A digital multirange indicator. (Honeywell Process Control Div./Fort Washington, Pa.)

Fig. 1-8. A portable digital temperature indicator. This meter includes a carrying case. (Ircon Inc.)

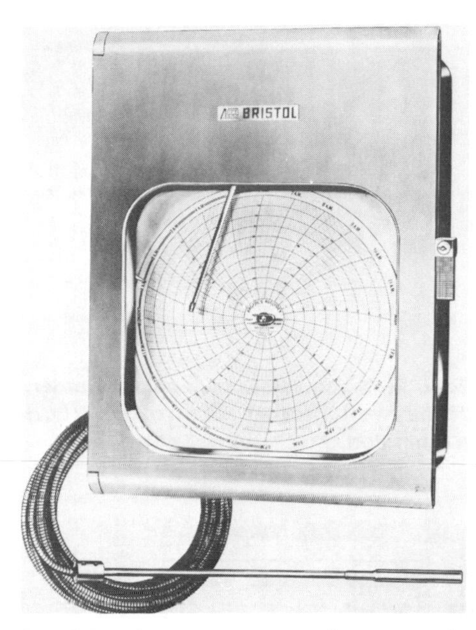

Fig. 1-10. A continuous, single-point strip chart recorder. (Bristol Division of Acco)

shows a portable temperature indicator.

Recorders and Monitors. Continuous recorders are available in several types. Fig. 1-9 shows a continuous strip chart recorder. Fig. 1-10 shows a single point circular recording chart used for remote readings. These recording instruments can be entirely mechanical, entirely electrical, or a combination of both.

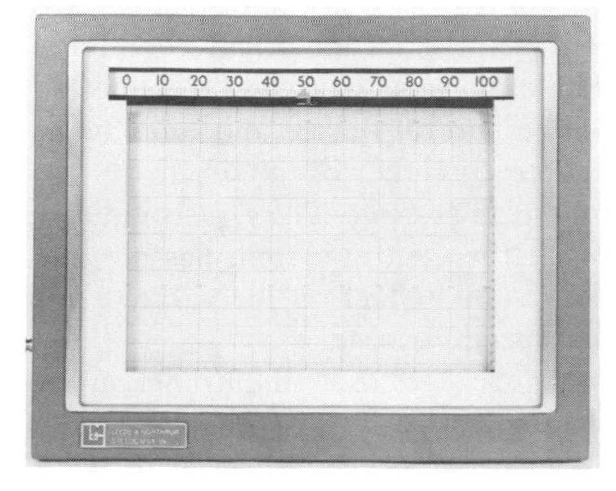

Fig. 1-9. A continuous strip chart recorder. (Leeds & Northrup)

There are two sets of units shown on most recording charts. One set indicates the value of the measured variable, for example temperature. The second set indicates the time. This allows for a close study of changes in the measured variable over any given period of time.

Some recording instruments only record the value of one measured variable. Other instruments can record many variables. The instrument shown in Fig. 1-11 is a multipoint recorder which has the capacity to record up to twelve measured variables connected to a maximum of three different, independent ranges.

Instruments that record the changes of only one measured variable are

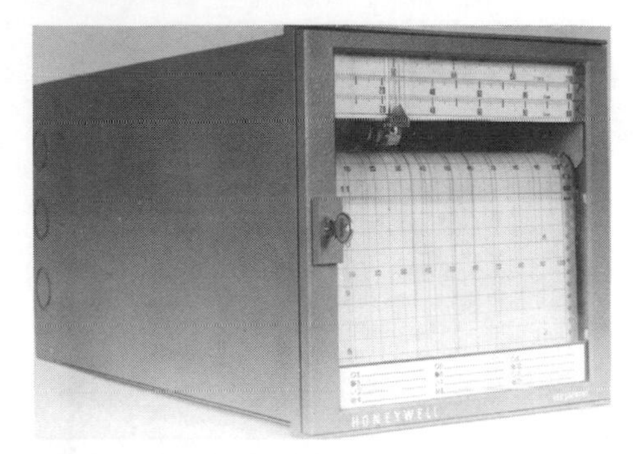

Fig. 1-11. A continuous multipoint recorder. (Honeywell Process Control Div./Fort Washington, Pa.)

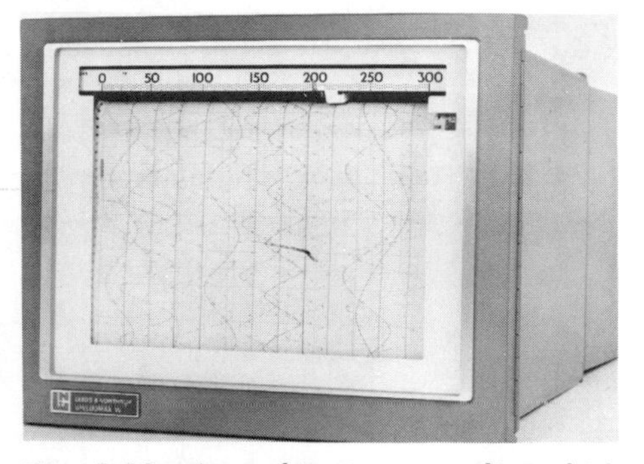

Fig. 1-12. A multipoint recorder which uses color to differentiate among the measured variables. (Leeds & Northrup)

called *single point recorders*. The trace on charts of such instruments is usually in the form of a single continuous line. Instruments which record more than one measured variable are called *multipoint recorders*. Multipoint recorders can have several continuous lines when only a few measured variables are recorded. A code, such as the use of numbers or color, is required when extensive variables are used. The instrument in Fig. 1-12 uses color and a code. The value of each variable is recorded at intervals. The exact interval depends on the operating speed of the instrument. Any one variable can be traced by checking the coded marks of all the variables recorded.

Mechanical recorders often draw continuous lines. The pen is positioned by a mechanical linkage, and the chart is driven by a spring-wound clock. In a combination mechanical-electric recorder, the chart can be driven by an electric clock.

Electric recorders draw continuous lines or print coded marks. The continuous line is drawn by an inking pen on the paper, or by a pressure stylus on carbon-coated film, or by a heated stylus on heat-sensitive paper. The instrument which uses a printed code ordinarily has an inked print wheel, which makes a mark on the paper chart. The print wheel mechanism is positioned on the chart by an electric motor. The motor receives its impulse for the measuring circuit. The chart is driven by a constant speed motor. Chart drives which are spring wound are likewise available for this type of recorder.

Status signals are used for monitoring control devices. (*Status* here means representing an abnormal condition.) These alarm signals can be connected to the control device and monitor its action directly, or attached to the measuring device. The signals indicate when the control system has not had the desired effect on the value of the measured variable.

Figure 1-13 is a multifunction instrument used for programmable data acquisition systems. It can be used for industrial, pilot-plant, and field use. The systems are able to be pro-

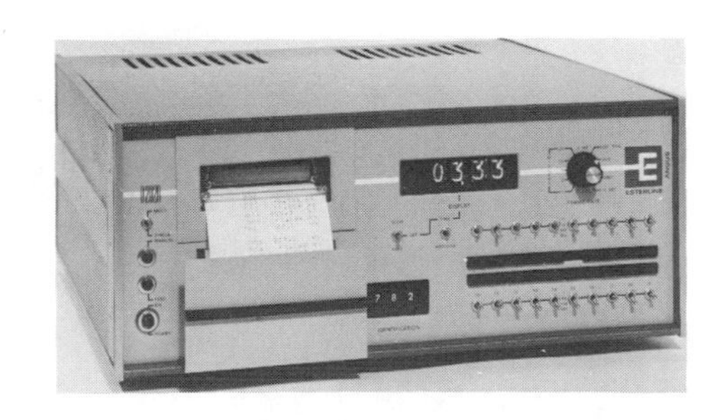

Fig. 1-13. A multifunction instrument capable of various programmable applications. (Esterline Angus)

grammed for each application. They rapidly provide digital printouts and/or tape information for various incoming analog signals. The self-contained printer prints channel number, data, time and data-frame identification. Programming is set by the switches shown. A solid state digital clock provides both visual and printed recording of hour and minutes. The systems can not only provide recordings but can also be programmed to serve as a monitoring system, which provides alarm settings to protect processes or instruments. These systems offer accuracy and versatility.

Manufacturers provide instruments which range from the simple, mechanical type to those involving the most sophisticated technology.

Panel Boards. Many indicating and and recording instruments can be mounted directly on plant equipment at the point of measurement and control. This type of mounting is referred to as *local mounting*. These instruments can also be centralized on panel boards located in control rooms or somewhat removed from the actual process. The three types of panel boards available are *graphic, semi-graphic,* and *non-graphic*. The graphic system includes a mimic display of the entire process system. The indicating or recording instruments are mounted on the panel in relation to the part of the process they measure or control. See Fig. 1-14. Special small size indicators and recorders are available for graphic panels. Fig. 1-15 shows a systems operations center. The panel is a dynamic schematic of bulk power loads, generation stations, and substations.

Semi-graphic panels include a miniature mimic display on the panels. The instruments can be identified on the graphic section by code numbers. Fig. 1-16 shows a semi-graphic master control panel for a turbogenerator. The top of the central section includes the mimic display which indicates the position of valves, breakers, etc. A semi-graphic panel can also be of the type shown in Fig. 1-17. All production processes are controlled from this instrument panel.

On non-graphic panels no attempt is made to provide a mimic display of the process layout. Fig. 1-18 shows an auxiliary control panel for a turbogenerator. The panel monitors water purity, soot cleaning systems inside the boilers, etc. A computer (not shown in the photo) provides details of panel information. Printers at right printout the

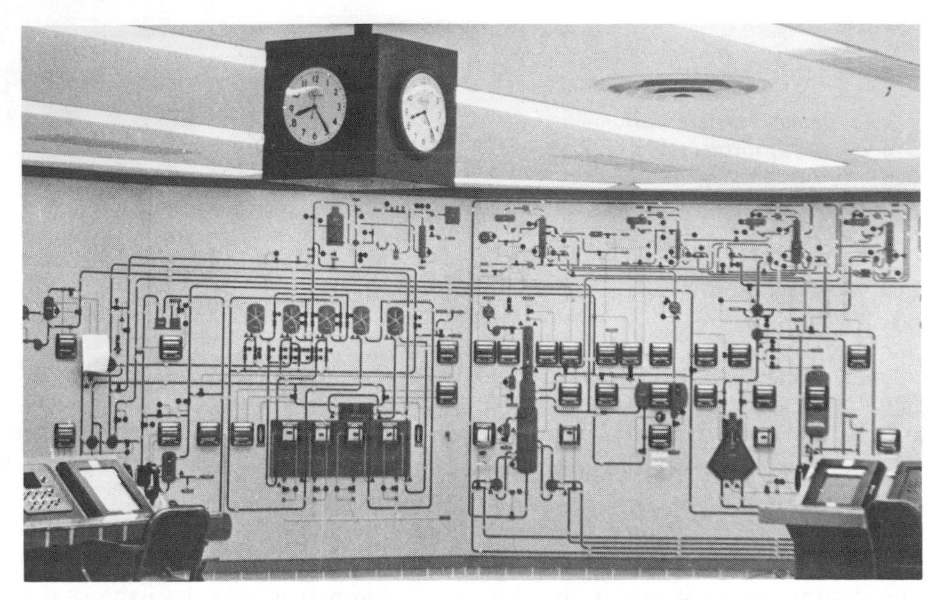

Fig. 1-14. A control panel for a refinery process. (Standard Oil of Ohio)

Fig. 1-15. This panel shows the network of main lines for a utility company. The computer monitors the board activity. (Cleveland Electric Illuminating Co.)

Fig. 1-16. A semi-graphic master control panel. The activity mimic display is shown at the top of the center section. (Cleveland Electric Illuminating Co.)

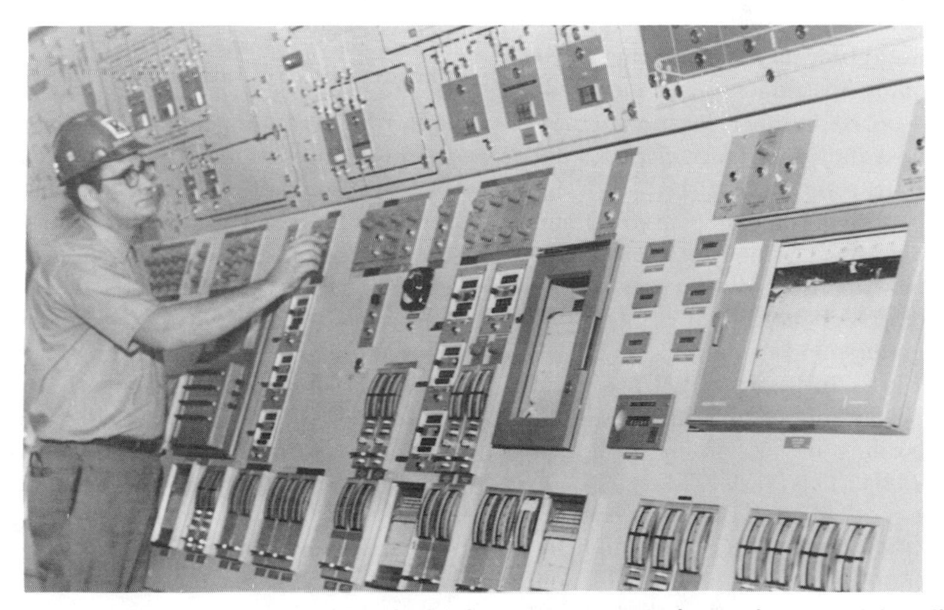

Fig. 1-17. This instrument panel overlooks the process area and provides control for all phases of the process. (Honeywell Process Control Div./Fort Washington, Pa.)

Fig. 1-18. Non-graphic control panel. (Cleveland Electric Illuminating Co.)

sequence of events should the monitoring system actuate the alarm. With non-graphic panels measuring and controlling instruments are mounted on the board and their location in the system is indicated on nameplates under the instruments. Fig. 1-18 is an example of recording all variables simultaneously at a single location. Such systems can use a display device or devices, such as digital printers, video monitors, and computers.

Many industries still use various types of indicating, monitoring, and recording devices singly or in combinations. This variety does not affect the fundamentals of measurement and control. Regardless of the simplicity or complexity of the industrial application, the primary task of the instrumentation technician is to understand the *fundamentals* of the devices used in process control.

Characteristics of Instruments

Before examining some of the common measured variables, such as temperature, pressure, flow, etc., encountered in process control, and some of the instruments used to measure and control these variables, it is well to become acquainted with the general characteristics of instruments. These characteristics are divided into two categories, *static characteristics* and *dynamic characteristics*. The static characteristics are those which refer to variables when they are not changing. The dynamic characteristics are those which apply when the variables are changing.

Static Characteristics. The static

characteristics of instruments are *accuracy, reproducibility,* and *sensitivity.*

Accuracy in a measuring instrument is the ability of the instrument to indicate or record the actual value of the measured variable. An accurate temperature indicator indicates the actual temperature of the process material, or other materials as required. The difference between the actual value of the measured variable and the reading on the instrument is called *static error.* Manufacturers specify the accuracy of their instruments to a certain percentage, plus or minus, of the total range or scale of the instrument or percentage of reading.

Accuracy in a control device is the ability of the device to achieve and maintain the true value of the measured variable required. An accurate temperature controller is able to achieve and maintain the substance at the temperature required by the process. Static error is the amount the controller deviates from the desired value.

Reproducibility in a measuring instrument is the ability of the instrument to indicate or record identical values of the measured variable each time the conditions are the same. If the desired temperature of the process substance is 210°F, an instrument with good reproducibility will indicate or record that value each time that temperature occurs. A gradual shift in the indicating or recording device, even though the true value of the measured variable has not changed, is called *drift.*

In control devices, reproducibility means the device is able to repeat an output signal each time the same input signal occurs. Drift is the gradual shift in the output signal, even though the input signal remains unchanged. If temperature of a substance is to be maintained at a certain value, the controller may indicate a change and take a control action, even though the temperature of the substance has not varied.

Sensitivity in a measuring instrument is the smallest change in the value of the measured variable to which the instrument responds. In a control device, sensitivity is the smallest change in input that causes a change in output. The *dead zone* is the range in which the measuring instrument or the controller does not respond. If the temperature of a substance varies a minute amount, the device can be incapable of response.

Dynamic Characteristics. The dynamic characteristics of instruments are *responsiveness* and *fidelity.*

Responsiveness in a measuring device is the ability of an instrument to follow changes in the value of the measured variable. The inability of the instrument to follow changes in the measured variable is called *measuring lag.* The period during which the instrument does not respond to a change in the value is called *dead time.* If the temperature of a substance varies, responsiveness defines how quickly the instrument indicates or records such a change.

Responsiveness in a controller is the ability of the controller to produce a changing output signal as the input signal changes. The inability of a controller to follow an input change is called *controller lag.* Dead time is the period during which the controller does not respond to a change in the input signal.

Fidelity in a measuring instrument is the ability of the instrument to correctly indicate or record a change in the value of the measured variable. In a controller, fidelity is the extent to which the output signal of the device correctly follows changes in the input signal. The difference between the changing value and the instrument reading or the controller action is called *dynamic error*.

The requirements of specific processes, of course, vary. One process may require close tolerances, another may not be as critical. Instruments vary in the degree they possess these general characteristics, but any useful instrument must have them.

Words to Know

process instrumentation
controlled process
control system
measured variable
manipulated variable
indicator
recorder
digital meter
readout
monitor
programmed system
panel boards
printout
static characteristic
dynamic characteristic
accuracy
static error
reproducibility
drift
sensitivity
dead zone
responsiveness
measuring lag
dead time
controller lag
fidelity
dynamic error

Review Questions

1. What is the difference between a *measured* and a *manipulated* variable?
2. What function does monitoring play in a control system?
3. What are some of the advantages of digital meters?
4. What is the difference between the use of local mounting and panel boards?
5. What is the difference between a graphic and a semi-graphic panel board?
6. What is a *mimic display?*
7. Name the three static characteristics of instruments.
8. What is *static error* as applied to a measuring instrument?
9. What is *drift* as applied to a controller?
10. What is meant by a *dead zone* in a controller?
11. Name the two dynamic characteristics of instruments.
12. What is *measuring lag?*
13. What is *dynamic error?*

Temperature

Chapter

2

There are changes in the physical or chemical state of most substances when they are heated or cooled. It is for this reason that temperature is one of the most important of the measured variables encountered in industrial processes.

Temperature is defined as the degree of hotness or coldness measured on a definite scale. Hotness and coldness are the result of molecular activity. As the molecules of a substance move faster, the temperature of that substance increases. *Heat* is a form of energy and is measured in calories or *BTU's* (British Thermal Units).

When two substances at different temperatures come into contact with each other, there is a flow of heat. The flow is away from the substance at a higher temperature toward the substance at a lower temperature. The flow of heat stops when both substances are at the same temperature.

Heat Transfer

The flow of heat is transferred in three ways: convection, conduction, and radiation.

Convection. Heat transferred by the actual movement of portions of a gas or liquid from one place to another is called *convection.* This movement is caused by changes in density due to rising temperature. For example, in a forced air heating system, the warm air entering the room through the supply duct is less dense, and therefore, lighter than the cooler air already in the room. As the warm air cools, it drops and moves through the cool air return and back through the heating system. See Fig. 2-1. Another example of convection is a water heating system. The heavier cold water moves down, forcing the heated water up through the pipes of the system. Convection takes place only in fluids (either a liquid or a gas).

Conduction. When heat is applied to one part of a substance, it is transferred to all parts of the substance. The movement is from molecule to molecule. Gases and liquids are poor conductors. The flow of heat by conduction takes place most effectively in solids. See Fig. 2-2.

Radiation. Heat energy is trans-

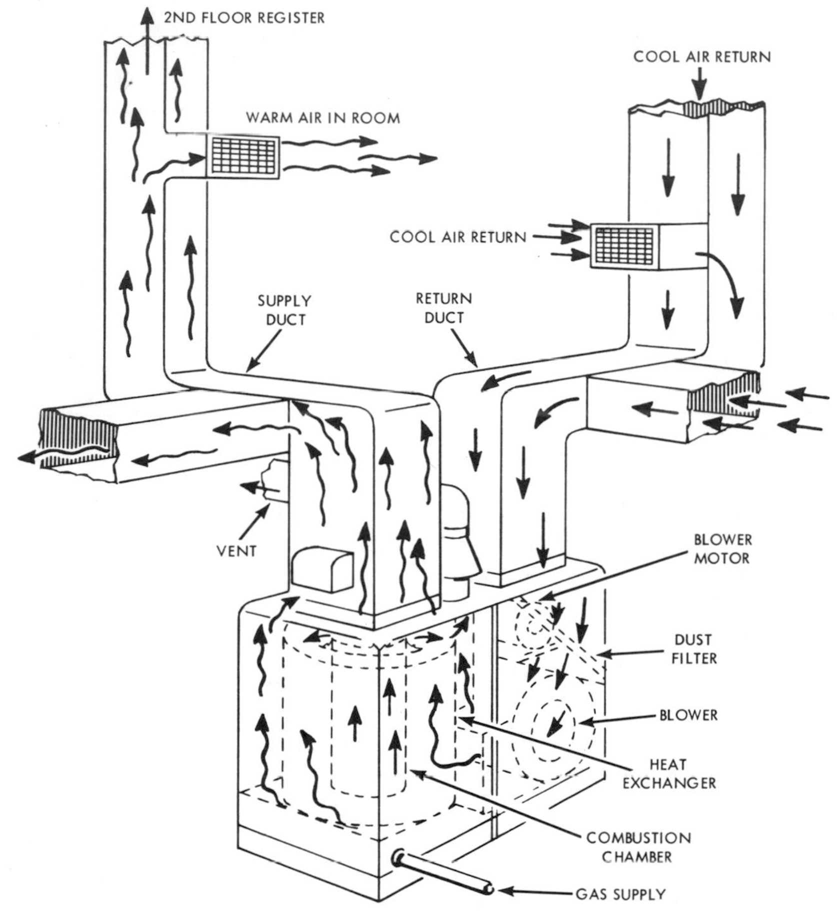

Fig. 2-1. This forced warm air system is an application of heat transferal by convection. The blower assists the circulation of air. (American Gas Assoc. Inc.)

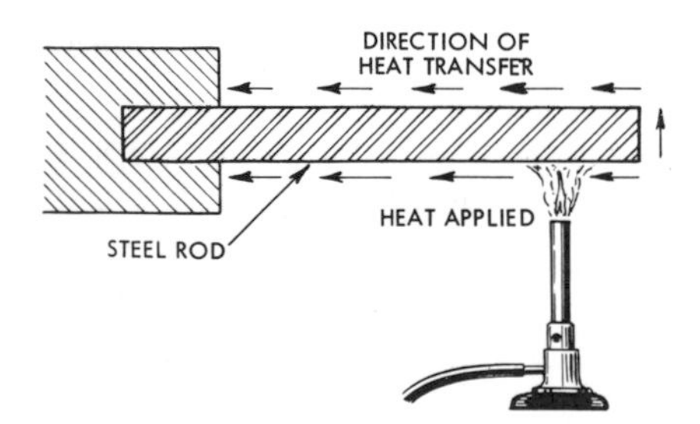

Fig. 2-2. In this simple application of conduction, the heat is transferred along the metal bar to the steel rod.

ferred in the form of rays sent out by the heated substance as its molecules undergo internal change. See Fig. 2-3. Only energy is transferred. The direction of the flow of heat is from the radiating source. The radiant energy is then absorbed by a colder substance or object. Radiation takes place in any *medium* (gas, liquid, or solid), or in a vacuum.

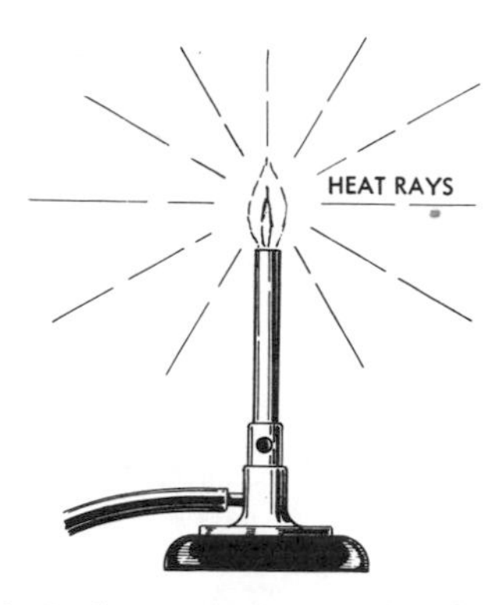

Fig. 2-3. Radiation is the transferal of heat by means of rays.

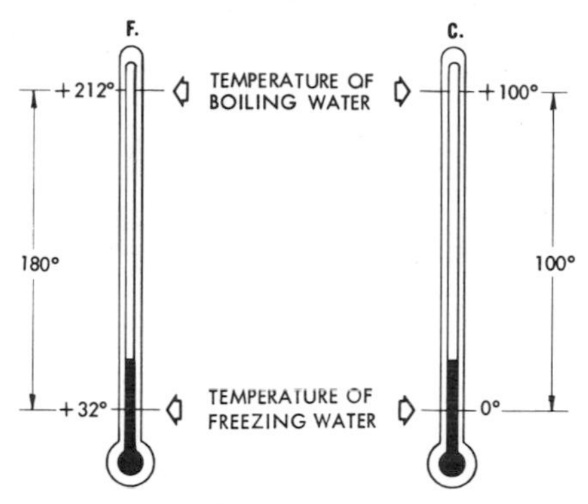

Fig. 2-4. The boiling point and the freezing point of water are used as fixed points on both the Fahrenheit and the Celsius (centigrade) scales.

Temperature Scales

The comparison of the heat content of one substance with another is made by referring to temperature scales. These scales also make it possible to measure and identify the heat level when there is a change in the state or condition of the substance, such as freezing, melting, or vaporizing (boiling).

The instrument used to measure temperature is the *thermometer.* The temperature scales most commonly used in calibrating thermometers are the *Fahrenheit scale* and the *Celsius* (or centigrade) *scale.* Fahrenheit and Celsius are the names of the men who invented the respective scales. The difference between these two scales concerns the temperature values assigned to their fixed points. On both scales the freezing temperature of water and the boiling temperature of water are used as fixed points. On the Fahrenheit scale the freezing temperature is set at 32°F and the boiling temperature at 212°F, with 180 degrees between the fixed points. On the Celsius (or centigrade) scale the freezing temperature is set at 0°C and the boiling temperature at 100°C, with 100 degrees between the fixed points. See Fig. 2-4.

Theoretically, there is a condition of no molecular motion, hence no heat energy. The temperature at this point is *absolute zero temperature,* or the lowest temperature possible. The Kelvin scale and the Rankine scale have their zero points at this absolute zero level. See Fig. 2-5.

On the *Kelvin scale* the freezing point of water is at + 273°K and the boiling point at + 373°K, with 100 degrees between the fixed points, just as on the Celsius (or centigrade) scale. The Kelvin scale is sometimes referred to as the *absolute centigrade scale.*

On the *Rankine scale* the freezing point of water is at + 491.7°R and the boiling point at + 671.7°R, with 180 degrees between the fixed points, just as on the Fahrenheit scale. The Rankine scale is sometimes referred to as

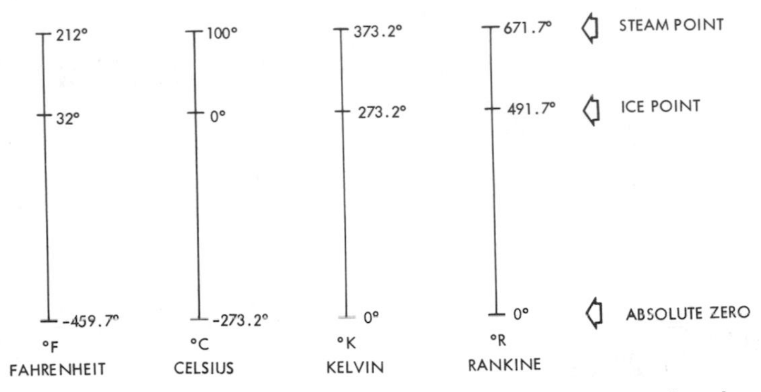

Fig. 2-5. Absolute zero is the temperature at which the movement of molecules (heat) completely stops. Absolute zero on the scale is −273.2°C. On the Celsius/Fahrenheit scale it is −459.7°F.

the *absolute Fahrenheit scale.*

The absolute scales are important because they are so often used in scientific and technical literature. The Kelvin scale is used in the study of physical chemistry. The Rankine scale is used in formulas for flow metering.

It is often necessary to convert temperature readings from one scale to another. Temperatures as read on the Fahrenheit scale are converted to the Celsius (or centigrade) scale, using the following formula:

$$°C = \frac{100}{180} (°F - 32°), \text{ or}$$

$$°C = 5/9\,(°F - 32°).$$

Celsius (or centigrade) readings are converted to Fahrenheit as follows:

$°F = (180/100°C) + 32°$, or

$°F = (9/5°C) + 32°$.

Examples:

Change 77°F to degrees Celsius (or centigrade).

$°C = 5/9\,(77° - 32°)$

$°C = 5/9 \times 45°$

$°C = 25°$.

Change 10°C to degrees Fahrenheit.

$°F = (9/5°C) + 32°$

$°F = (9/5 \times 10°) + 32°$

$°F = 18° + 32°$

$°F = 50°$.

Celsius (or centigrade) readings are converted to the Kelvin scale by this formula:

$°C = °K - 273°$

Example:

Change 100°C to °K.

$100°C = °K - 273°$

$100°C + 273 = °K$

$373 = °K$

Fahrenheit readings are converted to the Rankine scale by the formula:

$°F = °R - 459.7°$

Example:

Change 212°F to °R.

$212°F = °R - 459.7°$

$212°F + 459.7 = °R$

$671.7 = °R$

A variety of thermometers used for industrial and scientific applications are capable of measuring and controlling temperatures beyond the range of the thermometer based on the freezing and boiling points of water. Thermometers, depending on their fixed points, can be designed and calibrated to measure such temperatures as:

Boiling Point of Oxygen —183°C (—297°F)
Melting Point of Mercury —38.87°C (—37.96°F)
Melting Point of Tin +232°C (+449°F)
Melting Point of Cadmium +321°C (+610°F)
Melting Point of Zinc +419.5°C (+787°F)
Boiling Point of Sulfur +444.6°C (+832°F)
Melting Point of Antimony +630.5°C (+1176.9°F)

(**Note:** The melting point of Mercury is approximately the same on both the Fahrenheit and Celsius (or centigrade) scales because of the different amount of graduations or degrees on each scale. −40°F is the exact equivalent of −40°C.)

Temperature Measurement

Thermometers are used to measure temperature. But they vary greatly, depending on the requirements of the job they are intended for. In this section we discuss some of the more common types of thermometers, various principles and materials which go into their make-up, and how they are used.

Mercury-in-glass Thermometers. Mineral substances contract or expand a definite amount for each degree of temperature change. This is the principle of thermal expansion. When heat is applied to a mercury-in-glass thermometer, the mercury expands more than the glass bulb. This difference in expansion causes the mercury to rise in the small bore (capillary) glass tube. Because the mercury rises uniformly with temperature, the tubing can be

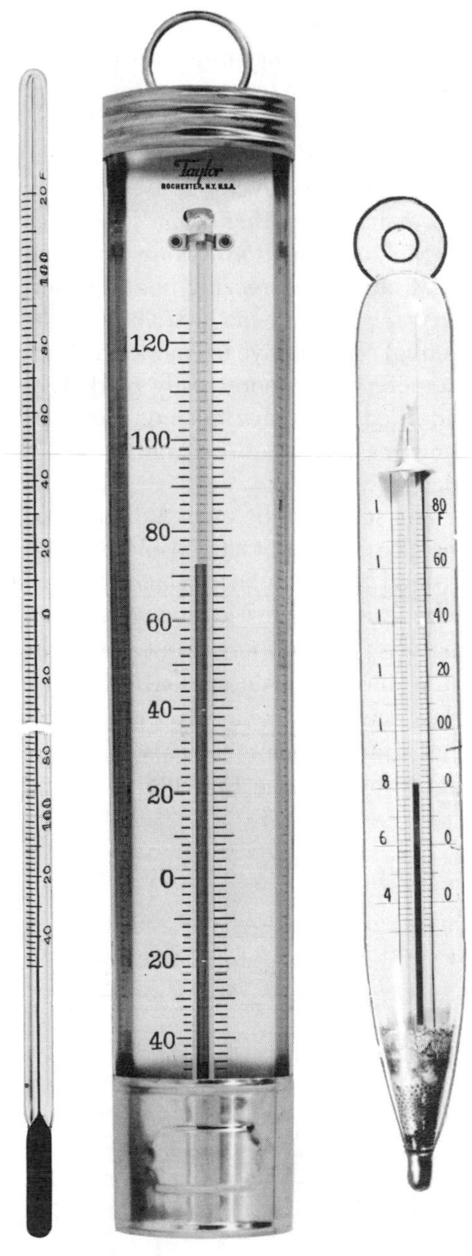

Fig. 2-6. The scale can be etched on the glass (left) or engraved on metal and directly attached to the thermometer (center). The thermometer and scale can both be enclosed in a glass envelope (right), which enables the instrument to float. (Taylor Instrument)

calibrated according to a temperature scale. The mercury-in-glass thermometer can be used for temperatures from −30°F to +800°F.

Liquid-filled Thermometers. Mercury is not the only liquid used in glass thermometers. Other liquids, such as alcohol, are used to measure temperatures below the freezing point of mercury (−38.87°C or −37.96°F). The alcohol contains dye to enable the thermometer to be more easily read. Liquid-filled glass thermometers can be used for temperatures from −300°F to +600°F.

The scale can be etched directly on the glass tube or it can be engraved on a metal plate or tube to which the glass tube is attached. Floating glass thermometers are also available. Both the tube and the scale are enclosed in a glass envelope, which is weighted so that the thermometer floats in an upright position. See Fig. 2-6.

Bimetallic Thermometers

The operation of bimetallic thermometers is based on the principle that different metals have different coefficients of thermal expansion. See Fig. 2-7. Two metallic alloys with different physical characteristics are fused together and formed into a spiral or *helix* (coil). This is the *bimetallic element.* When the bimetallic element is heated, it unwinds because there is a different thermal expansion for each alloy. A pointer attached to the helix by a shaft moves as the helix unwinds and indicates the temperature on a calibrated circular scale. See Fig. 2-8.

Bimetallic thermometers are available for industrial and laboratory use. The industrial type has a heavier construction which causes a slight loss of accuracy and speed of response. Processes, such as the refining of oil, use bimetallic thermometers.

Bimetallic thermometers are used to

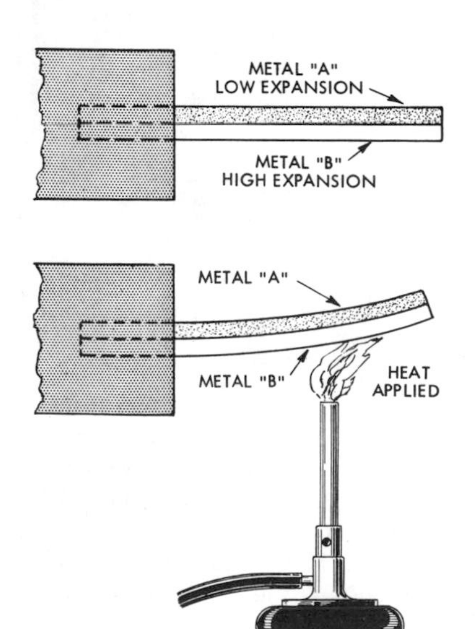

Fig. 2-7. The different rate at which metals expand when heated is the principle of the bimetallic strip.

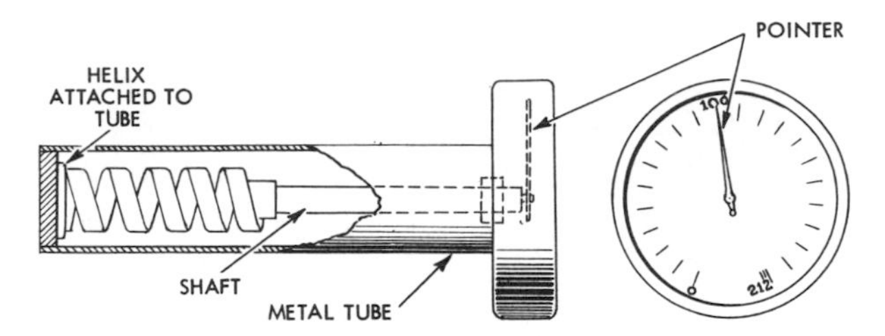

Fig. 2-8. This bimetallic thermometer has a helical element. When the thermometer is immersed in a hot substance, the helix unwinds which causes the pointer to move.

measure temperatures from −300°F to +800°F. They can indicate temperatures as high as +1000°F, but not on a continuous basis because the bimetallic element (helix) tends to overstretch at this temperature, causing permanent inaccuracy.

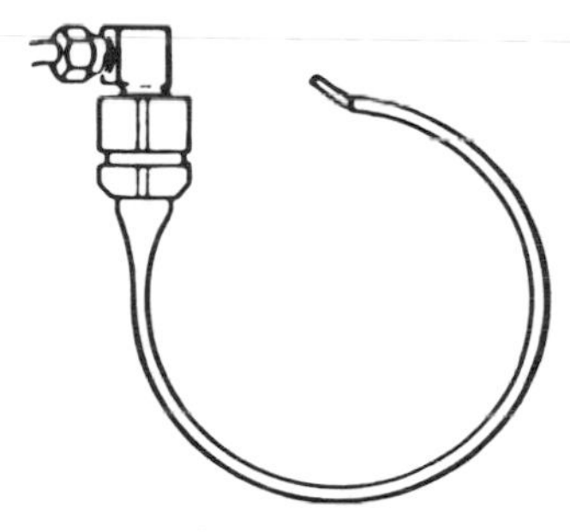

Fig. 2-9. A simple C-shaped Bourdon tube. (Bailey Meter Co.)

Pressure – Spring Thermometers

Both glass and bimetallic thermometers are made for *local readings* (the thermometer is placed in the substance at the point of measurement). But in industry it is often necessary to measure temperature at one location and record it at another. For this reason the pressure-spring thermometer was developed. This kind of thermometer can be used for continuous recording of temperature and also as the measuring element in control instruments.

Fig. 2-10. Spiral Bourdon tube. (Bailey Meter Co.)

The principal component of pressure spring thermometers is, of course, the pressure spring itself. This is a hollow spring that can be made in the C-shape of the original *Bourdon tube* (Fig. 2-9), but is more often used in the form of a spiral (Fig. 2-10), or helix (Fig. 2-11).

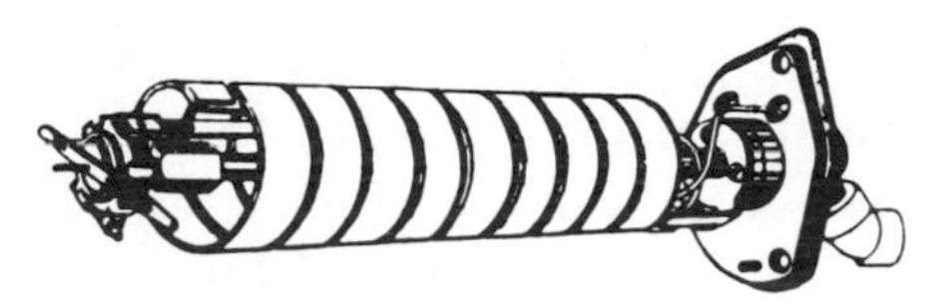

Fig. 2-11. Helical Bourdon tube. (Bailey Meter Co.)

One end of the capillary (small bore)

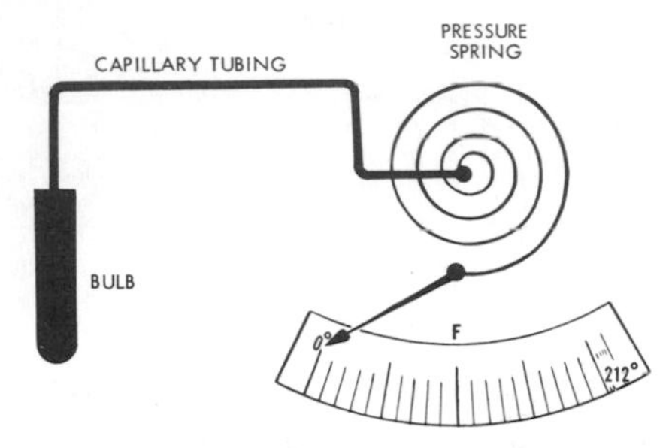

Fig. 2-12. When the temperature rises, the liquid in the bulb of the pressure-spring thermometer expands, increasing the pressure. This causes the spring to partially straighten and move the pointer on the scale.

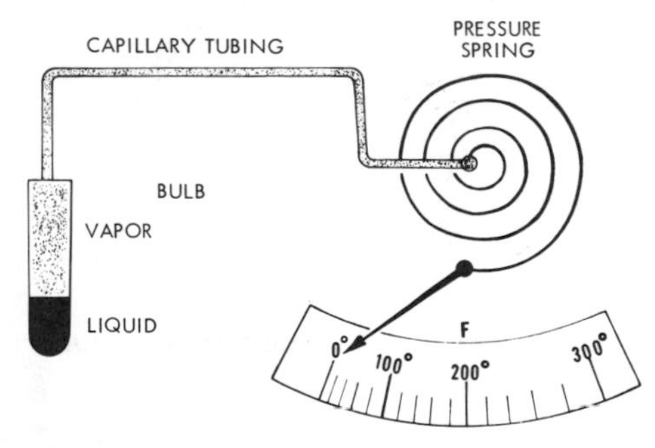

Fig. 2-13. A pressure-spring thermometer using a combination of liquid and vapor.

tubing is attached to the pressure spring. Joined to the other end of the tubing is a sensing bulb, which is the part of the thermometer that makes contact with the substance to be measured. When the spring, tubing, and bulb are filled with a suitable fluid, they form a measuring unit just as a glass tube, mercury, and the bulb of the glass thermometer do.

Pressure spring thermometers are classified by the type of fluid used. Mercury is used most often. Mercury is suitable for temperatures up to 1000°F. *Organic* (carbon based) *liquids,* such as xylene, are used for temperatures up to 600°F. An example of the liquid-filled type is shown in Fig. 2-12. Nitrogen is used in gas-filled systems for temperatures up to 1000°F. Some systems are partially filled with liquids which cause vapor pressure when heated. This type of pressure spring thermometer differs from the liquid-filled and gas-filled types. Vapor does not expand uniformly as liquids and gases do. See Fig. 2-13. The result is that vapor-actuated thermometers do not use a uniform scale. The graduations on the scale are more widely spaced apart at the higher readings than at the lower readings. This increases readability between the lines. Vapor-filled thermometers can measure temperatures from −300°F to +600°F.

Selecting the type of pressure spring thermometer for a particular applica-

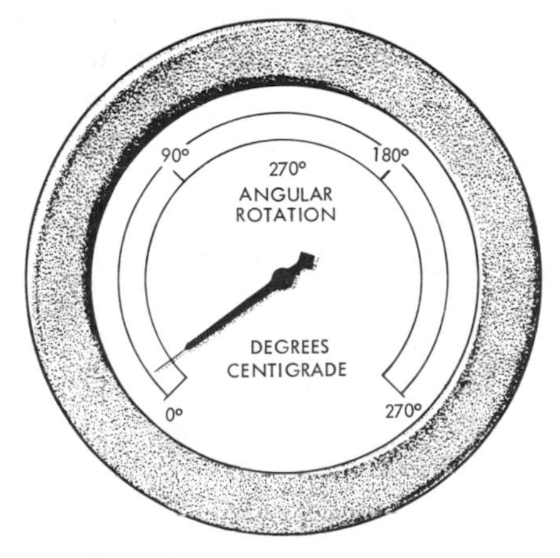

Fig. 2-14. The dial of a pressure-spring thermometer can be graduated to meet the needs of the application. This dial has a range of 270°.

tion depends on such factors as the useful temperature range and the length of the tubing required.

If the pressure spring thermometer is only used as an indicator, the simple dial type with a 270° scale (Fig. 2-14), or the rectangular case type with a 90° scale, can be used. If it is also used for recording, the thermometer can have a very short system, with the bulb attached to the recorder case, as shown in Fig. 2-15, or it can have a long tube system for more distant readings. See Fig. 2-16.

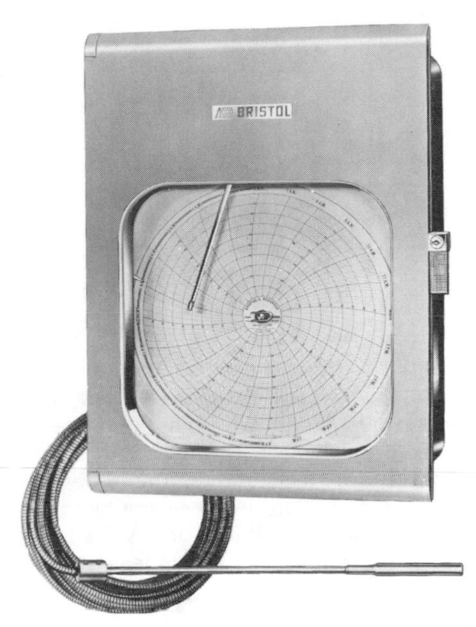

Fig. 2-16. A pressure-spring thermometer/recorder used for long distance (remote) readings. (Bristol Division of Acco)

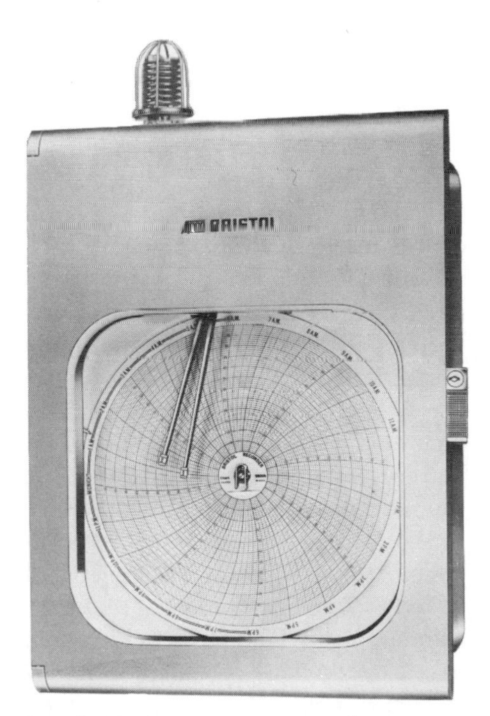

Fig. 2-15. A pressure-spring thermometer equipped with recorder. The sensing bulb is attached to the case. This thermometer is used for short range readings. (Bristol Division of Acco)

Thermocouples

The simplest electrical temperature-sensitive device is the thermocouple. Basically it consists of a pair of wires made from dissimilar metals. These wires are joined at one end. At the other end the wires are connected to a meter or a circuit. When the *hot junction* (the joined end of the wires) is heated, a measurable voltage is generated across the *cold junction* (the ends of the wires connected to a meter or circuit). The cold junction is also known as the *reference junction*. When the reference junction is kept at a constant temperature, the voltage does not vary. The voltage increases as the temperature being measured by the hot junction increases. See Fig. 2-17.

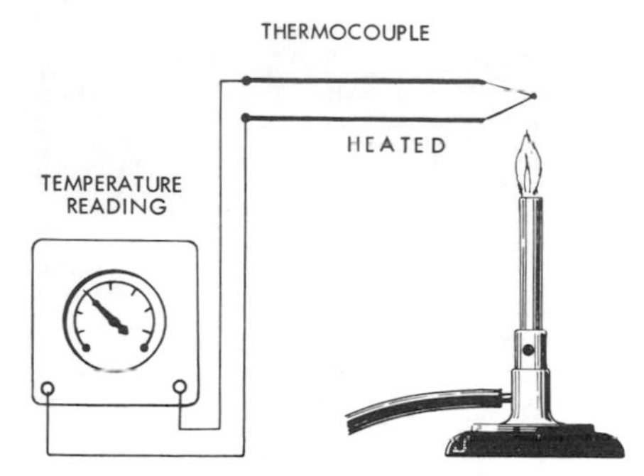

Fig. 2-17. Heat is applied to the joined ends of the thermocouple (the hot junction), while the opposite end (the reference or cold junction) is maintained at a constant known temperature. The difference in the temperature between the two junctions produces a voltage across the meter which is graduated in units of temperature.

TABLE 1 THERMOCOUPLE RANGES

TYPE OF THERMOCOUPLE	USEFUL TEMPERATURE RANGES (°F)
IRON - CONSTANTAN	0° to 1400°F
CHROMEL - ALUMEL	500° to 2300°F
PLATINUM / RHODIUM - PLATINUM	1000° to 2700°F
COPPER - CONSTANTAN	-300° to +700°F

The most common thermocouple wires are combinations of Iron-Constantan, Copper-Constantan, Chromel-Alumel, and Platinum/Rhodium-Platinum. One wire is positive and one wire is negative. The name of the metal appearing before the hyphen indicates the positive wire; the name of the metal after the hyphen indicates the negative wire. For example, in an Iron-Constantan thermocouple, Iron is used for the positive wire and Constantan for the negative wire. The characteristics and ranges of these wire combinations are shown in Table 1.

The meter used with a thermocouple is called a *millivoltmeter.* A millivoltmeter is an instrument made with a permanent magnet and a moving coil. It is extremely sensitive to changes in electrical voltage. (1 millivolt equals 1/1000th of a volt.) The coil moves with each minute change in voltage. A pointer attached to the coil registers these movements on a scale marked in voltage units. So when the millivoltmeter is connected to the thermocouple it doesn't actually measure temperature—it measures voltage. But since there is a definite relationship between the voltage generated by the thermocouple and the amount of heat detected by it, the scale of the meter can be graduated in units of temperature. See Fig. 2-18.

The circuit used with the thermocouple for more accurate voltage measurements is called a *potentiometer* circuit. At its simplest, it consists of a battery, a slidewire, the thermocouple, a standard cell, and a galvanometer. See Fig. 2-19. Some circuits can also include a recording meter. Part of the battery voltage is compared with the thermocouple voltage. Any difference in these voltages is sensed by the galvanometer. (The battery voltage is adjusted to 1 volt by comparing it with a standard cell which has a fixed voltage at all times. This is called *standardization.*) A slider moving along the slidewire alters the battery voltage to counteract the voltage produced by the thermocouple. This returns the galvanometer to its balance position,

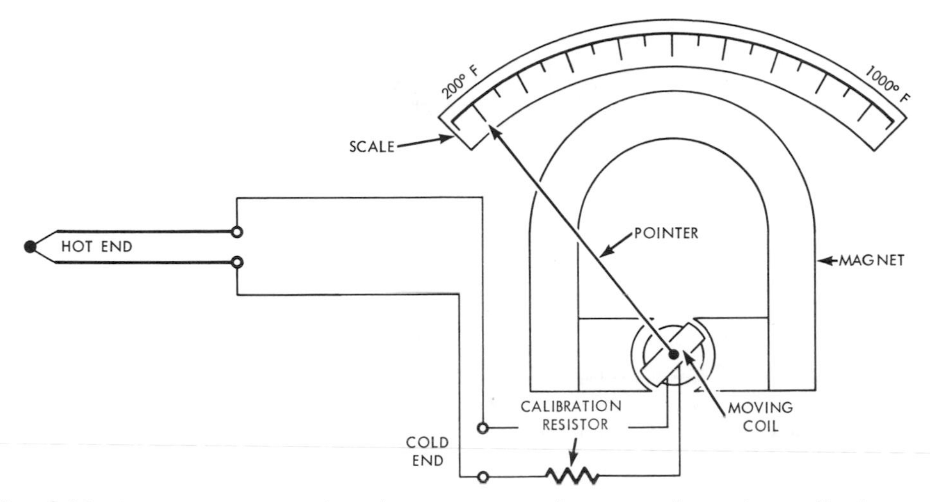

Fig. 2-18. A temperature-sensing device using a thermocouple and a millivoltmeter.

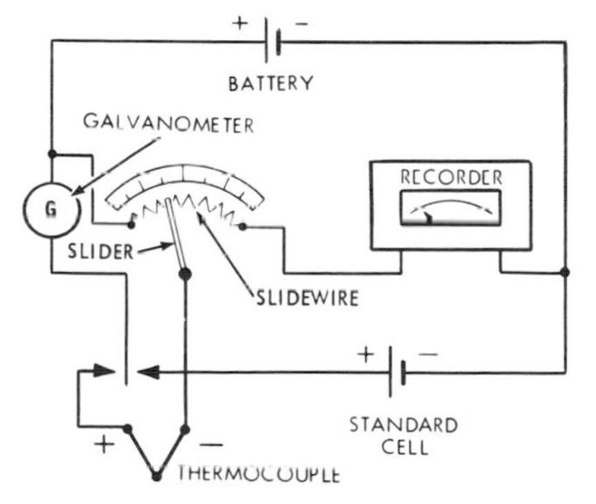

Fig. 2-19. A potentiometer circuit with a recording device used to measure temperature.

so that no current flows through it. This condition is called *null balance.* (A balance scale uses the same principle. When the weight placed on each arm of the scale is the same, the dial will indicate zero, or null balance.) A pointer attached to the slider indicates the degrees, Celsius (or centigrade) or Fahrenheit, on a scale graduated in units of temperature.

Resistance Thermometers

The operation of the resistance thermometer depends on an electrical circuit. In this instrument, the heat-sensitive element consists of a carefully made electrical resistor. Platinum, nickel, or copper wire, wrapped around an insulator, is most often used for the resistance wire of the element. See Fig. 2-20. A properly protected resistance thermometer increases its resistance when it is exposed to heat. A resistance thermometer is used as part of a circuit in which its electrical resistance is compared with a known resistance. This type of circuit is known as a *resistance bridge.* Unlike a thermocouple, the resistance thermometer does not generate its own voltage. An external source of voltage, like a battery, must be incorporated into the circuit.

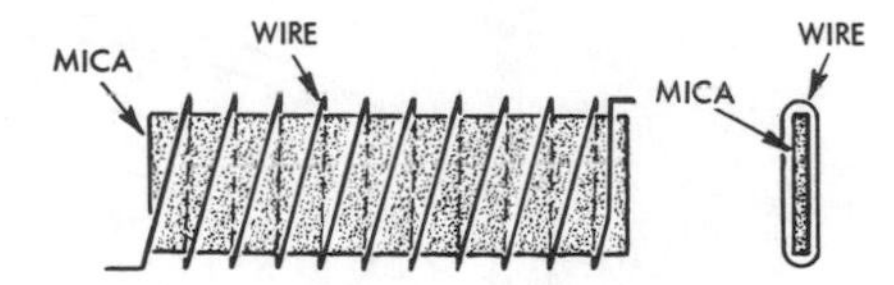

Fig. 2-20. Wire (platinum, nickel, or copper) wrapped around an insulator (mica) forms the resistance wire for the element of a resistance thermometer.

A millivoltmeter can be used to measure change of value in one of the resistances in a bridge circuit. The bridge is often shown schematically, as in Fig. 2-21. The battery is connected to two opposite points of the diamond shaped circuit, the meter to the other opposite points. The rheostat regulates the overall bridge current.

The regulated current is divided between the branch containing the fixed resistor and range resistor No. 1, and the branch containing the resistance themometer and range resistor No. 2. As the electrical resistance of the thermometer changes, the voltage at *X* and *Y* change. The millivoltmeter senses the difference in voltage caused by unequal division of current in the two branches. The range resistors establish the sensitivity of the bridge. If, for

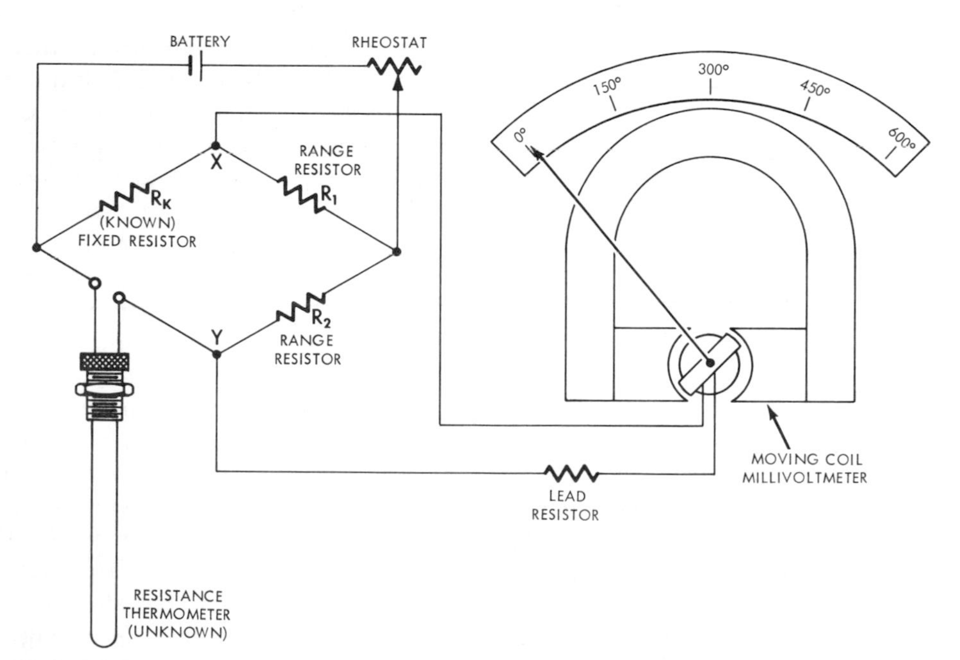

Fig. 2-21. A resistance thermometer using a bridge circuit. This particular bridge circuit is known as a Wheatstone bridge.

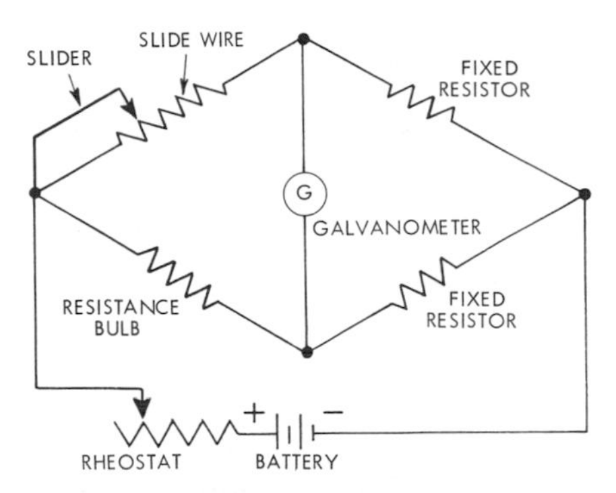

Fig. 2-22. Another arrangement for a resistance thermometer. The slidewire in this bridge circuit is the variable resistor.

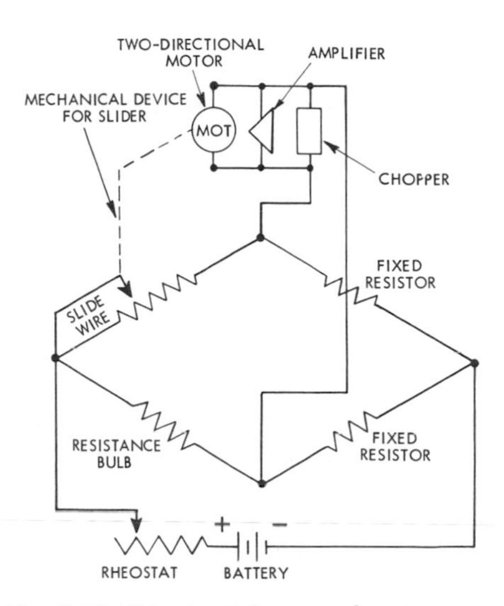

Fig. 2-23. This circuit has an electric, automatic self-balancing bridge which eliminates the need for a galvanometer.

example, range resistor No. 1 has an electrical resistance 100 times greater than range resistor No. 2, a fixed resistor with a resistance 100 times that of the thermometer is used. The sensitivity of the bridge is 1/100 or 1%. The meter can be calibrated in temperature units because the only changing resistance value is that of the resistance thermometer due to temperature change.

Another arrangement, as shown in Fig. 2-22, has the resistance temperature detector in a bridge circuit and uses a galvanometer to compare the resistance of the detector with that of a fixed resistor. A slider-slidewire combination is used to balance the arms of the bridge. The circuit is in balance whenever the value of the slidewire resistance is such that no current flows through the galvanometer. For each temperature change detected by the resistance thermometer there is a new value—and so the slider must take a new position to balance the circuit.

The null-balance type of instrument has been modified to eliminate the galvanometer, which is somewhat slow and delicate. An electronic instrument has been developed in which the dc voltage of the potentiometer or the bridge is converted to an ac voltage. The ac voltage is amplified in an electronic unit, and the stronger signal is then used to drive a two-directional motor that positions the slider on the slidewire to balance the circuit. See Fig. 2-23.

Pyrometers

Many special thermometers are required for industrial processes. The temperature of molten steel, for example, cannot be measured by a mercury thermometer. The glass would melt, and the mercury would, of course, boil away. The *pyrometer* is an instrument which measures temperatures, especially beyond the range of a mercury-in-glass thermometer, directly. For ex-

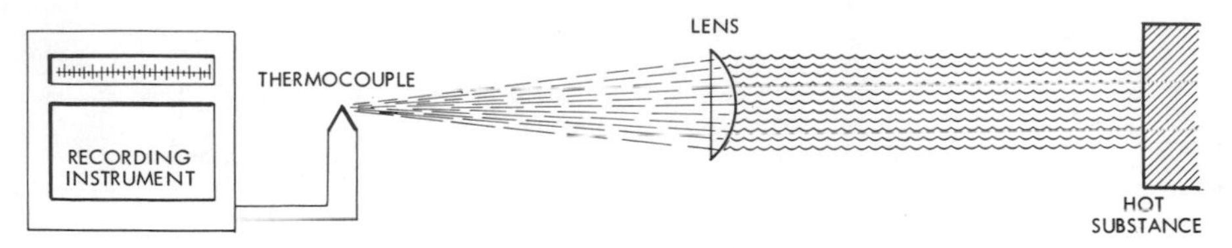

Fig. 2-24. A typical arrangement for a radiation thermometer.

ample, pyrometers can be made which use a comparison of electric currents produced by heating dissimilar metals (thermocouples). Since there is a direct relationship between voltage and temperature, the pyrometer is scaled in degrees Celsius (or centigrade) or Fahrenheit. Radiation thermometers can be used for measuring temperatures in the pyrometric range.

Radiation Thermometers. The radiation thermometer picks up the radiant heat by means of a lens and focuses the heat waves on a thermocouple. See Fig. 2-24. The potentiometer is most often used with the radiation thermometer, and it operates just as it does when connected to a thermocouple that makes direct contact with a hot substance.

Words to Know

temperature
British thermal unit
convection
conduction
radiation
fluids
medium
thermometer
Fahrenheit
Celsius
centigrade
absolute zero temperature
Kelvin scale
Rankine scale
bimetallic element
helix
Bourdon tube
organic liquid
pyrometer
thermocouple
hot junction
cold junction
reference junction
millivoltmeter
potentiometer
null balance
resistance bridge
range resistor

Review Questions

1. What is the difference between heat and temperature?
2. The flow of heat is transferred in what three ways?
3. What are the four principal temperature scales?
4. Convert the following temperatures from Fahrenheit to Celsius

(or centigrade): −40°, +41°, +176°, +255°.

5. Give the zero temperature values for the four temperature scales.
6. What principle causes the mercury in a glass thermometer to rise up the tube when the bulb is exposed to heat?
7. What method is used for checking the accuracy of a bimetallic thermometer after it has been exposed to a temperature beyond its range?
8. What are the three types of pressure spring thermometers?
9. What two types of devices are used to measure temperature electrically? Which of the two does not generate its own voltage?
10. A process requires the temperature to be measured at one point and indicated at a location 500 feet away. What thermometer would be suitable for this application?

Chapter

3

Pressure

Pressure is defined as force divided by area. The force of gravity on an object resting on the earth's surface is its weight. If an object weighs 100 pounds and rests on one square inch of the earth's surface, it is exerting a pressure of 100 pounds per square inch. See Fig. 3-1. All objects have weight and therefore exert a pressure on the earth that can be expressed in weight per unit of area, pounds per square inch or *psi*.

A column of liquid exerts a pressure on the earth just as a solid object does. This is known as *hydrostatic pressure. For example,* a column of water 34½ feet high exerts a pressure of 15 pounds per square inch. See Fig. 3-2. A column of mercury of the same height exerts a pressure of 204 pounds per square inch. This is because mercury is 13.6 times heavier than water. Therefore, it is possible to express pressure in inches or feet of a particular liquid. Inches (or millimeters) of mercury and inches of water are the common units of pressure measurement. One pound per square inch of pressure equals 2.04 inches of mercury or 27.7 inches of water. See Fig. 3-3.

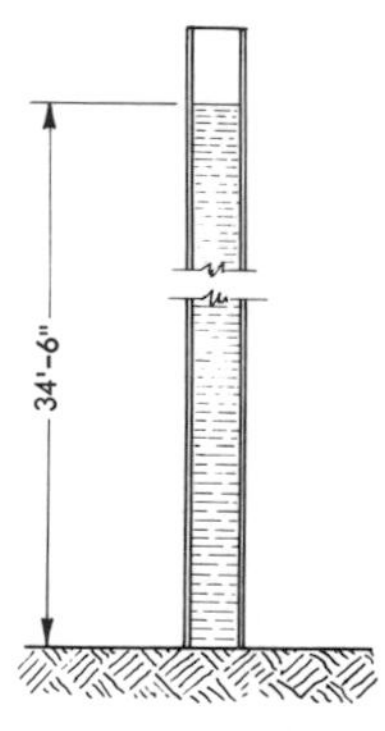

Fig. 3-2. A common example of hydrostatic pressure.

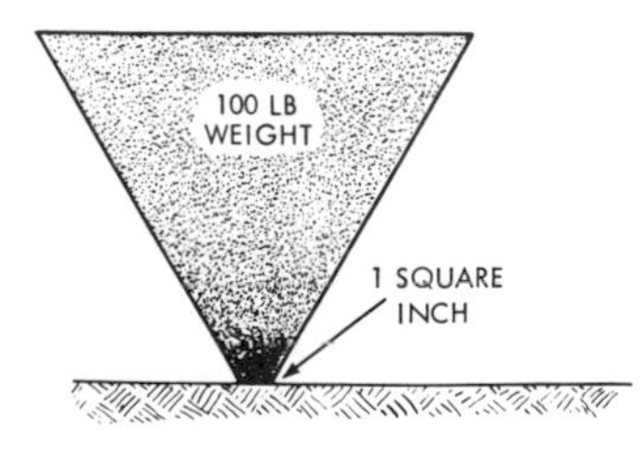

Fig. 3-1. A simple illustration of an object exerting a pressure of 100 pounds per square inch (psi).

To calculate psi, simply divide the

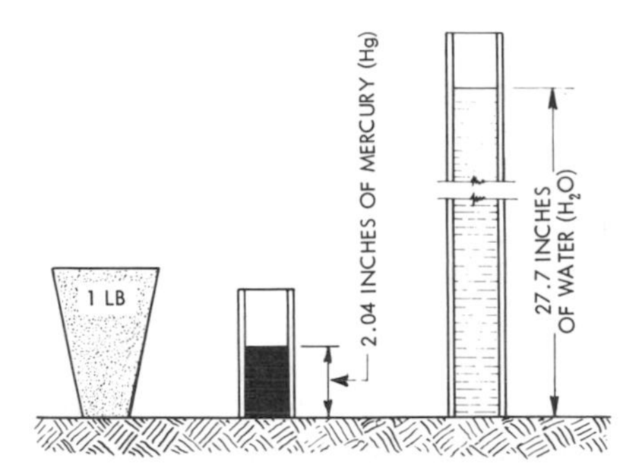

Fig. 3-3. A comparison of the inches of mercury and the inches of water needed to exert a pressure of 1 psi.

total number of inches by 27.7, if water is used, or 2.04, if mercury is used. *For example,* what is the psi, if the height of the column of fluid is 55.4 inches (approximately 4½ ft.)? If the fluid is water, the psi would be 2 (55.4 divided by 27.7). If the fluid is mercury, the psi would be 27.015 (55.4 divided by 2.04).

To express the psi in inches of the measuring fluid, simply multiply the psi by 27.7, if the fluid is water, or 2.04, if the fluid is mercury. *For example,* how many inches of the measuring fluid are required to obtain a psi of 4? If the measuring fluid is water, the amount required for a psi of 4 is 110.8 inches (approximately 9 ft. 3 in.), or 27.7 multiplied by 4. If the measuring fluid is mercury, the amount of fluid is 8.16 inches, or 2.04 multiplied by 4.

Manometers

The simplest device for measuring the pressure of liquids or gases is the *manometer.* The simplest form of the manometer is the *U-tube.* This consists of a glass tube, shaped like the letter *U,* and a scale marked in inches and tenths of inches. On the scale, the zero mark appears in the center. See Figs. 3-4 and 3-5. The manometer fluid (wa-

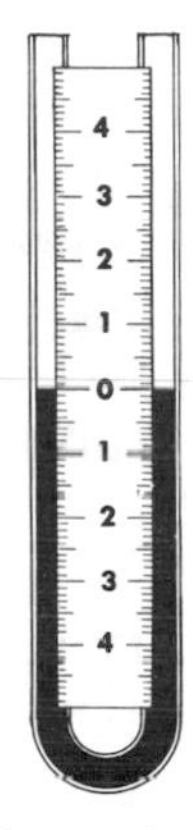

Fig. 3-4. A simple U-tube manometer.

Fig. 3-5. A typical industrial U-tube manometer. (Meriam Instrument Co.)

ter or mercury are the most widely used fluids) is poured into the tube until the level in both columns reaches the zero mark. With both columns open to the atmosphere, the level of the fluid will remain at zero. When a pressure line is connected to one column of the manometer, the fluid in that column will be forced down, and the fluid in the other column will rise. By measuring the difference in the height of the fluid in the two columns, the pressure of the inlet line can be expressed in inches of fluid. *For example,* say the manometric fluid is mercury as is the case in Fig. 3-6, and a pressure line is connected to the manometer, causing the mercury to be lowered one inch in one column and raised one inch in the other. The pressure in the inlet line is expressed as two inches of mercury, or 0.98 pounds per square inch (psi).

With water as the manometric fluid, the same pressure would cause the water to be lowered 13.58 inches in one column and raised 13.58 inches in the other column. The pressure could then be expressed as equal to 27.16 inches, the sum of the measurement of fluid in each column. See Fig. 3-7.

Two other types of manometers are the inclined-tube type and the well type.

In the inclined-tube type, the measuring column of the manometer is at an angle to the vertical. See Fig. 3-8. The angle serves to expand the scale of the instrument and increase readability. Since this type of manometer is

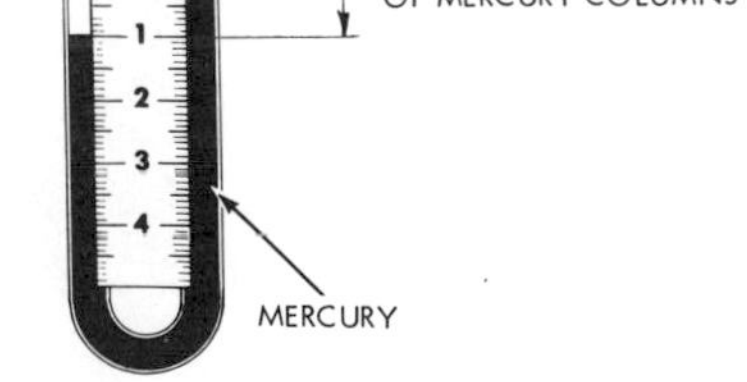

Fig. 3-6. The pressure at the inlet is expressed as 2 inches of mercury.

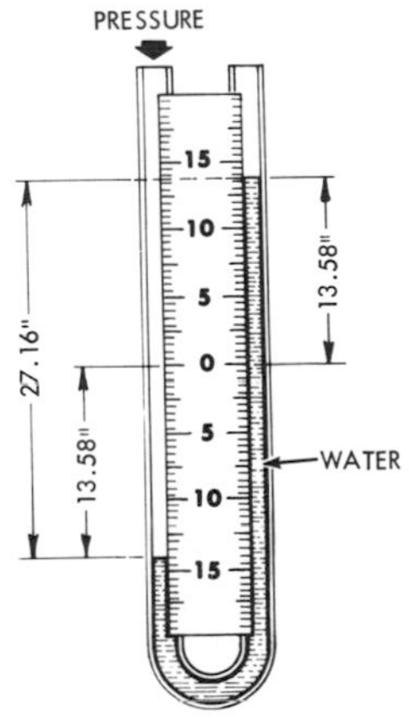

Fig. 3-7. In this application, which uses water, the same pressure shown in Fig. 3-6 is expressed as 27.16 inches of water.

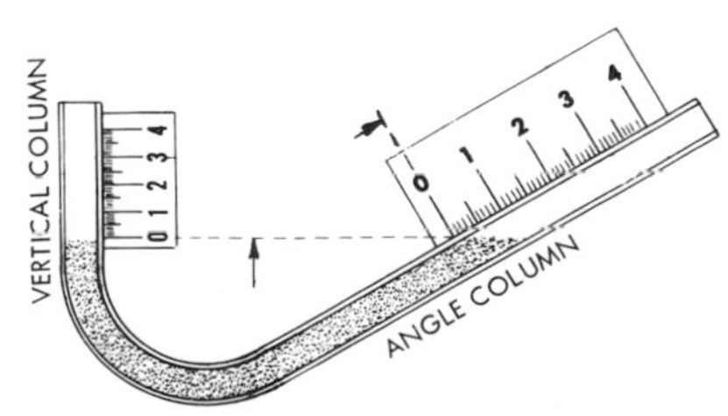

Fig. 3-8. An inclined-tube manometer.

used for low pressure measurement, the manometric fluid is water.

In the well type, one column is a large reservoir of fluid where the pressure is applied. Because of this design, the pressure can be measured simply by referring to the level of the fluid in the vertical tube. See Fig. 3-9.

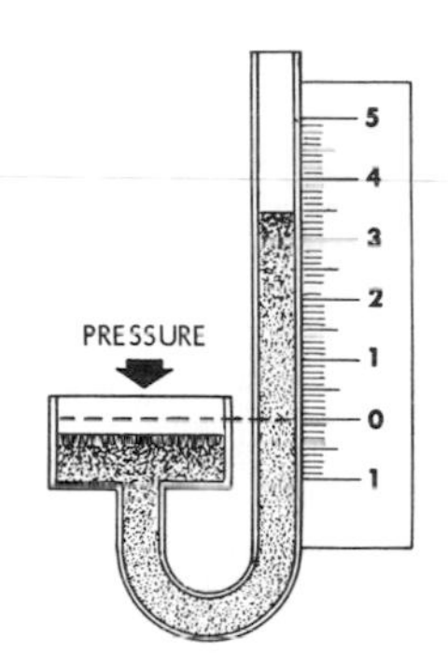

Fig. 3-9. A well-type manometer.

Pressure Elements

There is a series of mechanical devices which are designed to alter their shape when pressure is applied to them. These devices are called *elastic deformation pressure elements.* See Fig. 3-10. Each particular element responds to a different pressure range. Table 1 gives a list of such pressure elements. The upper and lower pressure limits of these devices are expressed in inches of water or pounds per square inch.

Fig. 3-11 shows an industrial pressure gage designed with a Bourdon pressure element. The pressure enters the Bourdon tube, causing the tube to unwind. This movement actuates a lever arm connected by gears to the pointer. A spiral spring connected to the shaft carrying the pointer takes up back lash in the gearing and maintains

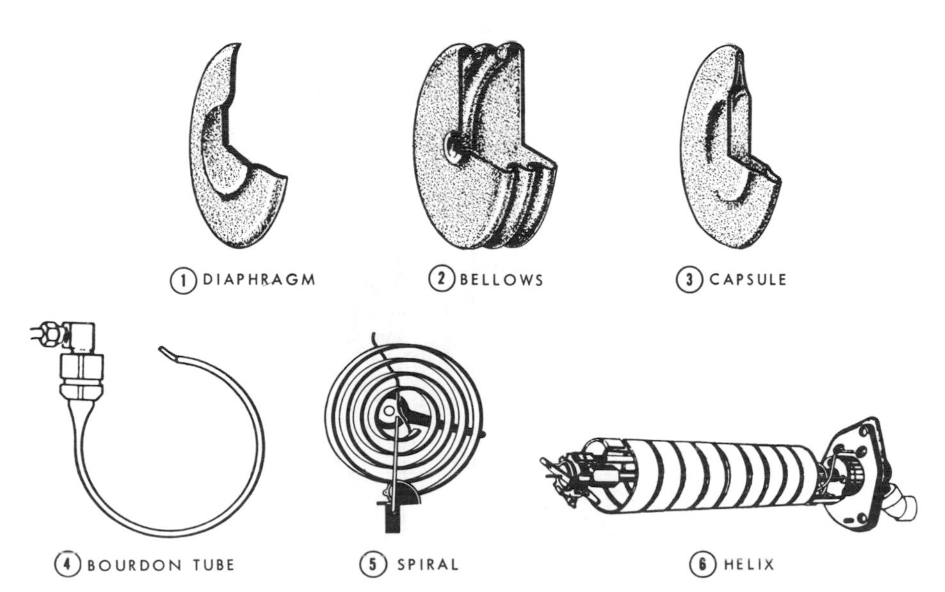

Fig. 3-10. Elastic deformation elements.

TABLE 1 PRESSURE ELEMENTS

PRESSURE ELEMENTS	MINIMUM RANGE	MAXIMUM RANGE
DIAPHRAGM	0" to 2" H_2O	0 to 400 psi
BELLOWS	0" to 5" H_2O	0 to 800 psi
CAPSULE	0" to 1" H_2O	0 to 50 psi
BOURDON TUBE	0 to 12 psi	0 to 100,000 psi
SPIRAL	0 to 15 psi	0 to 4,000 psi
HELIX	0 to 50 psi	0 to 10,000 psi

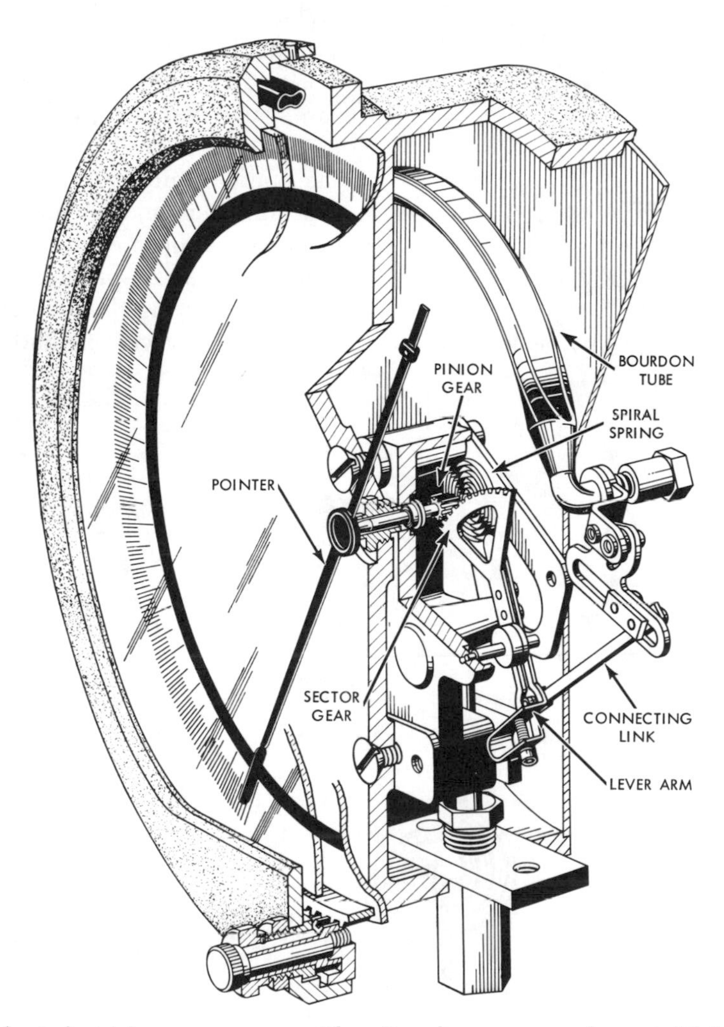

Fig. 3-11. An industrial pressure gage with a Bourdon pressure element. (Heise Bourdon Tube Co.)

Fig. 3-12. Typical industrial recording pressure gage. (Bailey Meter Co.)

Fig. 3-13. Interior of an electrical pressure gage. (Bailey Meter Co.)

the smooth sensitive action of the movement.

Recording pressure gages are frequently used in industry. See Fig. 3-12. Generally, they use elastic deformation elements of either the spiral or helical variety.

Pressure measuring instruments which provide an electrical output are available to be compatible with other equipment in various applications. These instruments combine an elastic deformation element (such as a diaphragm) with an electrical device. This device, when deflected by the pressure element, produces a change in an electrical characteristic, such as resistance. It produces a unit change in the electrical characteristic for each unit change of pressure applied. The conversion of one form of energy into another is produced by a *transducer*, a device, which in this application, changes pressure into an electrical variable, such as resistance or capacitance. Most of these electrical pressure transducers are used in bridge type circuits, similar to those used for the resistance thermometer. Fig. 3-13 shows a typical electrical pressure measuring device.

Differential Pressure Measurement

Frequently, it is important to measure the difference between two pressures. This is called *differential pressure measurement.*

Differential pressure measurement is similar to measuring one pressure. The same instruments are often used. The U-tube manometer is again the simplest instrument available. In single pressure measurement, one column is left open to the atmosphere. See Fig.

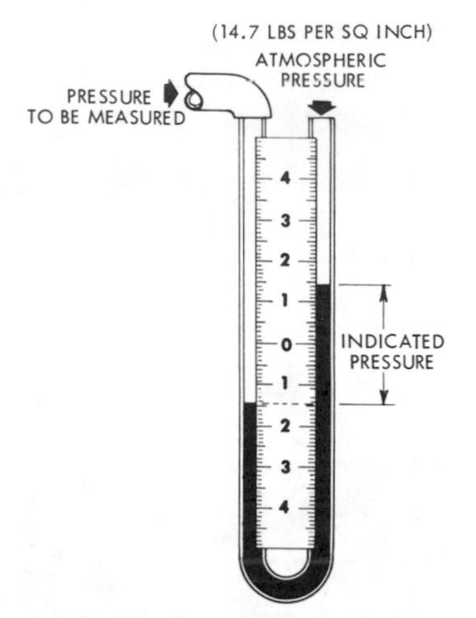

Fig. 3-14. Single pressure measurement with one column of the U-tube open to atmospheric pressure.

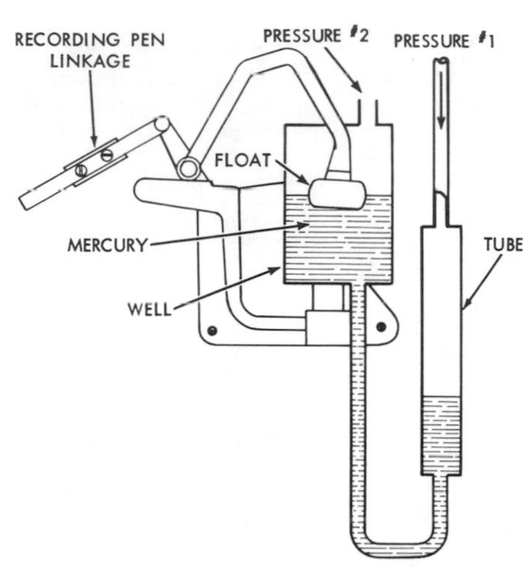

Fig. 3-16. A well-type mercury manometer for measuring differential pressure. (Bailey Meter Co.)

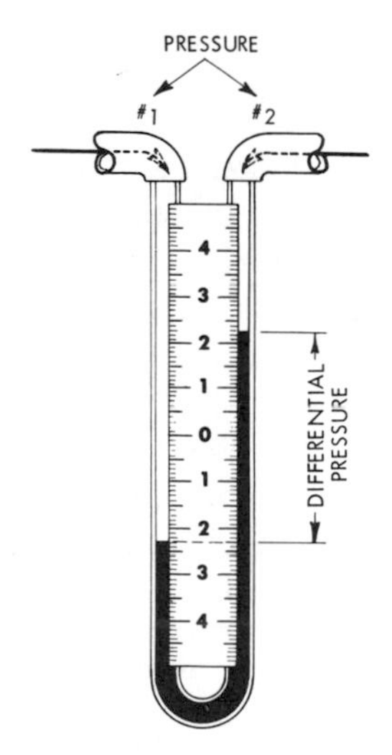

Fig. 3-15. Differential pressure measurement indicates the difference between two working pressures.

3-14. If both columns of the manometer are connected to separate pressures, the pressure indicated is the difference between them. See Fig. 3-15.

For recording differential pressures, a special well and tube manometer is used. See Fig. 3-16. In this type of instrument, the manometer is made of metal and is attached to, or enclosed in, the instrument case. The movement of the float is used to drive an indicating pointer or recording pen, rather than measure the difference in the height of the liquid in the column and the well. The float motion inside the manometer must be transferred to the outside linkage without losing the manometric fluid. The mechanism should be as friction free as possible. Manufacturers provide special bearings, torque tubes, or magnetic followers for this purpose. The float mechanism must be precisely manufactured to provide sensitivity and accuracy.

Another unit for measuring differential pressure is the liquid-filled bellows type. See Fig. 3-17. A pair of matched

#4 copper = 95A
#6 copper

Fig. 3-17. The bellows of a bellows-type differential pressure meter. The bellows are liquid-filled.

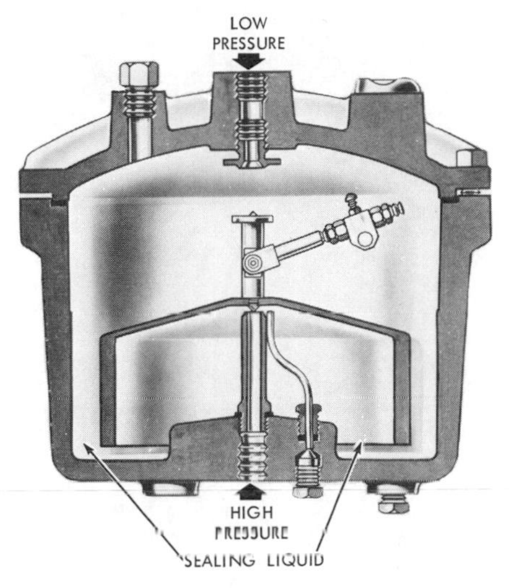

Fig. 3-18. Cutaway of a bell-type differential pressure meter. This type of meter is used when the difference between the pressures is very small. (American Meter Co.)

bellows is used in this instrument. The bellows are enclosed in separate pressure chambers. The pressure difference acting on the bellows actuates a shaft connected to them. This shaft, in turn, is connected to the indicating or recording mechanism by a torque shaft. Liquid in the bellows reduces the effect of rapid pressure changes. Valves control the flow of the liquid from one bellows to the other, thus preventing their rupture.

The bell manometer can be used to make differential pressure measurements. A bell (or inverted cup) is enclosed in a pressure housing. Two pressures are admitted to the bell, one to the inner surface and the other to the outer surface. See Fig. 3-18. A sealing liquid is required to separate the two pressure fluids. A higher pressure on the inner side causes the bell to rise; a higher pressure on the outer surface causes the bell to be lowered. This motion is transferred to the indicating or recording linkage through a pressure bearing, or other suitable device.

The weight-or-ring balance meter, shown in Fig. 3-19, measures differential pressure by displacing the liquid in the sensing device, causing one side of the ring to become heavier than the other. The mechanism is mounted on a pivot, which allows it to move. The weight acts as a balance, and the motion of the entire unit is transferred to an indicating pointer or recording pen.

There are many kinds of differential pressure measuring instruments, but almost all of them use the manometer, bellows, or bell as the actuating mechanism.

Several types of differential pressure sensors provide electrical outputs. These sensors are known as transduc-

Fig. 3-19. Ring balance meter for measuring differential pressure.

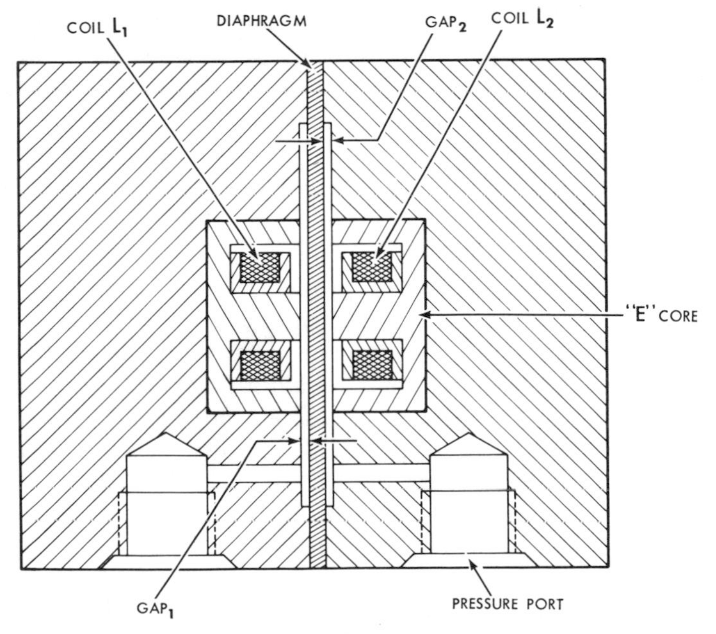

Fig. 3-20. Interior of a diaphragm-actuated transducer. (Validyne Engineering Corp.)

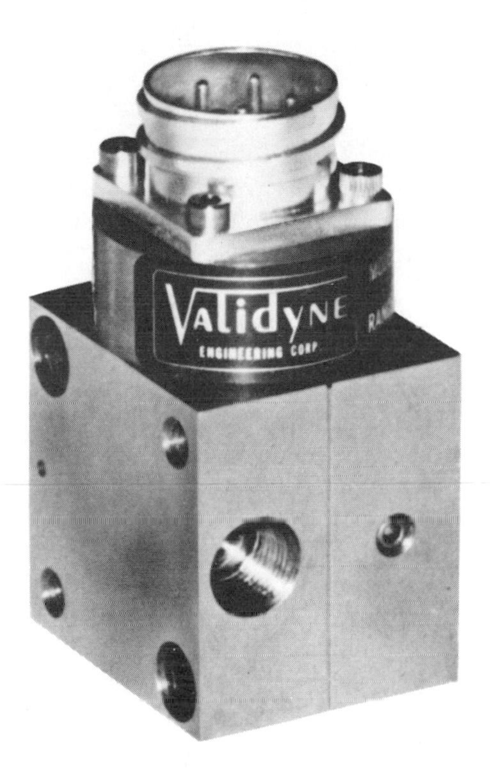

Fig. 3-21. Electrical transducer used in an electrical differential pressure meter. (Validyne Engineering Corp.)

ers. The motion of a float or bellows may be used to position a mechanical element attached to a slider, a transformer core, or a movable capacitor plate. In this way a unit change in differential pressure produces a proportional change in resistance, inductance, or capacitance. There are also diaphragm-actuated transducers in which the diaphragm is clamped between two blocks. Each block contains a core and a coil. See Fig. 3-20. Differential pressure deflects the diaphragm, causing a change in the electrical characteristics of the transducer, proportional to the differential pressure. A typical electrical differential pressure measuring transducer is shown in Fig. 3-21.

Words to Know

psi
hydrostatic pressure
manometer
U-tube
elastic deformation pressure element
transducer
differential pressure measurement

Review Questions

1. Convert the following pressures into inches of water: .7 psi, 3 psi.
2. Convert the following pressures into inches of mercury: 14.7 psi, 75 psi, 133 psi.
3. What is the simplest form of manometer?
4. What elastic deformation pressure elements would be found in gages with the following ranges: 0 to 1 psi; 1 to 10 psi; 0 to 1000 psi?
5. What is the difference in the height of the measuring columns, if a pressure of 60 psi is applied to one side of a mercury manometer and a pressure of 80 psi to the other side?
6. If one of the two pressures in problem 5 suddenly dropped to 0, a U-tube manometer should be used. Or is another type of differential pressure measuring device more suitable for this application? Explain.
7. In a bell type manometer, would oil be a good sealing fluid, if the fluids being measured contained oil? Explain.
8. Is the weight-balance differential pressure meter a form of manometer?
9. If only single pressure instruments were available, how would differential pressure be measured?
10. How can differential pressure be measured at one location and recorded at another point 100 feet away?

Level

Chapter

4

The vast amount of water, fuels, solvents, chemicals, and other materials used in manufacturing makes the measurement of level essential to modern process control. The level of material in a tank, bin, hopper, or other container indicates the amount available for processing or the quantity of finished product on hand. Instruments for measuring the level of liquids, slurries, and granular solids are necessary to determine the length of production time.

If the container has a uniform cross-section throughout, a change in level represents a uniform change in volume. See Fig. 4-1. When a horizontal cylindrical tank or a conical hopper is used, the level scale has non-uniform graduations, as shown in Fig. 4-2, so the level can be expressed in volumetric units.

Some phases of industrial processing may require a continuous measurement of the amount of material in a storage container, while others do not. In these latter instances it may only be necessary to know when the level reaches a particular point. Level

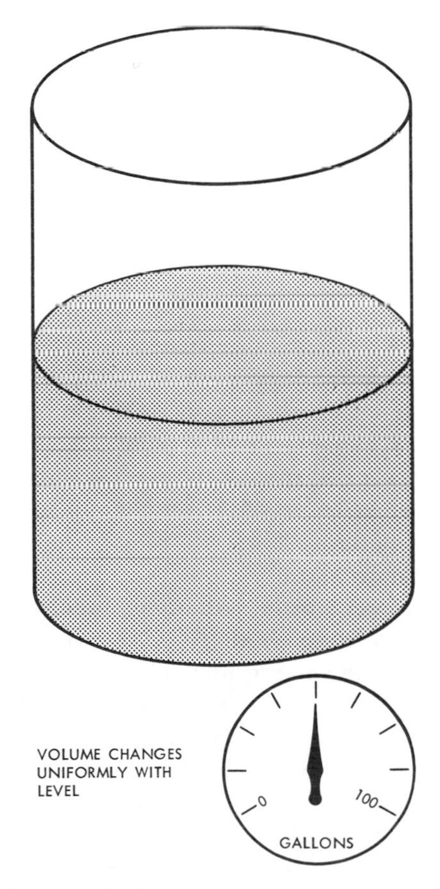

Fig. 4-1. Volume changes uniformly with level in containers having a uniform cross-section.

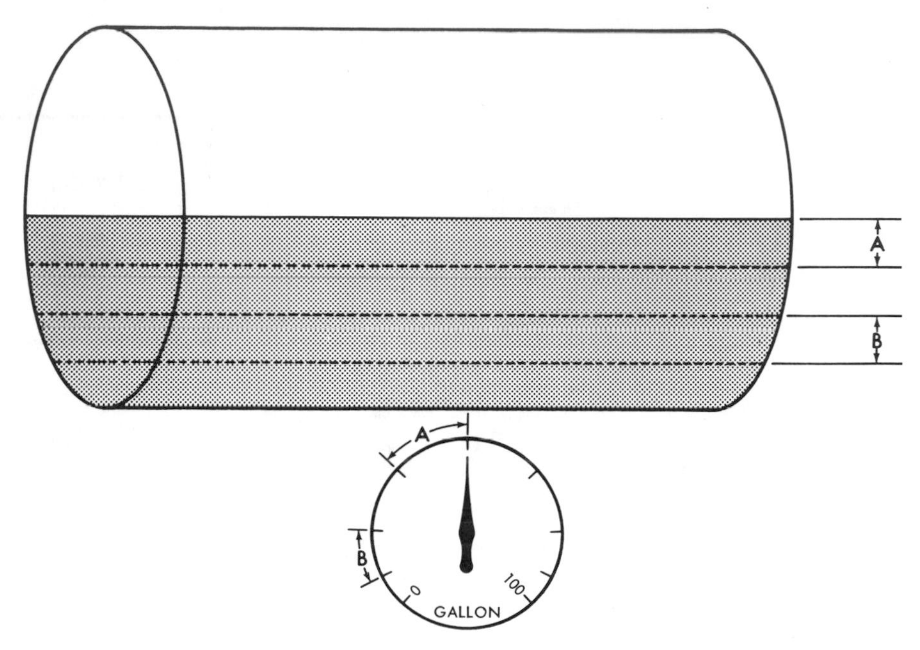

Fig. 4-2. A horizontal cylindrical tank does not have uniform volume. A change of level at point A above represents a greater volumetric change than a change of level at point B.

measuring devices are available which give either continuous or single point measurement. Single point level detectors are frequently used as control devices to prevent overflow or cut-off of material. These detectors can also actuate alarms.

Level can be measured directly or it can be inferred. The direct method uses the varying level of the material (liquids or granular solids) as a means of obtaining measurement. The inferential (indirect) method uses a variable, such as air pressure, which changes with the level of the material (liquids only) and actuates the measuring mechanism. Typical mechanisms for these applications are pressure gages and diaphragm boxes. Mechanical and electrical instruments are available for each method.

Direct Level Measurement

Floats. Floats, attached to a rod or a cable, are the common primary elements for liquid level measurement. The motion of the float as it follows the level of the liquid provides a sufficient measuring or controlling signal.

A simple single-point float-operated controller is shown in Fig. 4-3. It consists of a float attached to a lever-operated valve which supplies liquid to a tank. The float rises with the increasing liquid level. At a predetermined point, the float actuates the inlet valve which closes, shutting off the supply of liquid. The float may be lo-

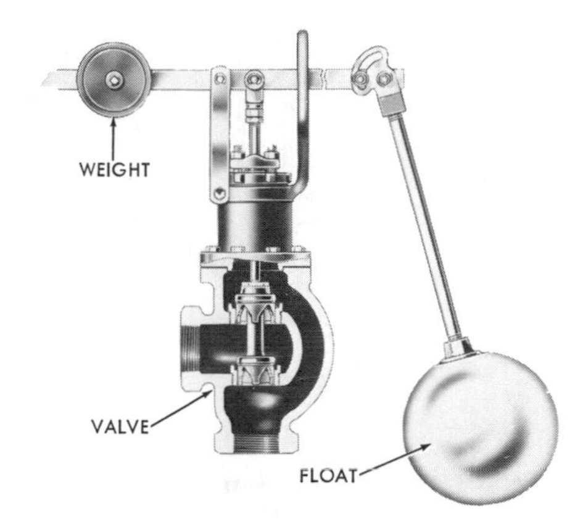

Fig. 4-3. This type of ball-float liquid level controller is used in open tanks. (Fisher Controls)

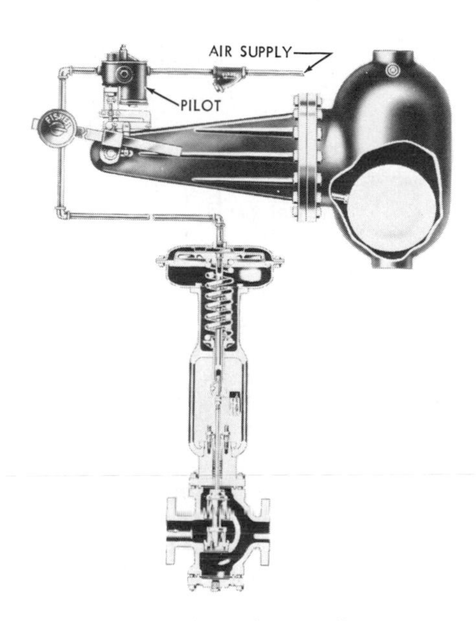

Fig. 4-5. A liquid level controller is operated by a pneumatic pilot to provide greater sensitivity. (Fisher Controls)

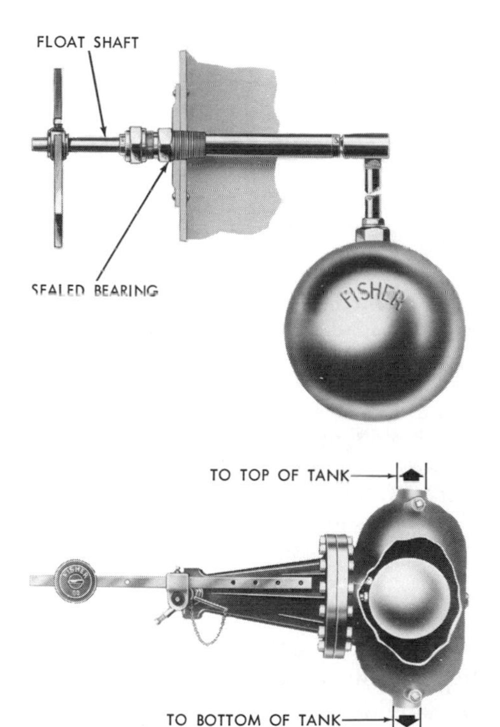

Fig. 4-4. A ball-float liquid level controller is used in closed tanks. A controller without float cage is shown in the top illustration. Bottom shows a controller with float cage. (Fisher Controls)

cated inside the tank or enclosed in a float cage outside the tank. See Figs. 4-4 and 4-5. The float cage is located at a height (level) suitable for the specific application. Pneumatic pilot-operated level controllers which use floats are also available. In these controllers, the float actuates a pneumatic relay which opens or closes the air supply to an air-operated valve.

A variation in float-operated level controls uses a toroidal (doughnut shaped) magnetic float. The float travels vertically along a dip tube in the tank which contains the liquid. A second magnet, inside the dip tube, follows the magnetic float and provides the liquid level signal. See Fig. 4-6. This type of system can be used to activate several pneumatic or electri-

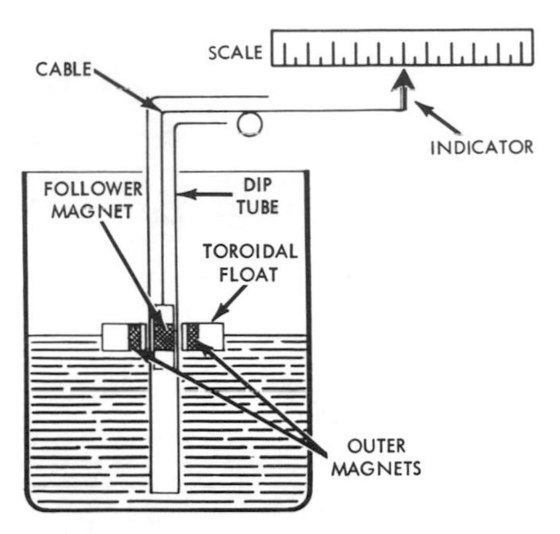

Fig. 4-6. A toroidal (doughnut shaped) magnetic float. The float responds to the level of the fluid. The outer magnets transfer the motion of the float to the follower (inner) magnet suspended in the dip tube by a cable which is also connected to an indicator. The movement of the follower magnet is thus transferred to the indicator.

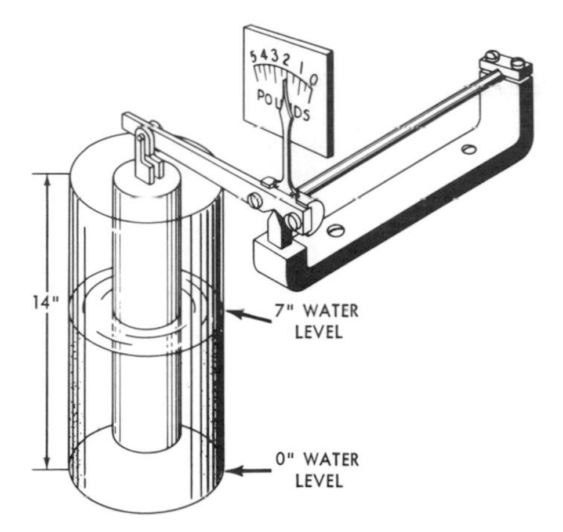

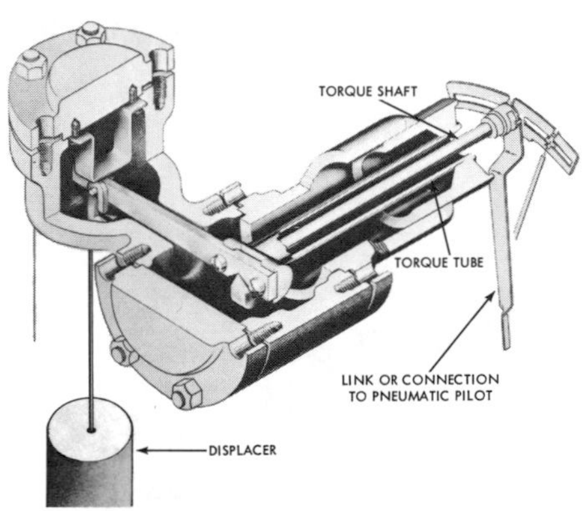

Fig. 4-7. The displacer in the top line drawing weighs 3 lbs. and is 14″ long. When it is placed in 7″ of water, its weight is 2 lbs. The weight of the displacer varies with the level of the liquid. These changes are converted into torque, as shown in the bottom cutaway of an actual unit. The torque or twisting force operates a pneumatic pilot. (Masoneilan International, Inc.)

cal switches as the float rises and falls. An advantage of this system is that it has longer float travel than single-point or ball-float devices.

Displacer. A displacer is similar in action to the buoyant float except its movement is more restricted. When the level of the liquid changes, the amount covering the displacer varies. The buoyant force acting on the displacer becomes greater as more of the displacer is submerged. This force is transferred to a pneumatic system through a twisting or bending shaft (torque). See Fig. 4-7. There is a different force affecting the shaft for each level of the liquid. This causes the shaft to assume a new position. The pneumatic system is designed so that there is a different air pressure to the indicator for each shaft position. The displacer float has the advantage over the single buoyant float in that it is more sensitive to small changes in level and is less subject to mechanical friction.

In some applications, several displacers are suspended on one cable,

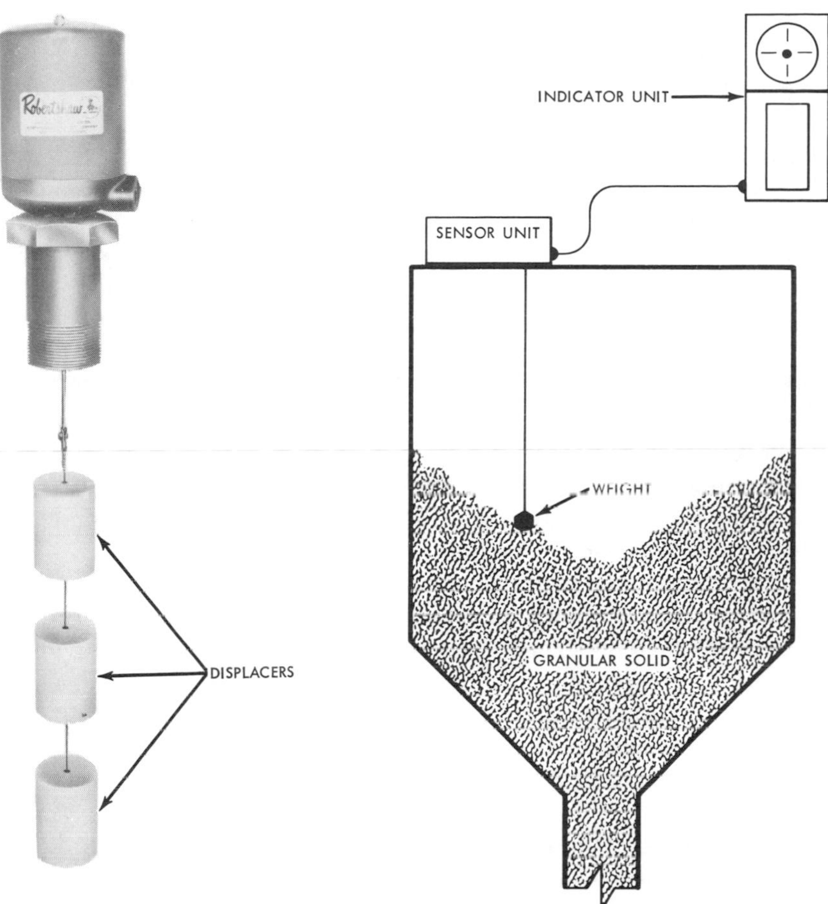

Fig. 4-8. The buoyant force in the cable changes as each of the displacers is covered by the level of the liquid. (Robert Shaw Controls)

Fig. 4-9. This unit uses a weight which comes into contact with the material to be measured. It is used for granular solids.

shown in Fig. 4-8. The buoyant force in the cable varies as any of the displacers is affected (covered or uncovered) by the level of the liquid. This force activates several pneumatic or electric switches.

Contact Method. The direct method of measuring level shown in Fig. 4-9 operates on a contact principle. The system can be adapted to use a float or a weight. The weight is used for granular solids, the float for liquids.

The cycle for measuring level is started by a push-button or a timer. At the beginning of each cycle, a weight, connected to a cable, is resting on the material to be measured. At this point there is no drag on the cable. After a preset interval, a motor is activated to draw up the cable and raise the weight. The increased force of the weight as it is lifted off the material is transmitted by the cable to an electrical sensor. This sensor then causes the motor to drive the cable and lower the weight until it is again in contact with the material. This completes the measuring cycle. An electrical indicator in the motor circuit provides the level measurement. The time interval between raising and lowering the weight can be adjusted for the specific application.

If, at the beginning of the cycle, the weight is covered by the material that is being measured, the motor maintains its drag on the cable until the weight is free of the material. Then the cycle continues as before. When the material to be measured is a liquid, a float replaces the weight. Otherwise, the system and the cycle are basically the same.

Electric Probes. Conductivity, capacitance and ultrasonic probes are several methods of electrically measuring liquid level. Most often these are single-point devices, designed to measure a single predetermined level.

Because of their extended center electrode, conductivity probes somewhat resemble an automotive spark plug. These probes are used with liquids that conduct an electric current. The electrodes are supplied with a dc voltage. Two or more of the probes are mounted on the top of the storage vessel. The electrodes extend into the vessel. See Fig. 4-10. When the liquid makes contact with any of the electrodes, an electric current flows between the electrode and ground (the vessel itself).

The current energizes a relay coil which causes the relay contacts to open or close, depending on which stage of the process is involved. The relay contacts actuate an alarm, a pump, or both. A typical system has three probes: a low level probe, a high level probe, and a high level alarm probe. The alarm probe has the shortest electrode and extends the shortest distance into the vessel. When the level of the

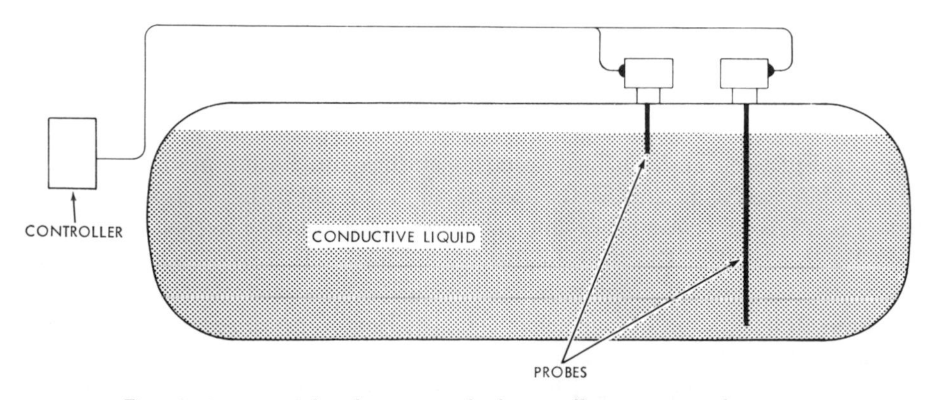

Fig. 4-10. Liquid level measured electrically using conductivity.

liquid drops so that it is only in contact with the low level electrode, a circuit forms and actuates the filler pump. During the filling cycle, the level of the material rises until the high level probe is reached. At this point a second relay opens and stops the pump. If the pump continues in operation beyond the full point and the liquid reaches the alarm probe, a circuit is

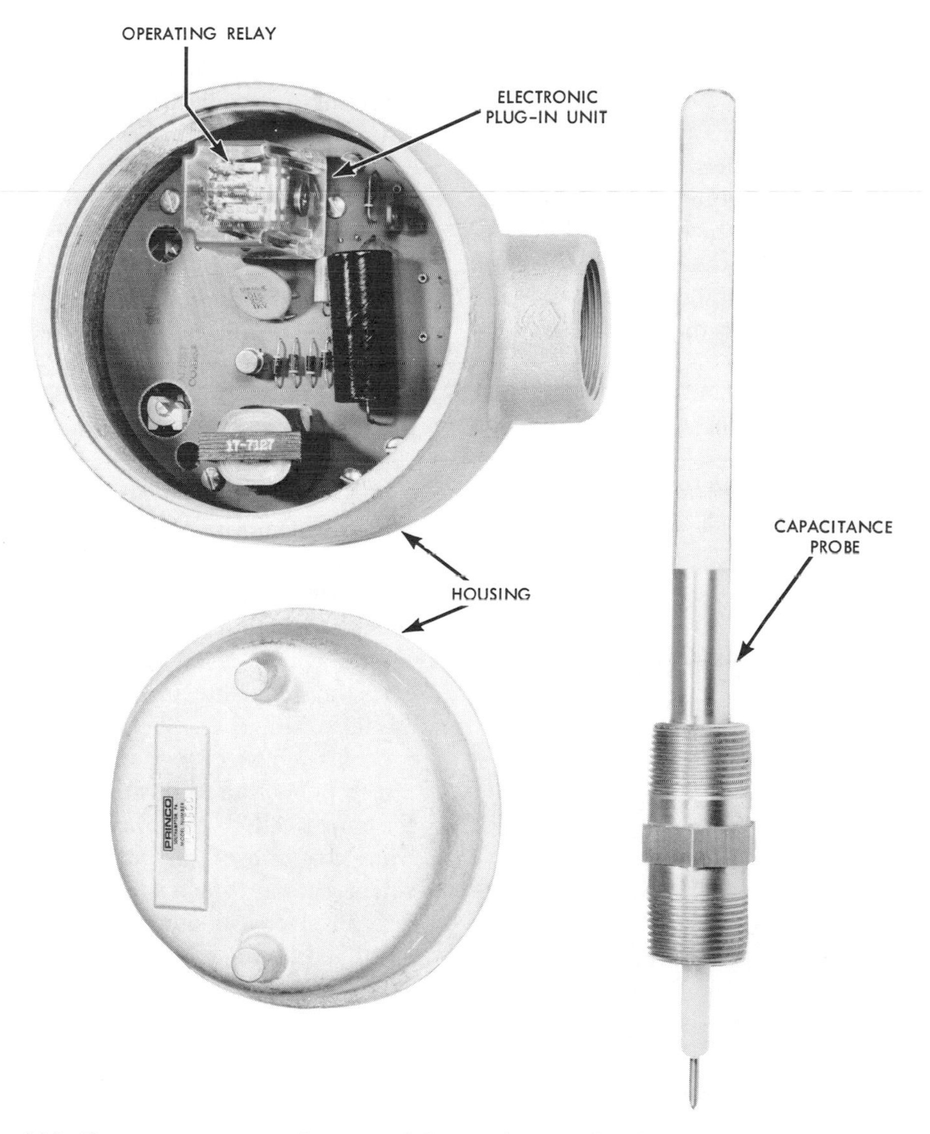

Fig. 4-11. This capacitance probe is used for single-point level detection. (Princo Instruments, Inc.)

formed which closes the relay. This actuates a visual or audible alarm.

Capacitance probes are used in similar applications as conductivity probes. The tubular metal shell and inner rod serve as the electric capacitor plates. See Fig. 4-11. When the liquid level rises above the probe, the liquid acts as the dielectric of the capacitor. When the level drops and exposes the probe, the air or vapor above the liquid becomes the dielectric. This alters the capacitance of the probe and changes the frequency of the oscillation circuit in which it is used. A sufficient change in the frequency actuates a relay. Capacitance probes are also used to measure the level of granular solids or slurries.

The electrical characteristic which expresses the dielectric quality of a material is called the *dielectric constant.* The dielectric constant of air is 1. Other materials are generally compared with air. For example, the dielectric constant for carbon tetrachloride is 2.4 and for pure water it is 80. The higher the number of the constant, the more conductive the material is. Water is more conductive than either carbon tetrachloride or air. Capacitance probes cannot be used for materials with dielectric constants higher than 80 or lower than 2. In addition to detecting a specific level for one material, capacitance probes are used to detect the interface between two materials. See Fig. 4-12. When equipped with a long probe, they are used for continuous level measurement.

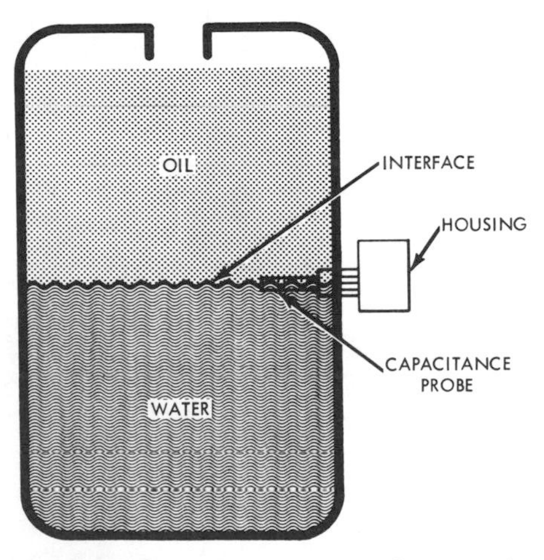

Fig. 4-12. A capacitance probe can be used to detect interface between two materials.

An ultrasonic probe basically consists of a transmitter and a receiver mounted in a probe with a ½″ gap between them. See Fig. 4-13. Both the transmitter and the receiver are piezoelectric crystals. The gap is an important component of the sensing system. All three comprise a single unit which is connected to a control unit by means of a cable. The control unit supplies an electrical signal to the transmitter crystal which is energized and converts the signal into an ultrasonic transmission. However, there can be no transmission of energy unless the liquid to be measured fills the gap. Once the gap is filled, the ultrasonic signal from the transmitter reaches the receiver crystal which converts it back into an electrical signal. The signal is amplified by the control unit and actuates an electrical relay. The relay can operate an indicator, recorder, alarm, pump, or whatever is needed for the particular application. These probes are especially suitable for measuring level in processes which involve sticky liquids, such as syrup.

Fig. 4-13. The major components of an ultrasonic probe are the transmitter, the receiver, and the gap. (National Sonics Corp.)

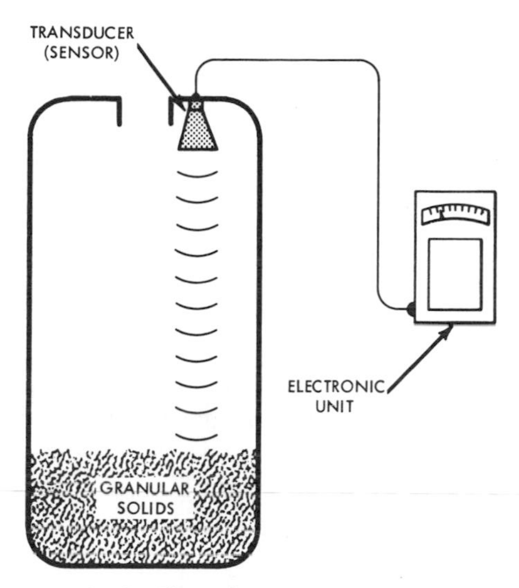

Fig. 4-14. This ultrasonic transmitter-receiver is used for continuous level measurement.

When the ultrasonic probe is strictly used as a single-point level sensing device, it is installed at the low level or high level point of the tank, as with capacitance probes. For continuous level measurement, the ultrasonic transmitter-receiver is located at the top of the vessel. See Fig. 4-14. In this particular application, involving granular solids, the transmitted ultrasonic waves are reflected like an echo off the surface of the material in the container. The time interval between the transmission and the return of the signal varies with the level.

Indirect Liquid Level Measurement

Hydrostatic Pressure. Several types of indirect level measuring devices are actually hydrostatic pressure sensors. The simplest is a pressure gage located at the zero level in a liquid container. See Fig. 4-15. Any rise in the level causes an increase of hydrostatic pressure which is measured by the gage. The gage scale is graduated in feet or inches.

Certain liquids, because of their nature (corrosive, hot, etc.), must not come into direct contact with the pressure gage. In these applications, a transmitting fluid is used between the liquid and the gage. A common fluid used is air because it is inexpensive and readily available. The *air trap* and the *diaphragm box* are two devices

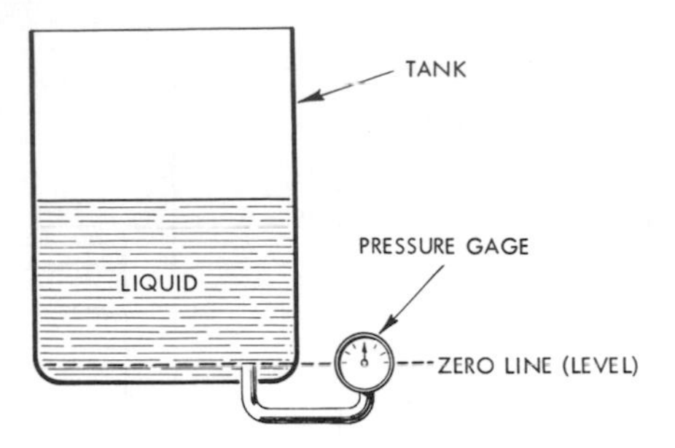

Fig. 4-15. As the tank fills, the pressure of the liquid increases. This increase of pressure can be read on the gage and in feet, inches or gallons.

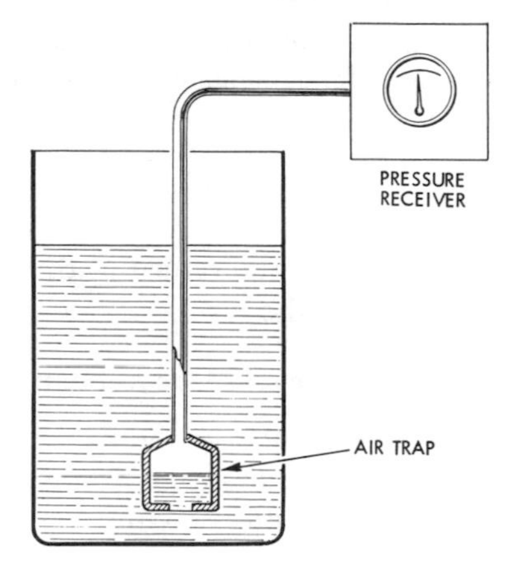

Fig. 4-16. The air pressure in the air trap is expressed on the scale in units of level.

employed when air is the transmitting fluid.

The air trap is a box which is lowered into the liquid to be measured. See Fig. 4-16. As the level of the liquid rises, the pressure on the air trapped inside the box increases. The air pressure is sent through a tube to a pressure gage which has a scale for indicating the level.

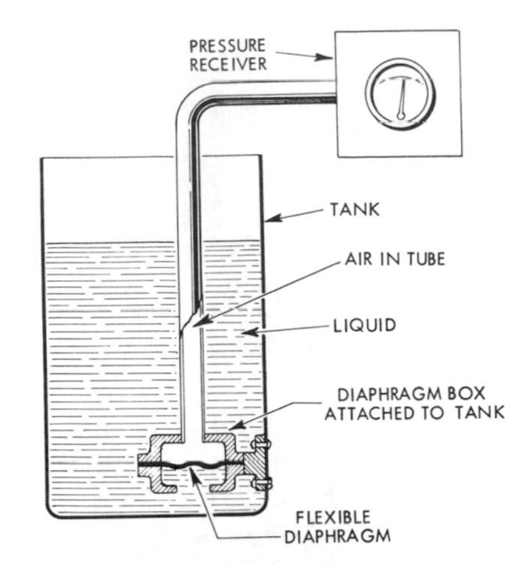

Fig. 4-17. The gage in this unit responds to the deflection of the flexible diaphragm. The deflection is caused by a rise in liquid level.

The diaphragm box, like the air trap, transmits air pressure to a gage. See Fig. 4-17. Here, however, the air is trapped inside by a flexible diaphragm covering the bottom of the box. As the level of the liquid rises, the pressure on the diaphragm increases. This pressure acts on the air in the closed system and is piped to the pressure gage where a reading can be taken.

The bubbler method of level measurement also uses the variation of hydrostatic pressure caused by the liquid column. A pipe is vertically installed in the container. The open end of the pipe is placed at the zero level of the container. The other end of the pipe is connected to a regulated air supply and to a pressure gage. When a level measurement is to be made, the air supply is adjusted so that the pressure is slightly higher than the pressure caused by the height of the liquid. This

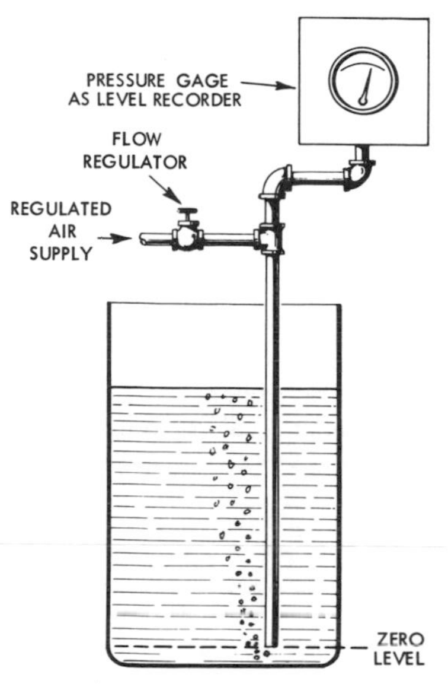

Fig. 4-18. The air pressure in the pipe is slightly more than the liquid pressure in the vessel, so that the air pressure becomes the measure of level in the tank.

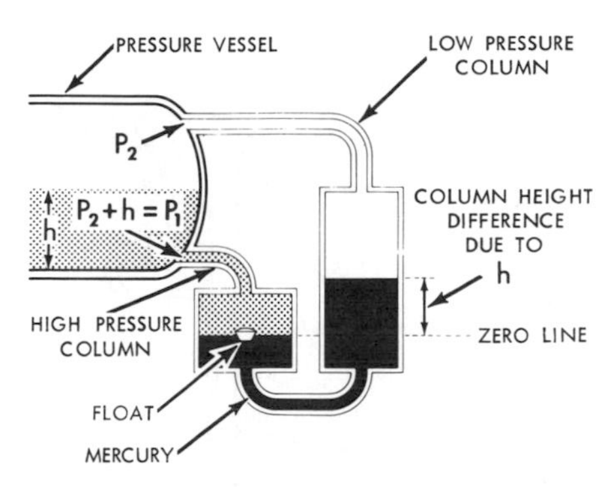

Fig. 4-19. When the liquid is under pressure, level can be measured using a well-type mercury manometer.

is accomplished by regulating the air pressure until bubbles are observed slowly leaving the open end of the pipe. See Fig. 4-18. The gage then measures the air pressure needed to overcome the pressure of the liquid. The level gage is calibrated in feet, inches, or gallons.

The methods described above can only be used when the liquid container is uncovered, or has an opening to the atmosphere. A differential pressure meter is used to determine the level of a liquid under pressure. Fig. 4-19 shows a well-type mercury manometer. Tubing or pipe is used to connect the upper tap of the vessel to the low pressure column of the differential pressure meter. The lower tap of the vessel is connected to the high pressure column of the manometer. The vessel pressure enters both sides of the meter and is cancelled out. The difference in pressure detected by the meter is due to the pressure created by the liquid in the vessel. Most of the differential pressure devices described in Chapter 3 can be used in this manner for level measurement.

Radioactive Devices. Radioactive devices can also be used for single-point and continuous level measurement. They are primarily used when the material to be measured is too corrosive, or the temperatures at that point in the process are too hot, or, in general, when the situation makes it unsuitable to install primary elements inside the storage vessel.

In an application where only single-point level detection is required, the radioactive material is mounted on one side of the storage vessel. A detector is mounted on the opposite side. See Fig. 4-20. The mounting in this system is at the high level point. The system is installed at the level required by the

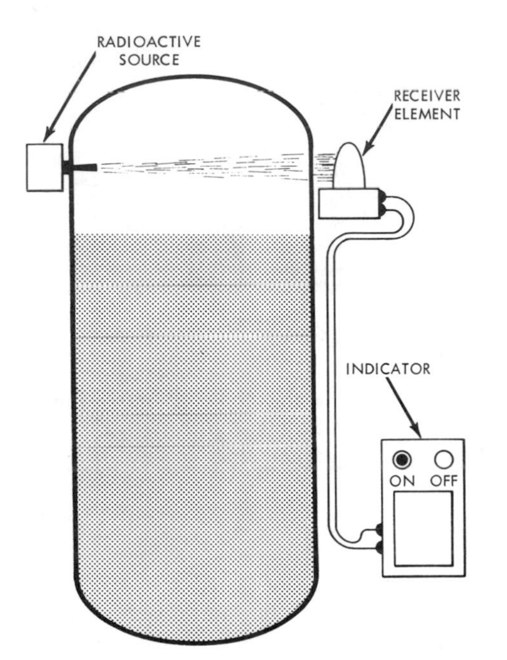

Fig. 4-20. This radioactive system is used for single-point level control.

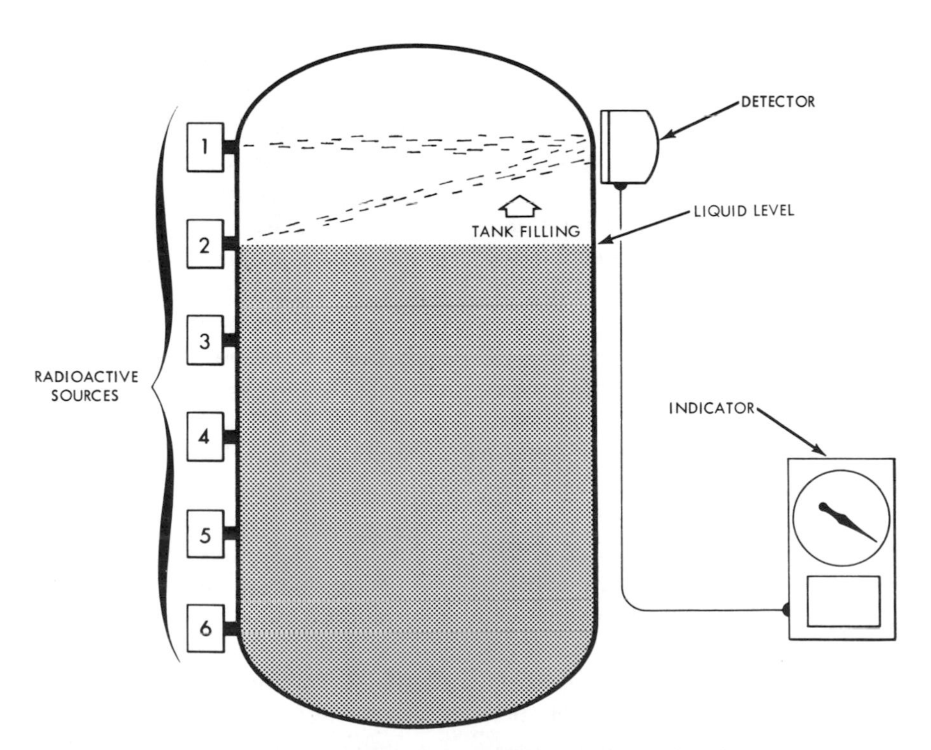

Fig. 4-21. Radioactive sources can be used for continuous level measurement.

process. When the material in the storage vessel rises to the level of the radioactive rays, or above, it cuts off the rays to the detector. A relay in the detector closes a contact which stops the supply of material to the vessel.

One or more radioactive sources are used in an application requiring continuous level measurement. See Fig. 4-21. These radioactive sources are mounted on one side of the storage vessel and a detector on the other side. Some systems use multiple sources and some use multiple detectors. As the level in the storage vessel varies, it affects the amount of radioactive rays which reach the detector. The energy of the sources is sized to make the change in radioactive energy uniform as the level changes. The electronic circuitry further linearizes the signal.

Loss of Weight Devices. Another method of determining the level of materials is to weigh the entire storage vessel, since the weight changes as the level of the material changes. The vessel can be weighed mechanically or electrically. Fig. 4-22 shows a mechanical scale. The vessel can be weighed electrically, using *load cells.* See Fig. 4-23. Load cells are specially constructed mechanical units which contain *strain gages.* Strain gages provide a measurable electrical output proportional to the stress applied by the weight of the vessel on the load cells. As the pressure on the cell changes, the electrical resistance of the strain gage changes. The strain gage is connected into a bridge circuit that contain an electrical meter graduated in feet or inches.

It should be noted that the weigh-

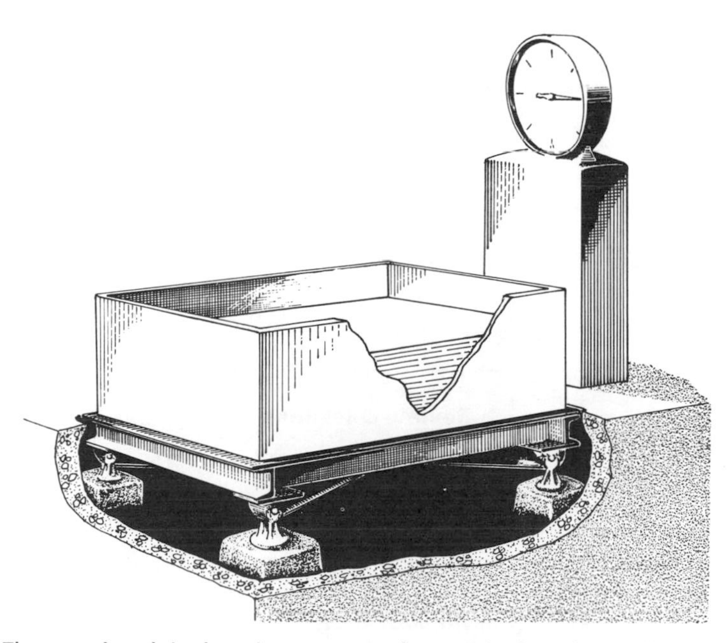

Fig. 4-22. The vessel and the liquid content can be weighed mechanically using this scale.

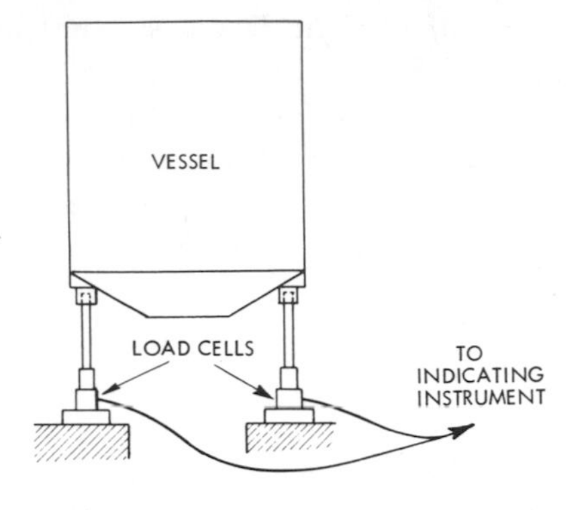

Fig. 4-23. Load cells are used to weigh the vessel and the content electrically.

ing method is accurate only if the density and particle size of the material being measured are uniform, and the moisture content remains constant. The change in weight must be entirely due to the change in level.

There are many other devices used for determining liquid level. In this chapter, we have been concerned with continuous level measurement. Many of the devices described can also be used for *discrete level measurements.* Discrete level measurement determines liquid levels at particular points.

Words to Know

slurry
granular solid
float
single-point level measurement
displacer
electric probe
dielectric constant
load cell
strain gage

Review Questions

1. What is the basic difference between a *direct* and an *indirect* level measuring device?
2. What force acts upon floats and displacer elements?
3. What is meant by the term *dielectric constant*?
4. Where is the ground connection for conductivity probes?
5. What are the three main components of an ultrasonic level detector?
6. What level measuring device should be used when the process

material is too corrosive?

7. If a pressure gage mounted at the base of an open water tank indicates a pressure of 5 psi, what is the height of the water above the pressure gage connection?
8. What is the basic operating difference between an air trap and a diaphragm box?
9. What air pressure is required in a bubbler system installation when the maximum level to be measured is 40 feet?
10. When installing a differential pressure-type liquid level meter, how should one connect the high pressure and the low pressure sides of the meter?

Chapter 5

Flow

In industrial applications, many different methods are used to measure flow. This chapter will outline the fundamental differences among these methods. In Part Two, particular instruments will be described, along with their principles of operation.

There are two kinds of flow measurement: *Flow Rate* and *Total Flow*. Flow rate is the amount of fluid that moves past a given point at any given instant. Total flow is the amount of fluid that moves past a given point during a specified period of time.

Flow Rate Meters

Meters designed to measure flow rate include differential pressure meters, variable area meters, and weir, flume, and open nozzle meters.

Differential Pressure Meters. The flow rate of a fluid in a pipe is related to the differences in pressure across a restriction inside the pipe. Forcing the fluid through a reduced area (restriction) causes the pressure on the upstream side of the restriction to be greater than the pressure on the downstream side.

The simplest pipe restriction used for measuring flow is the *orifice plate*. An orifice plate is a thin, circular metal plate with a sharp edged hole in it. It is held in the pipe between two flanges which are called orifice flanges. The shape and location of the hole are the distinguishing feature of the three kinds of orifice plate: The *concentric plate* has a circular hole located in its center, Fig. 5-1; the *eccentric plate* has a circular hole located below its center, Fig. 5-2; and the *segmental plate* has a hole that is only partly circular located below its center, Fig. 5-3. The kind of orifice plate used depends on the characteristics of the fluid to be measured.

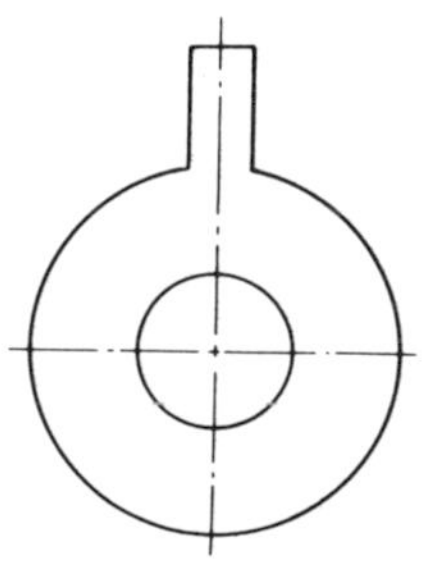

Fig. 5-1. Concentric orifice plate.

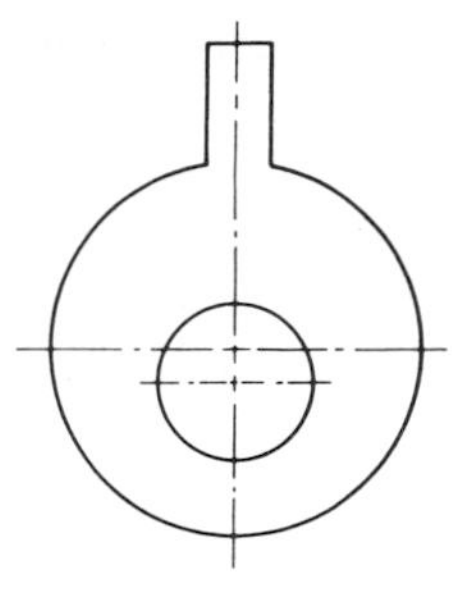

Fig. 5-2. Eccentric orifice plate.

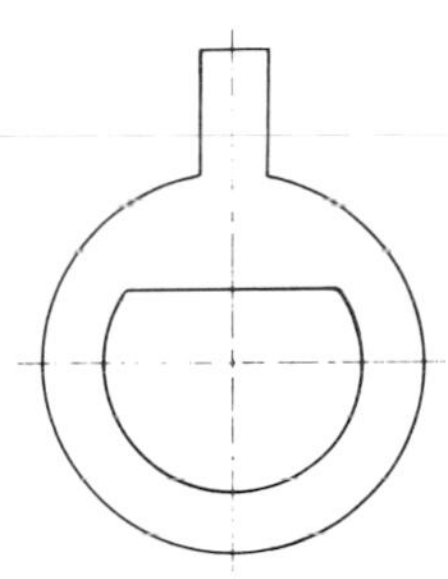

Fig. 5-3. Segmental orifice plate.

Another pipeline restriction for flowmetering is the *Venturi tube,* which is a specially shaped length of pipe resembling two funnels joined at their smaller openings. See Fig. 5-4. The Venturi tube is used for large pipes. It is more accurate than the orifice plate, but considerably more expensive, and more difficult to install. A compromise between the orifice plate and the Venturi tube is the *flow nozzle,* which resembles the half of the Venturi tube

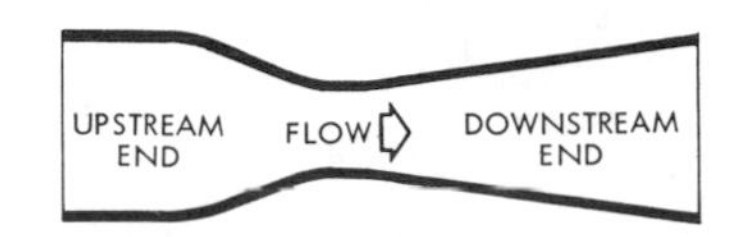

Fig. 5-4. Venturi tube.

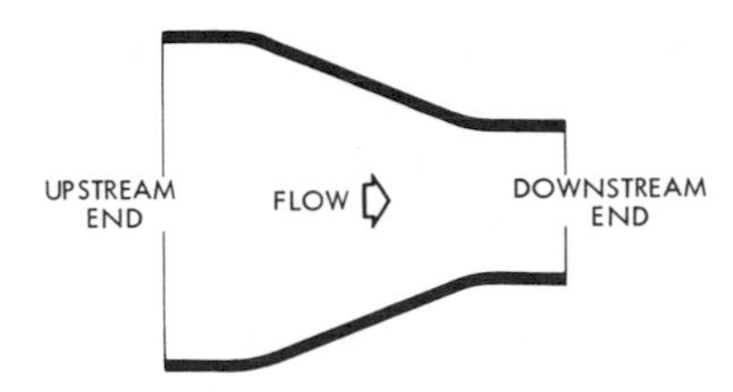

Fig. 5-5. Flow nozzle.

where the fluid enters. See Fig. 5-5. The flow nozzle is almost as accurate as the Venturi tube and it is not as expensive or as difficult to install.

To obtain the pressures upstream and downstream of the primary element requires taps on both sides of the restriction. The location of these pressure taps varies. Three kinds of taps can be used with the orifice plate: *flange taps, pipe taps,* and *vena contracta taps.* Flange taps are located on the flanges which hold the orifice plate in the pipe. See Fig. 5-6. Pipe taps are located at fixed distances upstream and downstream of the orifice plate. The

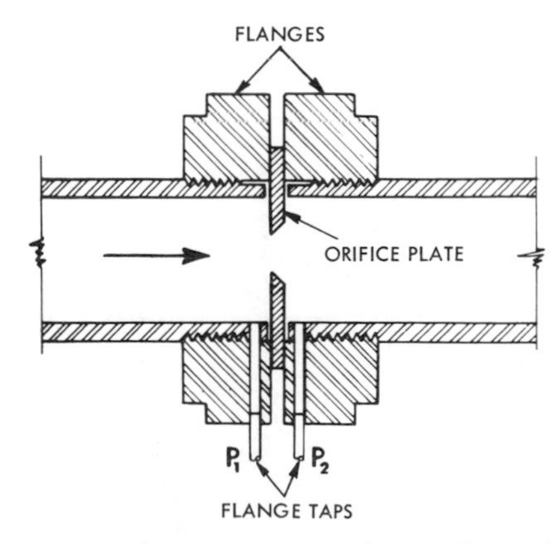

Fig. 5-6. Flange taps are located on the bottom of the flanges which hold the orifice plate in place.

upstream tap is placed 2½ pipe diameters from the plate, and the downstream tap 8 diameters away. See Fig. 5-7. Vena contracta taps are located 1 pipe diameter from the plate on the upstream side, and at the point of minimum pressure (determined by calculation) on the downstream side. See Fig. 5-8.

The pressure taps used with the Venturi tube are located at points of maximum and minimum pipe diameter.

The pressure taps used with the flow nozzle are located at distances upstream and downstream of the nozzle as designated by the manufacturer. This location is critical, and the manufacturer's recommendations must be followed.

Any of the differential pressure instruments described in Chapter 3 can be used, along with these primary flow elements, for flow rate measurement. Since the desired measurement is flow rate, differential pressure must be converted to flow rate. This conversion can be made on the scale or chart of the instrument itself.

Variable Area Meters. There are two types of variable area meters: the *rotameter* and the *valve type area meter.* In the rotameter, the area is varied by a float in a tapered tube. See Fig. 5-9. In the valve type area meter, the

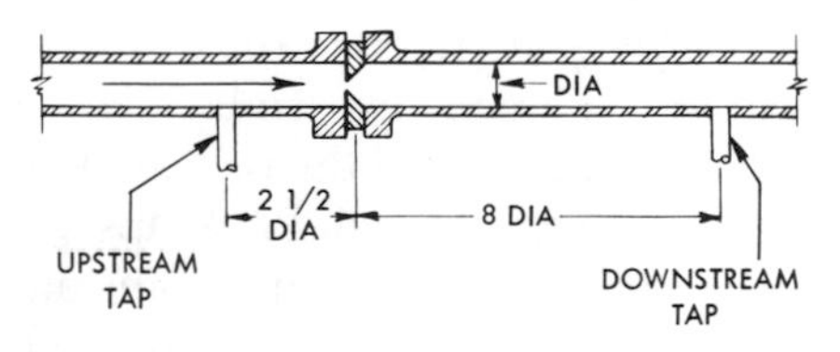

Fig. 5-7. The relative positioning of the upstream and downstream pipe taps.

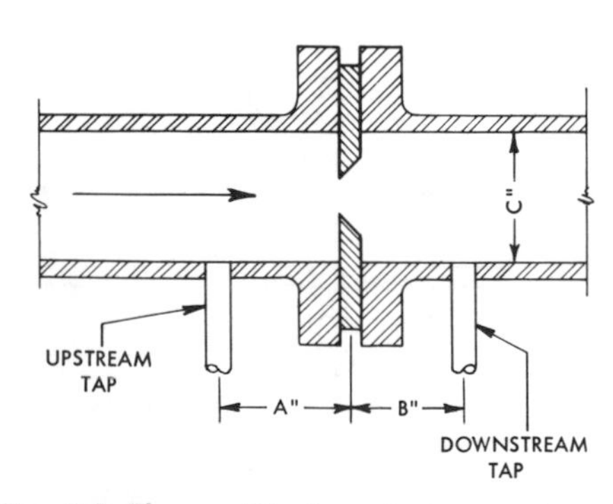

Fig. 5-8. The positioning of vena contracta taps. A" equals C". B" is calculated from application data.

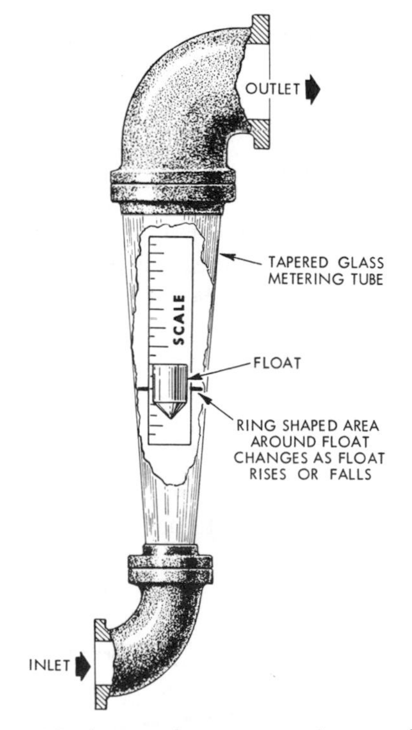

Fig. 5-9. A typical rotameter, showing the principal of operation.

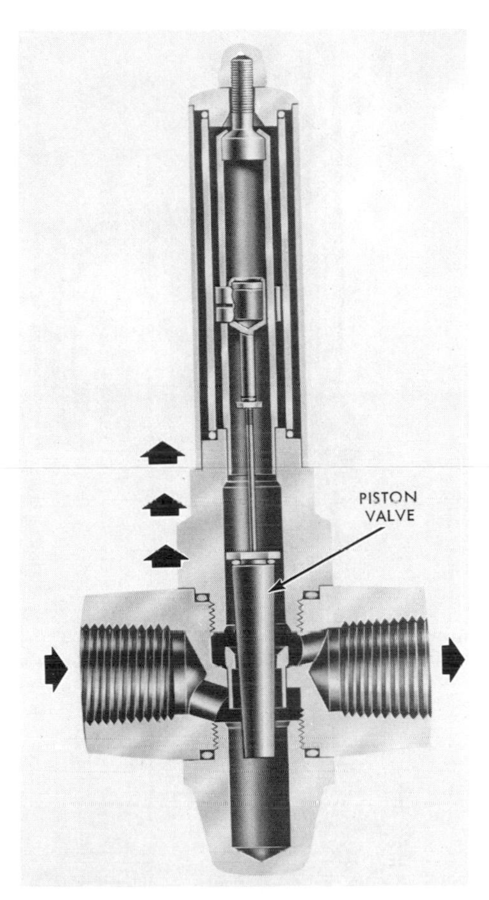

Fig. 5-10. In the area meter, the piston valve moves to a new position for each flow rate. (Wallace & Tiernan, Inc.)

movement of a self-positioning piston valve varies the area. See Fig. 5-10.

If we compare the differential pressure instruments with the variable area meters, we find that the difference lies in the quantities they maintain at a constant. In the differential pressure meter, the area remains constant, while the differential pressure varies with the flow rate. In the variable area meters, the area is varied the amount necessary to maintain the pressure differential at a constant as the flow rate changes.

The float of the rotameter adjusts the size of the area by rising and falling in the tapered tube. Depending on the rate of flow, the float takes a position in the tube that increases or decreases the size of the area, and thus keeps the differential pressure constant. In the valve type meter, a specially shaped plug or piston moves to a new position to keep the differential pressure constant for each rate of flow.

Rotameters are used for measuring the flow of gases or liquids. They are available with glass, metal, or plastic tapered tubes. The floats can also be made of glass, metal, or plastic. Proper selection depends on the nature of the fluid, its temperature, and its pressure. When glass tubes are used, the flow rate can be measured directly by noting the position of the float on a scale attached to the glass, or on a metal scale alongside it. See Fig. 5-11A.

Metal and plastic tubes are required for applications where the fluid is too corrosive or too dark in color to enable the scale to be read. A magnetic follower is used with such tubes because the float position cannot be observed directly. Both the float and the indicator contain a magnet. When the float rises or falls, whether with increasing or decreasing flow, the magnetic coupling causes the indicator to change its position on the scale. See Fig. 5-11B.

Open Channel Measurement. The primary elements used in applications where accurate measurement is not of critical importance, and the fluid flow is in an open channel, are *weirs, flumes,* and *open nozzles.*

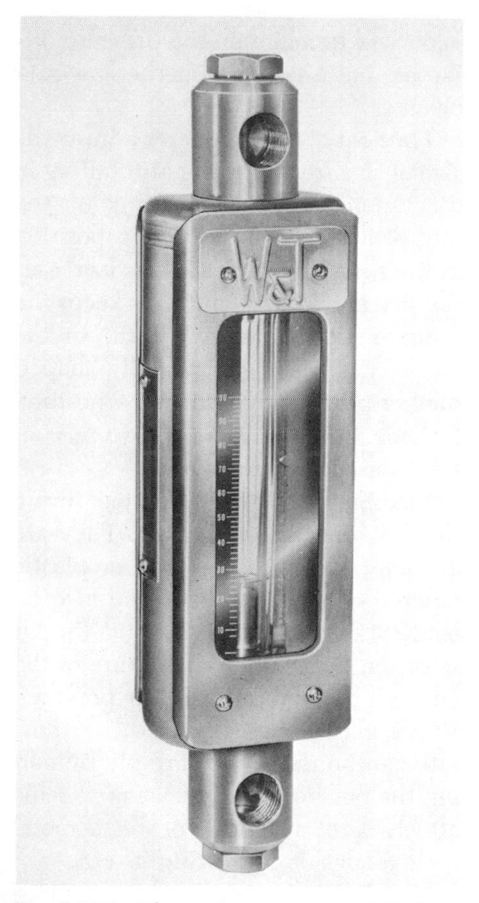

Fig. 5-11A. Glass tube rotameter. (Wallace & Tiernan, Inc.)

Fig. 5-11B. Metal tube rotameter. (Wallace & Tiernan, Inc.)

A weir is a flat bulkhead with a specially shaped notch along its upper edge. It is installed across the open fluid stream, forcing the fluid to rise up the notch as the flow rate increases. See Fig. 5-12. A flume is a formed structure installed in the open fluid stream, forcing the fluid to rise within it as the flow rate increases. See Fig. 5-13. The open nozzle is shaped so that the level of the fluid in the nozzle rises uniformly as the flow rate increases. See Fig. 5-14.

When weirs, flumes and open nozzles (the primary elements), are used the flow rate is measured using a level sensor placed in a still well which is adjacent to the flow channel. As the flow through the channel increases, the level rises simultaneously in both the channel and the still well. Hence the float position in the still well reflects

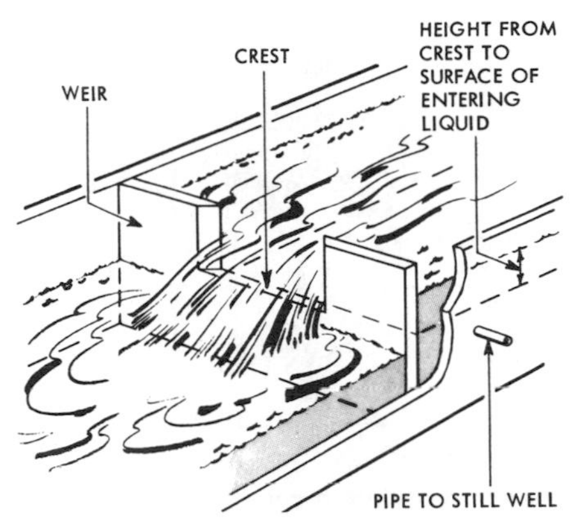

Fig. 5-12. The weir forces the fluid up the crest as flow rate increases.

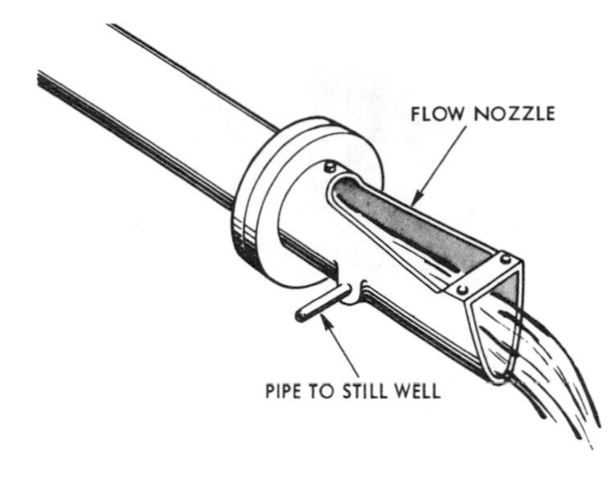

Fig. 5-14. An open nozzle used for open channel measurement.

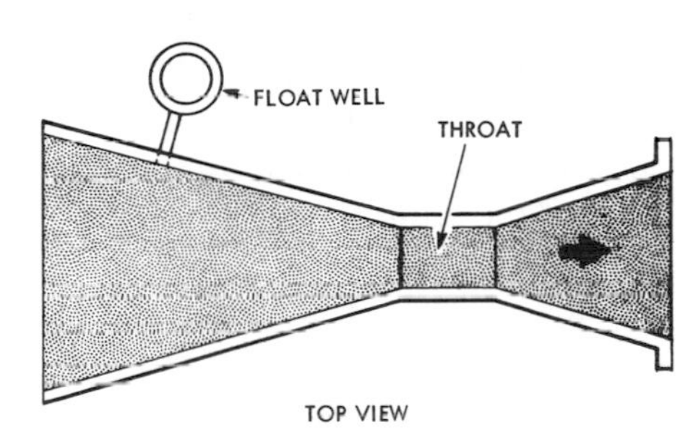

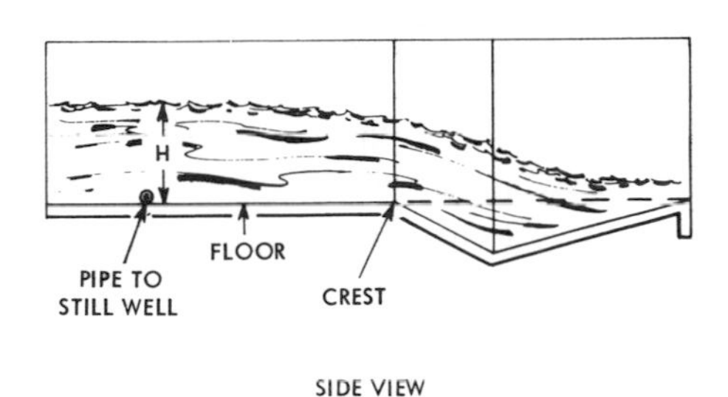

Fig. 5-13. Top view and side view of a flume.

the flow rate. The accompanying scale is calibrated in flow units (usually millions of gallons per day).

Total Flow Meters

Positive Displacement Meters. Meters operating by positive displacement admit fluid into a chamber of known volume and then discharge it. The number of times the chamber is filled during a given interval is counted. Two chambers are used. They are arranged so that as one is filling, the other is being emptied. This provides continuous flow. The total flow for a given interval can be determined by multiplying the number of times the chambers are filled by the volume of the chambers. The average flow rate is computed by dividing this total flow by the time units in the period.

One type of positive displacement flow meter is shown in Fig. 5-15A. The fluid moves through the chambers, causing the disc to rotate and wobble (nutate) as one chamber fills and the other is emptied. The movement resembles that of a slowly spinning top. The rotary motion is transferred to a gear assembly which drives the counter. As the disc nutates, the cham-

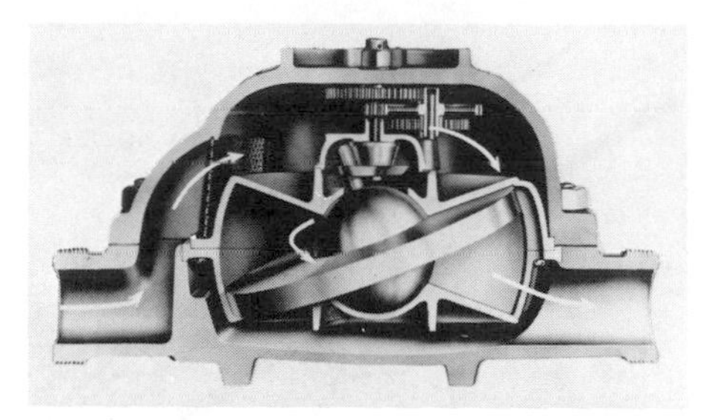

Fig. 5-15A. Nutating disc-type of positive displacement flow meter.

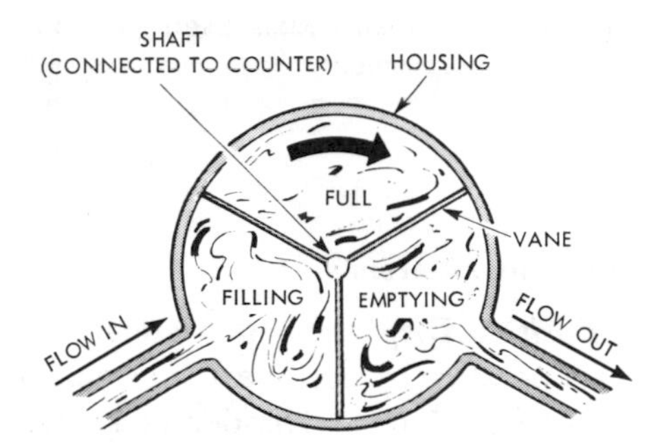

Fig. 5-15B. A positive displacement flow meter using three chambers.

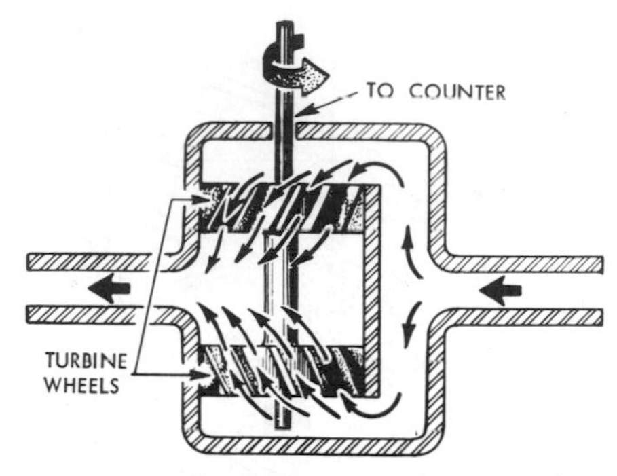

Fig. 5-16. A velocity meter with turbine wheels to drive the counter.

bers alternate functions. The inlet chamber becomes the outlet chamber, and vice versa.

Other positive displacement flow meters contain internal mechanisms similar to those found in pumps. These can have three chambers. See Fig. 5-15B. Among these are oscillating pistons, rotary vanes, and oval gears. Though they all differ in details of design, they operate in a similar fashion in that they are driven by the rate of flow and measure the quantity of fluid flow.

Velocity Meters. A velocity meter converts the velocity of a flowing liquid to a rotary shaft motion. The liquid enters the meter and drives a turbine wheel or propeller at a speed that varies with the flow rate. The turbine wheels drive a gear train which is connected to a counter. See Fig. 5-16. The counter then registers the total quantity of the liquid which has passed through the meter.

Some meters of this type contain secondary metering mechanisms and are called *compound meters.* For example, a positive displacement type mechanism might be included to provide better measurement of a low flow rate than would be possible with a simple velocity meter.

Integrators. Total flow can be measured with flow meters that incorporate integrators. An *integrator* is a calculating device that combines multiplication and addition. The changing flow rate is multiplied, and the product is added continuously to the previous totals on the counter. If differential pressures are used as input, a means of converting differential pressure to flow rate must be included in the device.

Such integrators can be mechanical or electrical, continuous or intermittent. An example of a continuous mechanical integrator is the ball and disc type shown in Fig. 5-17. The disc is rotated at a constant speed to provide the time input. Two steel balls, one on top of the other, are held in position by a carriage which moves radially across the face of the disc. The radial position of the carriage is determined by the flow rate. As the bottom ball moves from the center of the disc toward the outer edge, its speed increases, so that the roller driven by the ball rotates faster. The roller drives a counter which totals the flow by continuously multiplying the rate of flow by an increment of time.

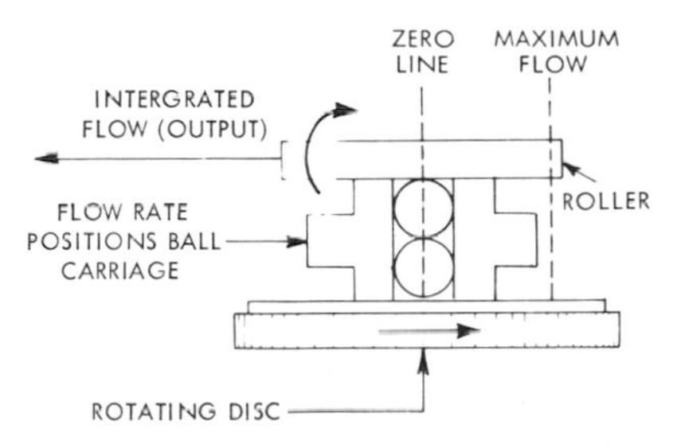

Fig. 5-17. A ball and disc-type integrator.

In flowmetering, when the flow rate is not a rapidly changing factor, the intermittent mechanical integrator can replace the continuous type. In this device, as shown in Fig. 5-18, a constant speed motor drives a counter for a period of time, depending on the duration of the flow. When the duration of the flow is longer, the motor operates for a longer period of time, and there is greater counter movement. Generally, such an integrator includes a mechanism that converts the measured differential pressure to flow rate. The counter is then driven by the product of flow rate and time and indicates total flow.

Continuous electrical flow integrators that use electrically-operated mechanical devices to regulate the rotation of the counter are available. The intermittent electrical flow integrators

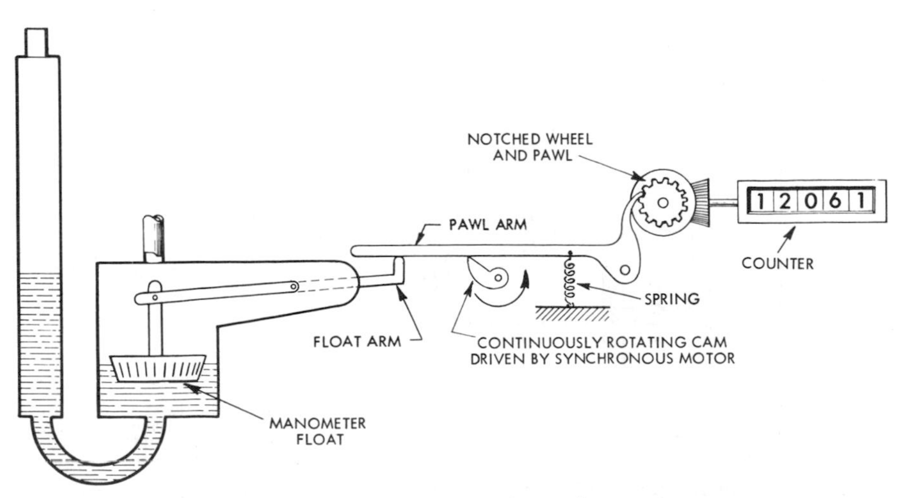

Fig. 5-18. An intermittent mechanical integrator.

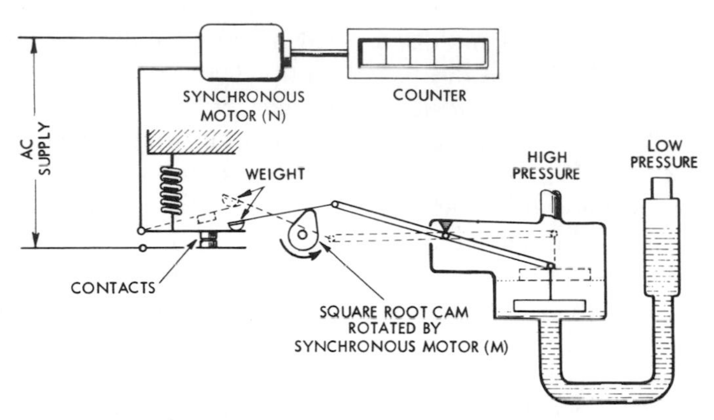

Fig. 5-19. An intermittent electrical integrator. Synchronous motor (N) drives the counter. Synchronous motor (M) drives square root cam. (The Foxboro Co.)

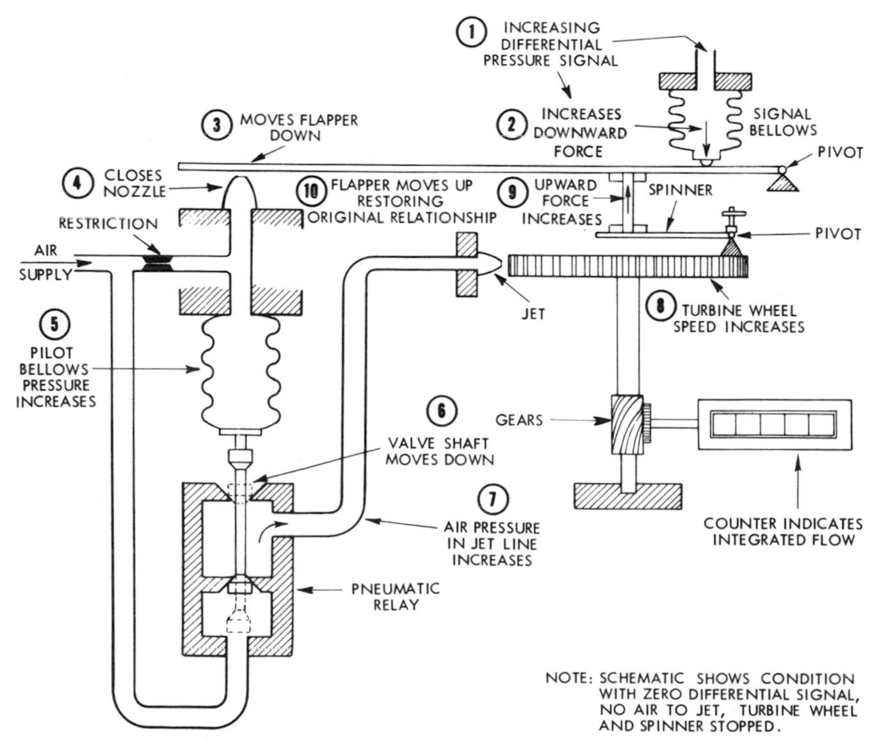

Fig. 5-20. A continuous pneumatic flow integrator. The numbers indicate sequence of events. (The Foxboro Co.)

contain mechanically-operated contacts to control the number of impulses to the counting mechanism. See Fig. 5-19.

Also available are pneumatic integrators, developed to satisfy a demand for non-electrical devices in applications where explosives or other hazards are present. The pneumatic integrator operates on the force-balance principle. A force is produced by the spinner as a result of the speed of its rotation. The flow rate causes a second force to be produced by the signal bellows. These two forces act on the flapper in opposition to one another. See Fig. 5-20. The amount of air that leaves the jet is determined by the position of the flapper. Depending on the quantity, the air causes the turbine wheel to move faster or slower. As the turbine wheel moves faster, the spinner applies a greater force on the flapper, and the counter increases speed. The relationship between the speed of the turbine wheel and the air pressure from the jet is such that the counter directly indicates flow rate. See Fig. 5-20.

Except for their source of power and the method of converting differential pressure measurement to flow rate, all integrators resemble one another and perform the same function. The principal requirements are an accurate measurement of the differential pressure, a near perfect conversion to flow rate, a constant time input, and an easy-to-read counter.

Words to Know

flow rate
total flow
orifice plate
venturi tube
rotameter
weir
flume
flow nozzle
open nozzle
positive displacement
velocity meter
compound meter
integrator

Review Questions

1. What are the two kinds of flow measurement? What is the difference between them?
2. What are the three primary flow elements used with differential pressure flowmeters?
3. Which of the primary elements in Question 2 is described as concentric, eccentric, or segmental?
4. What is the relationship between *flow rate* and *differential pressure?*
5. Why is the rotameter known as a variable area flowmeter?
6. What are the three primary elements used for measuring flow rate in open channels? How is flow rate measured with each of these elements?

7. How is the average flow rate measured when using a positive displacement type flowmeter?
8. What is the function of the turbine wheel in a velocity flowmeter?
9. A flow integrator is a calculator. What calculation does it perform?
10. What is meant by the terms *intermittent flow integrator* and *continuous flow integrator?*

Humidity

Chapter

6

Humidity is the measure of the amount of water vapor in a given volume of air, or other gases. The terms used to describe humidity are *absolute humidity* and *relative humidity*. Absolute humidity is the actual amount of water present in a specified volume of air. It is expressed in grains of water per cubic foot of air, or grams per cubic centimeter. (1 grain equals 1/7000 of a pound.) Relative humidity is the ratio of the actual amount of water vapor in the air to the maximum amount the air would contain at the same temperature, if saturated. Relative humidity is also defined as the ratio of the actual water vapor pressure in the air to the water vapor pressure in saturated air (containing the maximum amount) at the same temperature. Humidity expressed as 100% relative humidity indicates the air is saturated.

Another measure of humidity is *specific humidity*. Specific humidity is the ratio of the mass of water vapor to the mass of dry air and moisture. It is expressed as grains of water vapor per pound of the mixture of dry air and water vapor.

Fig. 6-1 shows the amount of water vapor contained in air at various temperatures, when at atmospheric pressure (29.97 inches of mercury). Fig. 6-1 also illustrates the *psychrometric chart*, which is the key to the measurement of humidity. The psychrometric chart graphically describes the properties of moist air. Fig. 6-1 is an example of how the variables for humidity measurement are plotted on a typical psychrometric chart.

On the typical psychrometric chart, the dry bulb temperature lines are vertical, and the dry bulb temperatures, in degrees Fahrenheit, are indicated at the bottom of the chart. The wet bulb and dew point temperature lines run diagonally downward toward the right. The temperature values are read at the point where these lines intersect with the 100 percent relative humidity line. The lines indicating the percent of relative humidity curve upwards to the right, and the percentage values are shown directly on the lines themselves. The absolute humidity, measured in grains per pound of air, is read on the vertical scale at the right

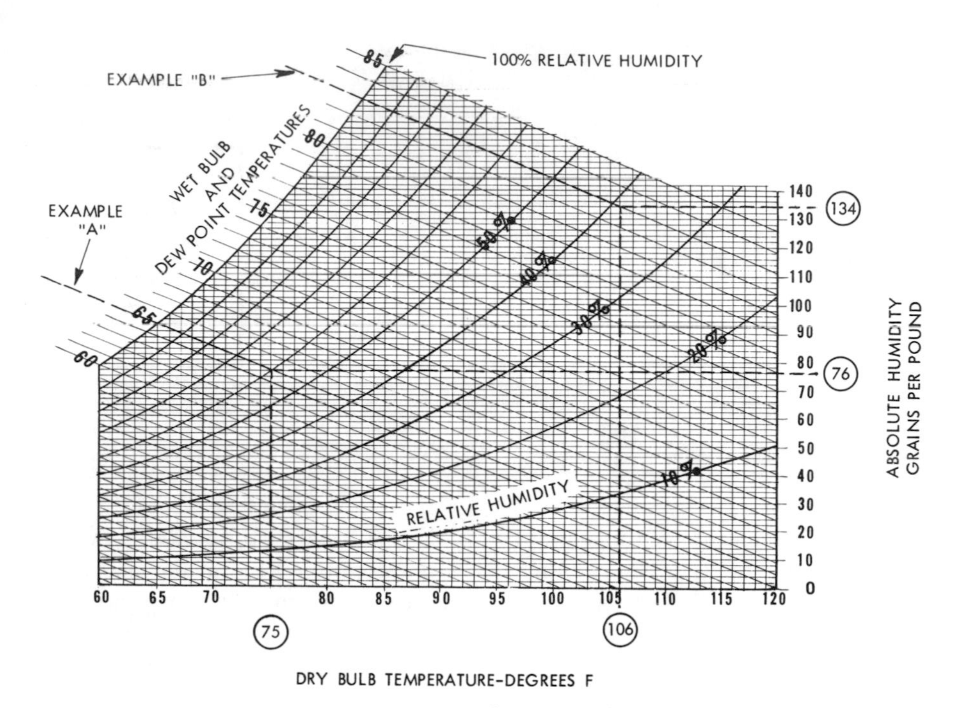

Fig. 6-1. A psychrometric chart.

of the chart, using the horizontal line that leads from the intersection of a wet bulb (or dew point) temperature line with a dry bulb line. (The terms *wet bulb* and *dry bulb* will be explained later in this chapter.)

Several examples can show how this somewhat complex chart reads in actual usage.

Example A: To determine both relative humidity and absolute humidity, given a dry bulb temperature of 75°F and a wet bulb temperature of 65°F, first locate the 75°F dry bulb temperature reading at the bottom of the chart. Then locate the 65°F wet bulb reading along the 100 percent relative humidity line. The relative humidity is the point where these two lines intersect. In this example, the relative humidity is approximately 60 percent. A horizontal line, drawn from the point of the intersection to the right hand scale of the chart, shows the absolute humidity. In this case, it is 76 grains per pound.

Example B: To determine both relative humidity and absolute humidity, given a dry bulb temperature of 106°F and a dew point temperature of 83°F, use the same procedure as in Example A above. The relative humidity is 39 percent, and the absolute humidity is 134 grains per pound.

See Fig. 6-1, on which special guidelines have been drawn to aid your more complete understanding of this procedure.

Measuring Relative Humidity

Relative humidity is measured with either a *hygrometer* or a *psychrometer.* The hygrometer works on the principle of certain physical or electrical changes which occur in various materials as they absorb moisture. The psychrometer (we have already discussed the *psychrometric chart*) registers the temperature difference between two primary elements, one of which is kept moist so that water continuously evaporates from its surface.

Hygrometers. Hygrometers which depend on physical changes employ human hair, animal membrane, or other materials that lengthen or stretch when they absorb water. Fig. 6-2 shows a typical recording hygrometer.

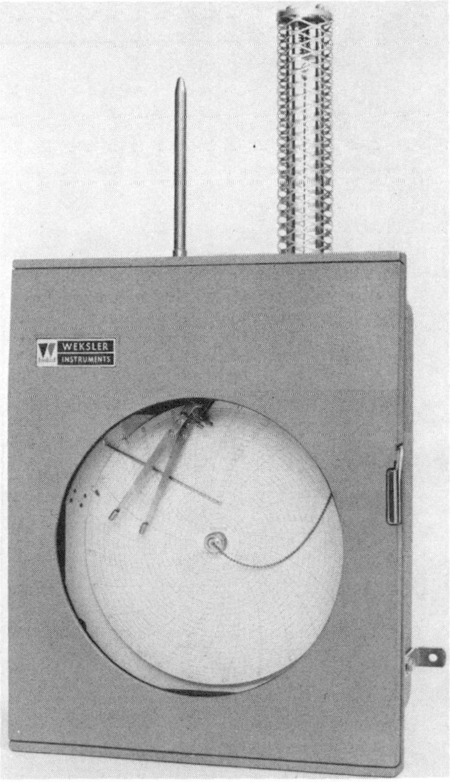

Fig. 6-2. A typical recording hygrometer. (Weksler Instruments)

Electrical hygrometers employ transducers to convert humidity variations into electrical resistance changes. The type of transducer shown in Fig. 6-3 consists of an insulating plate bearing two interleaved grids coated with *hygroscopic material.* (Hygroscopic materials readily take up and retain moisture.) The electrical resistance of the hygroscopic material depends on the amount of water it absorbs from the air at any given moment. The transducer can be used with a bridge-type instrument which measures resistance changes and indicates them on a scale calibrated in percent of relative humidity.

Another type of electrical hygrometer uses a capacitance probe as its primary element. See Fig. 6-4A. An anodized aluminum strip provides a por-

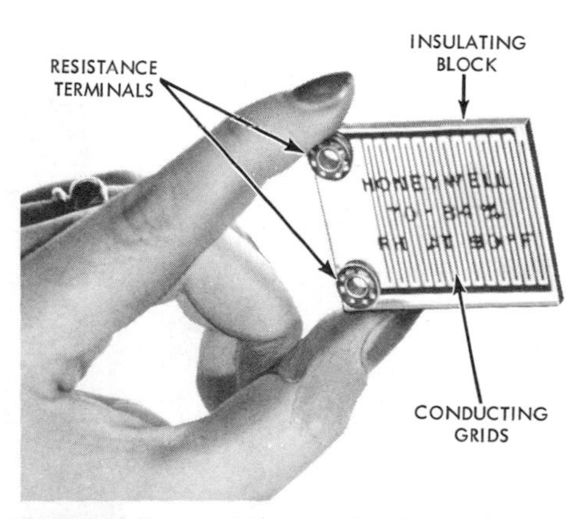

Fig. 6-3. A transducer used with an electrical hygrometer. (Honeywell Process Control Div./Fort Washington, Pa.)

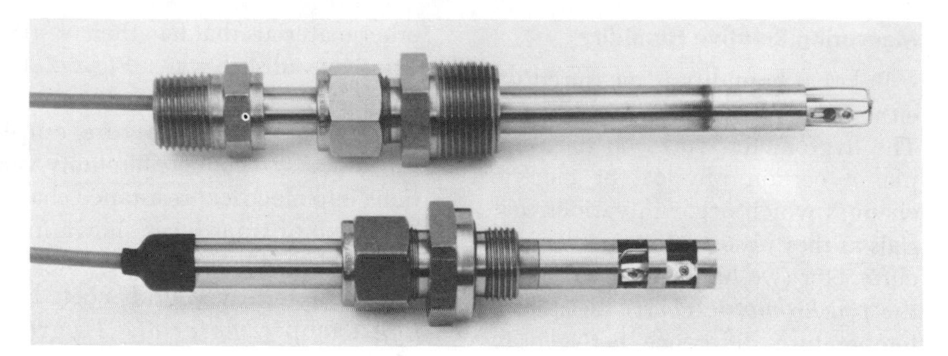

Fig. 6-4A. Various capacitance probes used with an electrical type hygrometer. (VEEKAY Limited)

Fig. 6-4B. Interior view of an electrical hygrometer, showing the electrical circuit. (VEEKAY Limited)

Fig. 6-4C. Electrical hygrometer with digital readout. (VEEKAY Limited)

ous aluminum oxide layer. A thin film of gold is deposited on the oxide layer. The gold and the aluminum serve as the plates, and the oxide layer as the dielectric of an electrical capacitor. When the probe is installed within the environment of a process containing water vapor, molecules of water condense and collect on the surface of the aluminum oxide layer. This alters the electrical capacitance of the probe. The amount of water collected varies with the humidity of the process environment. The instrument can be calibrated for digital readout in relative humidity units. Fig. 6-4B, on the previous page, shows the electric circuit used with the hygrometer probe. Fig. 6-4C is an electrical hygrometer with digital readout.

Psychrometers. The psychrometer, as already discussed, is an instrument for determining atmospheric humidity, using two thermometers. The bulb of one thermometer is kept moist (wet bulb), and the other is dry. Several types of psychrometers are available. Two of the most common are the sling psychrometer and the recording psychrometer.

The simple sling psychrometer, shown in Fig. 6-5, consists of two glass thermometers attached to an assembly that permits the two thermometers to be rotated through the air. One thermometer has a moistened cloth cover (or wick) over the sensing bulb to maintain a moist condition. The thermometers are rotated. The wet bulb will show a lower temperature than the dry bulb because of the evaporation of water from the wick. These temperature readings, along with a psychrometric chart, are used to determine the relative or absolute humidity of the specific environment.

Recording psychrometers use the same principle as the sling psychrometer. Two separate pressure spring thermal units or two separate resistance

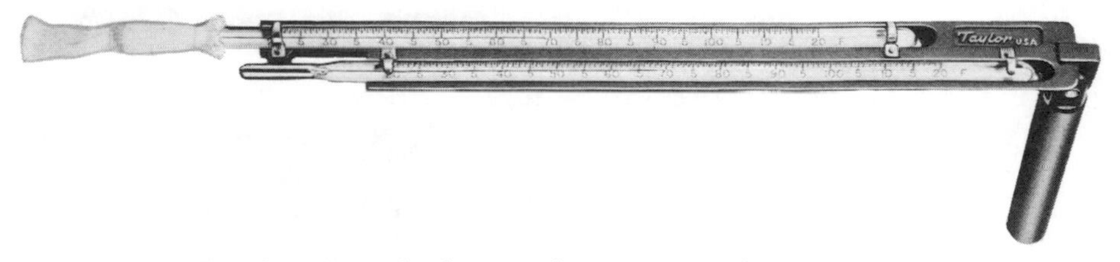

Fig. 6-5. A simple sling psychrometer. (Taylor Instruments)

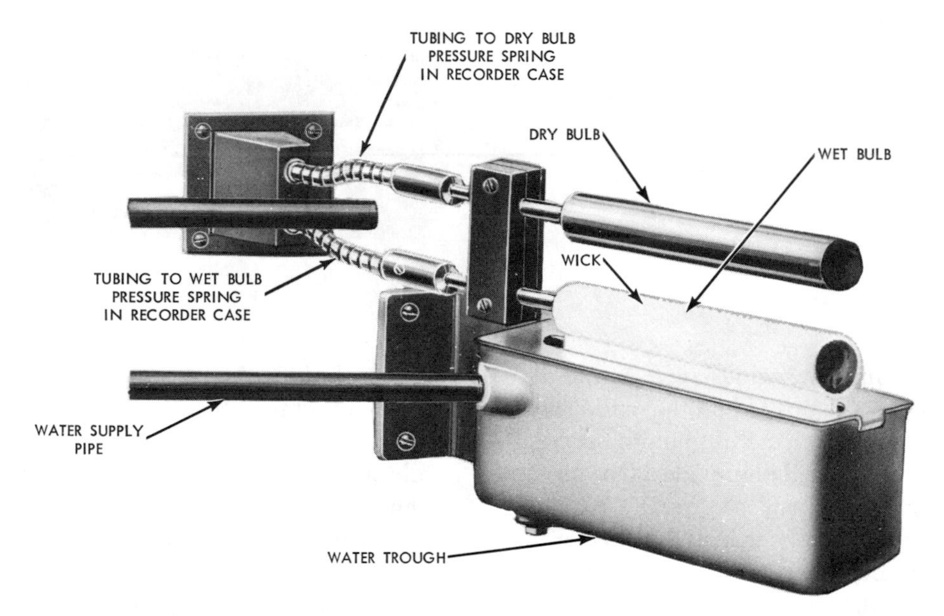

Fig. 6-6. Typical pressure-spring psychrometer, showing the installation of the wet bulb and dry bulb elements.

thermometers bulbs are used. Fig. 6-6 shows a typical pressure spring psychrometer. Psychrometers are generally used in process applications for determining the humidity of air at temperatures between 32°F and 212°F.

Measuring Dew Point

In many industrial processes, the *dew point* is a more significant measurement than relative humidity. The dew point is the temperature at which a given sample of moist air is fully saturated and begins to deposit dew.

The typical instrument for measuring dew point is shown in Fig. 6-7. It uses a gold-plated mirror surface which is bonded to a copper thermistor holder. This assembly is chilled by a Peltier effect thermoelectric cooler. (The Peltier effect is discussed in

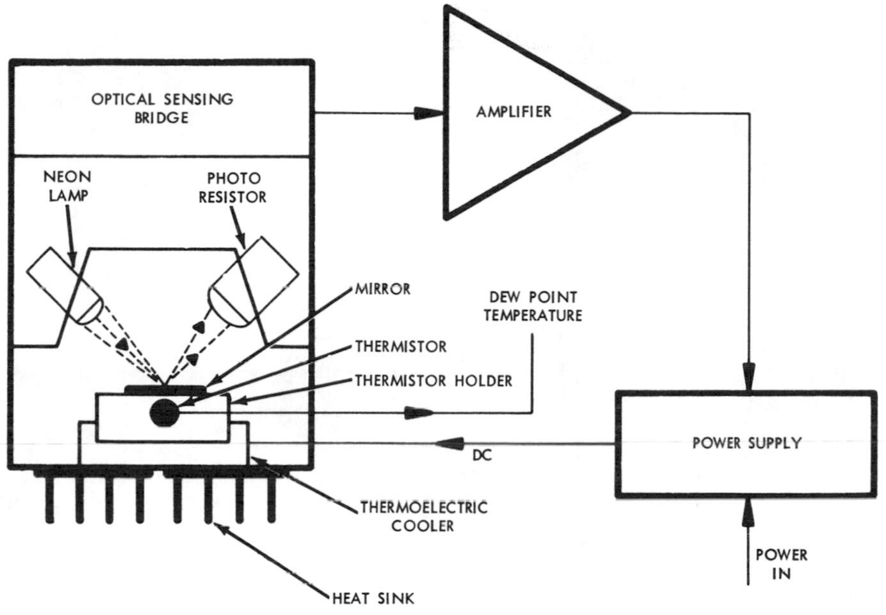

Fig. 6-7. A dew point hygrometer using a system of primary measurement. (EG&G, Environmental Equipment Div.)

Chapter 10.) The air or other gas being measured for dew point is passed by the mirror. A neon lamp is beamed on the mirror which reflects the beam toward a photoelectric resistor. As dew forms on the mirror and clouds it, there is a change in the amount of light reflected. This change is detected by an optical sensing bridge. The result is a change in the electrical resistance which becomes the input for an amplifier. This effects a proportional change in the power supplied to the cooler. The thermistor in the cooling assembly measures the dew point temperature. The heat sink serves as a thermal capacitance.

Fig. 6-8 shows a continuous dew point recorder. This instrument uses a specially constructed sensing element, consisting of a thin-walled metal tube coated with insulating varnish, and wrapped with glass fibre cloth. See the left view of Fig. 6-8. This assembly is then covered with the double winding of gold wire shown in the center view. A solution of lithium chloride is used to saturate the glass cloth. Lithium chloride is a hygroscopic chemical salt which absorbs moisture from the air surrounding it. As the moisture absorbed by the lithium chloride increases, the current passing through the gold wire increases. The increase of current through the wire creates heat which passes through the tube wall to the temperature sensing bulb. The sensing bulb can be a type of resistance thermometer or a filled system. The temperature of the cell is the dew

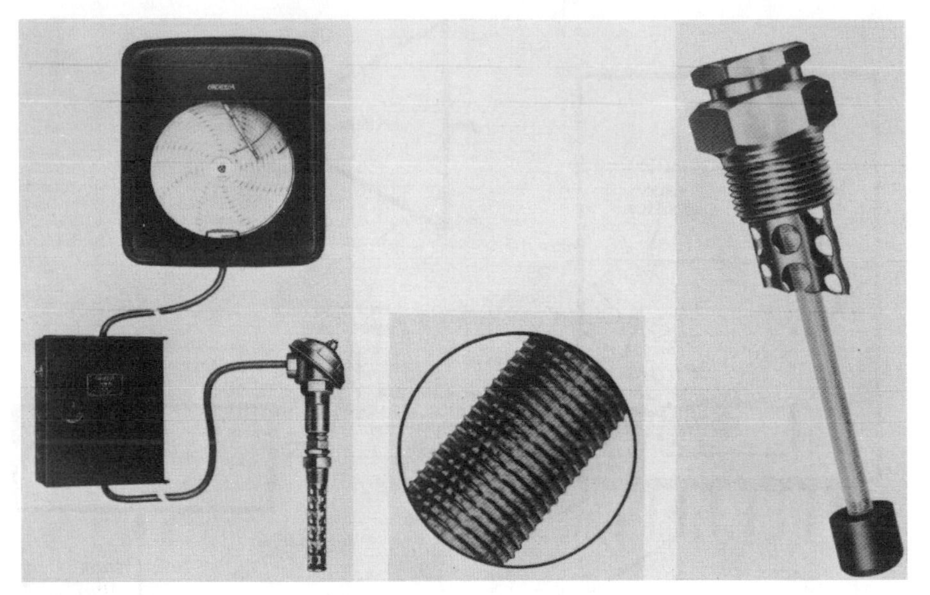

Fig. 6-8. A continuous dew point recorder, showing, (left) the hygroscopic element, (center) a close-up of the wire, and (right) the complete unit. (The Foxboro Co.)

point. The right view of Fig. 6-8 shows the complete continuous dew point recording system.

Measuring Moisture

The continuous measurement of moisture content is crucial to the mass production of strip materials, such as paper and textiles, and granular materials, such as chemicals, plastics, grains, powders, and minerals. Strip materials are processed as continuous webs. Granular materials are carried through the process in pipes, on conveyor belts, or through chutes and hoppers. Determining the moisture content of materials of these types requires a non-contact detection method. Samples can be taken and analyzed in a laboratory. But two methods more suited to continuous high speed processing are based on the use of microwaves (radar waves) and infrared waves.

Microwaves. One of the unique characteristics of water (moisture) is its ability to absorb microwaves. As they are absorbed by the moisture, the microwaves alter in amplitude and phase. The measurement of these variations in the microwaves is the method used to determine the number of water molecules present in the processed material, or its moisture content.

The microwave moisture gage, shown in Fig. 6-9, consists of a transmitter, a receiver, and two oval horns. The horns are used for directivity. Using the principle of microwave transmission through the process material, the sensors are mounted directly on

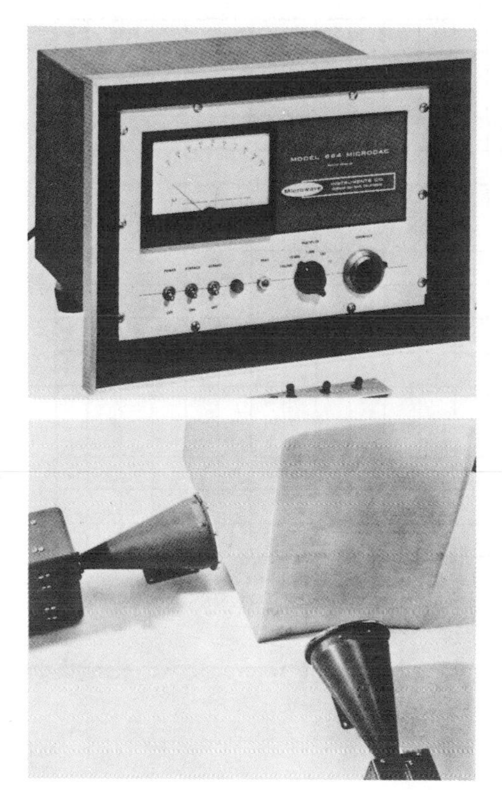

Fig. 6-9. A microwave system for detecting moisture content. Top view shows the transmitter-receiver and the bottom view shows the oval horns used for detecting moisture content. (Microwave Instruments Co.)

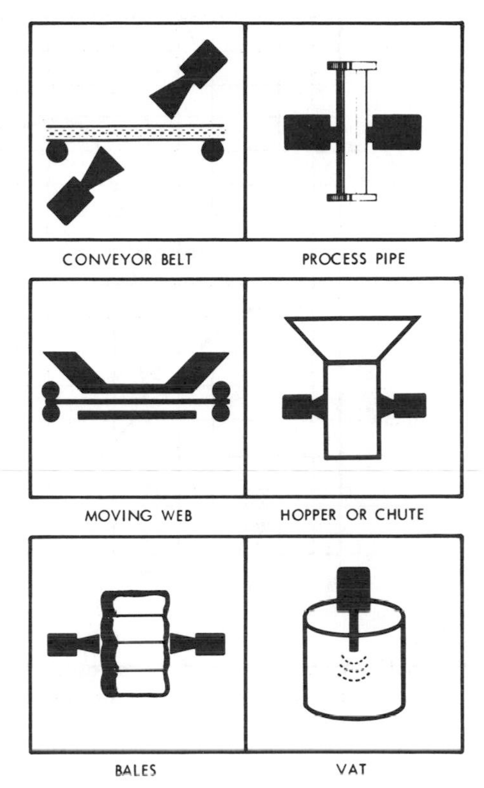

Fig. 6-10. The positioning of the detector horns for various applications. (Microwave Instruments Co.)

webs, pipes, chutes, belts, tanks, etc. The method of mounting and the specific type of unit selected depend on the requirements of the application. See Fig. 6-10. In all units, the transmitter produces the microwaves, and the receiver measures the change in their amplitude and phase after passing through the process materials. The amount of change detects the moisture content which is recorded on a dial indicator. For strip materials it is necessary to scan the width of the strip to determine the over-all moisture content.

Infrared Waves. Measuring moisture content with infrared waves is based on the *reflectance* of the process material. Reflectance is the ability of a material to absorb and scatter such radiation. This property of the material depends on its chemical composition, including its moisture content. The moisture analyzer, an instrument used to detect moisture content, mea-

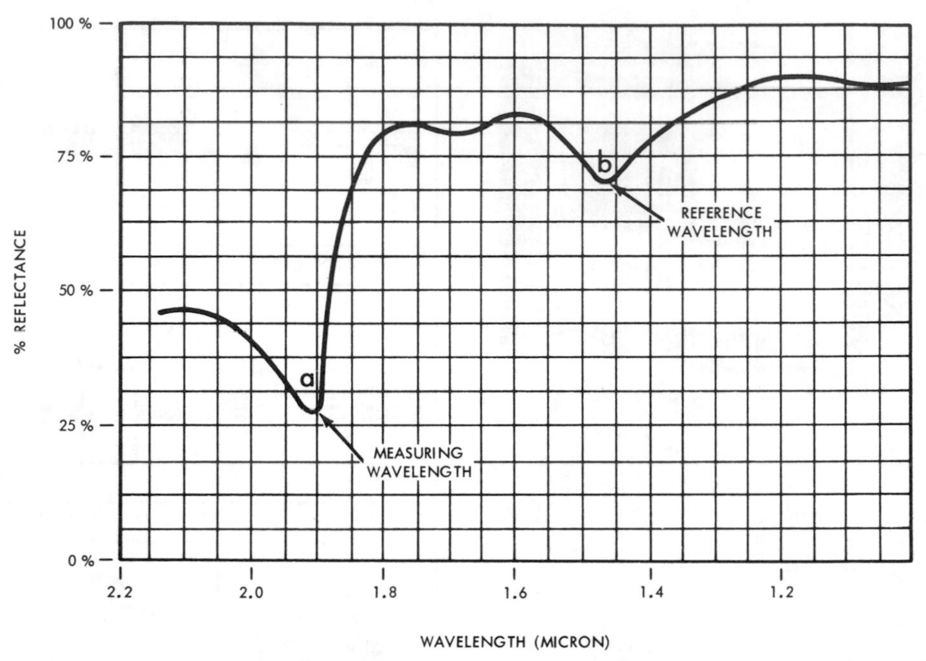

Fig. 6-11. A comparison of the measuring wavelength and the reference wavelength in a system using infrared waves. (Anacon, Inc.)

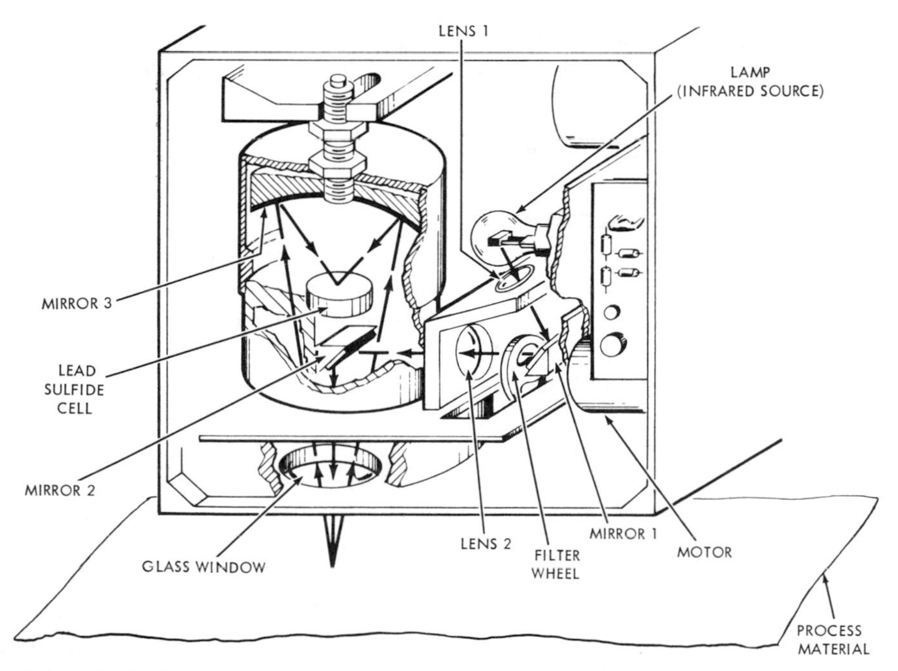

Fig. 6-12. Method of detecting moisture content using infrared waves. (Anacon, Inc.)

sures changes in the material's reflectance. Any other factors, such as surface variations, which affect reflectance must be compensated for .This is accomplished by selecting two wavelengths, a measuring wavelength and a reference wavelength, and comparing the amount of energy reflected by the process material for each wavelength. See Fig. 6-11. The changes due to secondary factors are cancelled by using the ratio of the two reflected signals. Because of the wavelength selected, the analyzer responds only to the changes in moisture content. The filter wheel, driven by a synchronous motor, carries two optical filters. One of the filters transmits light at the measuring wavelength, and the other at the reference wavelength.

The infrared waves are projected on to the surface of the process material indirectly, using a system of lenses and mirrors. See Fig. 6-12. The function of the lenses is to provide a uniform distribution of the infrared waves on the mirrors. The energy reflected from the process material is directed by a concave mirror (Mirror #3) to the lead sulfide cell, as shown in Fig. 6-12. The cell converts the radiant energy into electrical energy. A signal is produced in the cell by alternating the measuring and reference wavelengths. The amplitude of one series indicates the reflectance at the measuring wavelength, and the amplitude of the second series indicates the reflectance at the reference wavelength. These pulses are separated and converted into two dc currents, each corresponding to the amplitude of one of the series. The difference between the two currents, indicated on a meter or a remote recorder, is proportional to the moisture content.

Words to Know

absolute humidity
relative humidity
specific humidity
psychrometric chart
dry bulb
wet bulb
dew point
hygrometer
psychrometer
hygroscopic materials
reflectance

Review Questions

1. What two terms are used to describe humidity measurement? How do these measurements differ?
2. Given the wet and dry bulb temperatures, what two measurements can be obtained from a psychrometric chart?

3. How does a *hygrometer* differ from a *psychrometer*?
4. Describe a sling psychrometer.
5. What is the operating temperature range of a psychrometer?
6. What is the function of the hygroscopic coating used on a dew point element?
7. Name some processes which require moisture measurement.
8. What unique characteristic of water makes possible the microwave measurement of moisture?
9. What is the advantage of measuring moisture content at two different infrared wavelengths?

Transmission

Chapter

7

The transmission of a measured variable from the point of measurement to a remote point is an important function in process instrumentation because of the size and complexity of modern industrial plants. Temperature, pressure, level, and flow are the variables most commonly transmitted. The measuring elements can be pressure springs, thermocouples, bellows, floats, or any of the other elements we have discussed in previous chapters. Fig. 7-1 shows graphically how the transmission of process variables takes place.

Transmitting transducers convert the measured variables into a signal (pneumatic or electrical in this discussion). This enables the measured variable to be received by a remote indicating, recording, or controlling device. When the signal is transmitted to the receiver, it is reconverted to a measurement which is indicated on a gage or recorded on a chart. An elec-

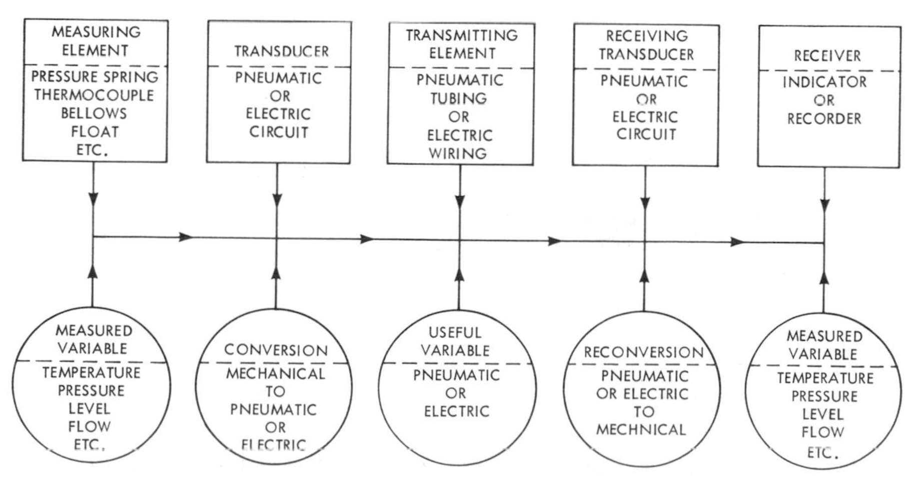

Fig. 7-1. The above diagram shows the transmission of process variables.

trical signal, for example, can be converted to mechanical energy which positions a pointer or a recording pen. The selection of either a pneumatic or an electrical transmitter in process instrumentation, depends on the nature of the variable and the distance the signal must travel.

Pneumatic Transmission

The first requirement for pneumatic transmission is a constant, steady air supply. The most common supply pressures are 20 psi and 35 psi. A means of changing the air pressure is needed so that a definite pneumatic pressure signal is produced for each value of the measured variable. The signal pressure varies between 3 psi and 15 psi when the air supply pressure is 20 psi. When the air supply pressure is 35 psi, the variance is between 6 psi and 30 psi. See Fig. 7-2. The receiver converts the signal pressure into a value of the measured variable. Basically, the receiver is an indicating or recording pressure gage with its scale or chart graduated in units of the measured variable. Pneumatic transmission is limited to distances of 600 feet or less.

Many models of pneumatic transmitters are available. They vary in details, but many of them operate on the flapper-nozzle principle shown in Fig. 7-3. The main components of this mechanism are the nozzle, supplied

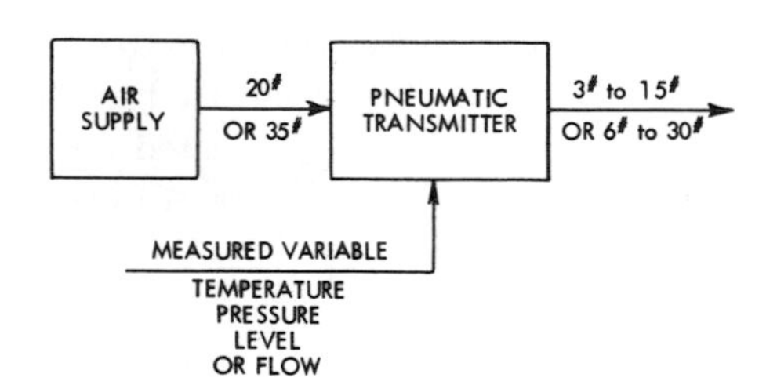

Fig. 7-2. Diagram showing the proper air supply required for pneumatic transmission.

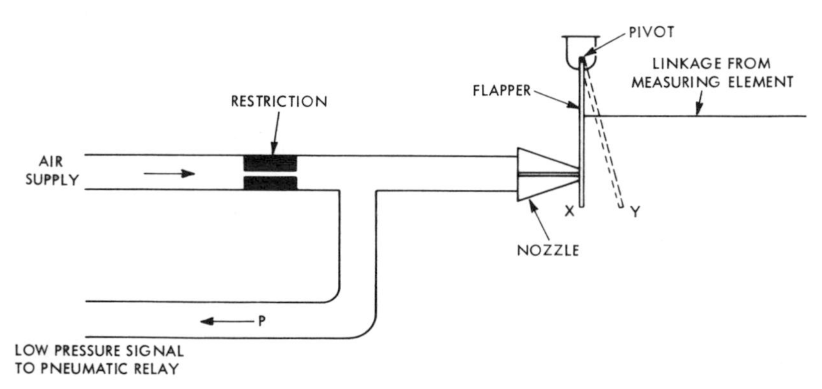

Fig. 7-3. Flapper-nozzle pneumatic transmitter. When the flapper is at X, low pressure (P) increases. When the flapper is at Y, low pressure decreases.

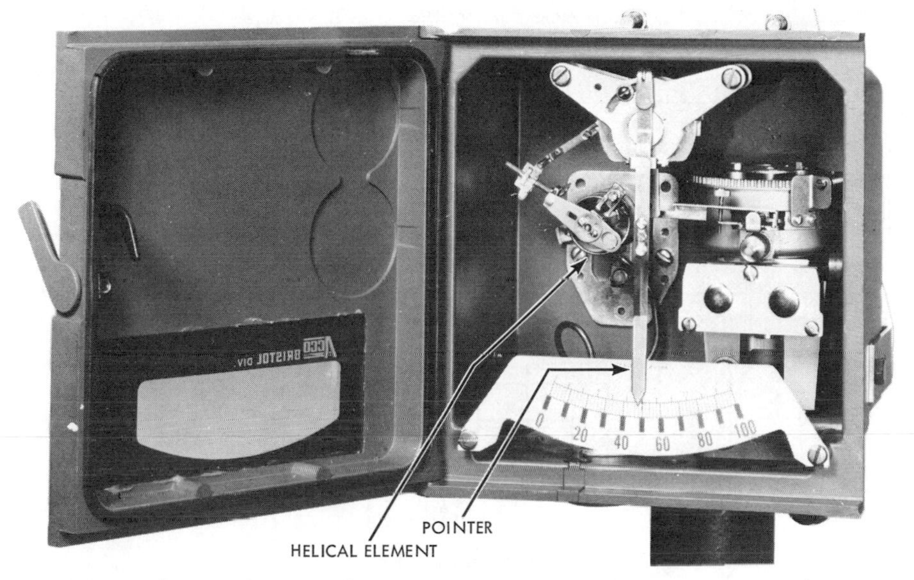

Fig. 7-4. Internal view of a typical pneumatic transmitter. The transmitter and pointer are directly linked to the helical element. (Bristol Division of Acco)

with air through a restriction, and the flapper positioned by the measuring element. See Fig. 7-3.

When the flapper is positioned against the nozzle, air cannot escape, and maximum air pressure passes to the amplifier. If the flapper is positioned away from the nozzle, air escapes, reducing the amount of air pressure to the amplifier. The flapper positions control the amplifier. The amplifier produces an air pressure proportional to the measured variable. It provides a signal strength sufficient for transmission over the required distance. Fig. 7-4 shows a typical pneumatic transmitter.

Electrical Transmission

There are two classes of electrical transmitters. One is used for applications in which the transmitted signal is not required to travel more than 1500 feet. They are frequently used in industrial process plants. The second is primarily designed for applications which require the transmitted signal to travel distances greater than 1500 feet. Examples are gas and oil pipelines and water and sewage systems. This long distance transmission is called *telemetering*. Telemetering is discussed below. In transmitters designed for industrial process plants, the transmission of the measured variable is a function of electric current or voltage.

Fig. 7-5 is a diagram of a simple current transmitter used to transmit pressure. The instrument represented by the diagram employs a pressure spring sensor. Whenever the pressure changes, the pressure spring moves the

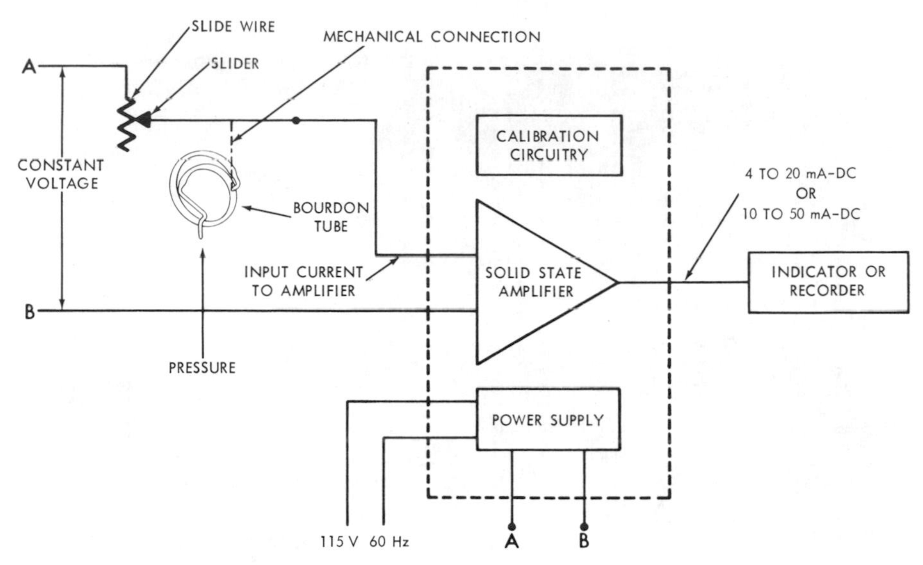

Fig. 7-5. A simple current transmitter.

slider along the slidewire, altering resistance in proportion to the pressure. If a constant dc voltage is supplied to the slidewire, the current changes according to Ohm's Law. (Current $(I) = \frac{\text{Voltage } (E)}{\text{Resistance } (R)}$ is a statement of Ohm's Law.) The current is supplied to an electronic amplifier that converts it to an electric current value in one of the standard ranges (4 to 20 milliamperes or 10 to 50 milliamperes). A solid-state amplifier has several advantages. It can permit isolation of the output from the input signal. This eliminates the interference in the input signal, and the interference is not transmitted as part of the measured variable. The calibration circuitry is located in the amplifier. The solid-state amplifier also provides a constant voltage. Similar amplifiers are available. They are used with differential pressure sensors, resistance thermometers, thermocouples, and other process measuring devices. Reference junction compensation circuitry is provided in the amplifier when thermocouples are used.

As shown in the diagram, Fig. 7-6, the voltage transmitter is similar to the electric current-type transmitter shown in Fig. 7-5. The only difference between the two is that the output of the amplifier in the voltage transmitter can be a voltage or a current signal. The standard output voltage ranges are 0 to 5 volts, or 0 to 10 volts. Fig. 7-7 is a photograph of a solid-state voltage transmitter.

There are current-to-current, current-to-voltage, voltage-to-voltage, and voltage-to-current transmitters. Selection depends on the output of the measuring transducer and the type of signal required by the indicating or recording instrument.

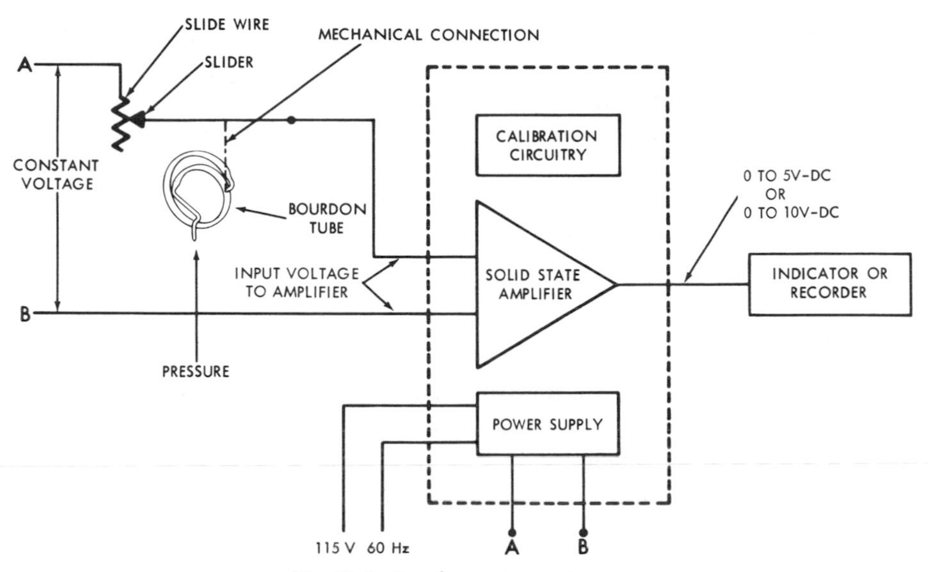

Fig. 7-6. A voltage transmitter.

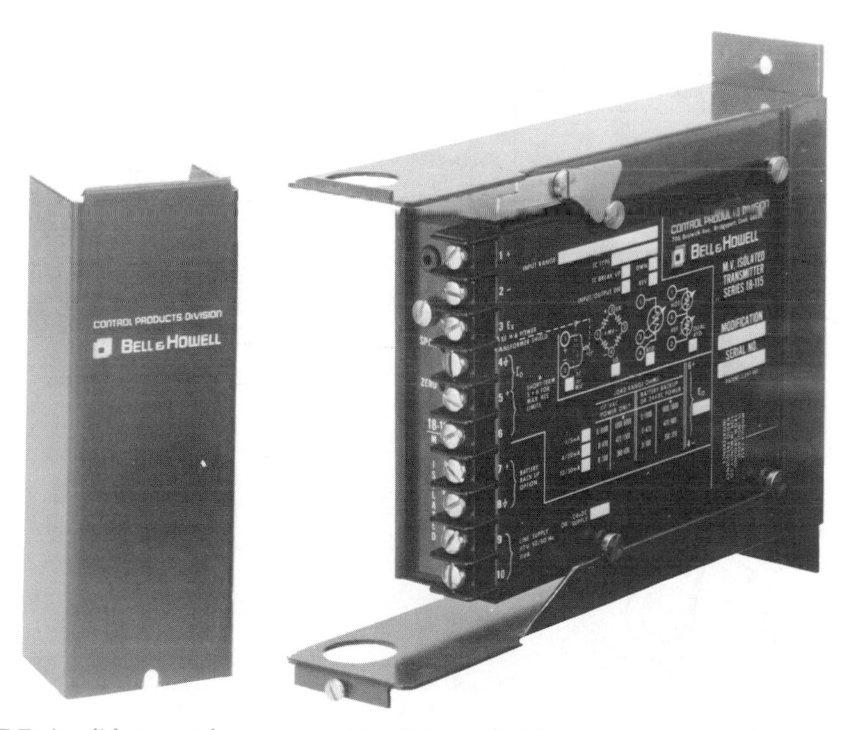

Fig. 7-7. A solid-state voltage transmitter. It is used with a strain gage, a thermocouple, or other millivolt source. (Bell & Howell)

Telemetering

The measurement in telemetering is transmitted as a pulse, a frequency, or a tone.

Pulse Telemetering. When pulse telemetering is used, the measurement is transmitted in terms of time rather than the magnitude of any electrical value. In the simple pulse duration pressure transmitter shown in Fig. 7-8, the cam follower (rider) is positioned by the Bourdon pressure spring. The cam itself, on which the follower rides, is driven by a constant speed (synchronous) motor. The switch is closed or open, depending on whether or not the follower is resting against the cam. When the follower is resting against the cam, the switch is open. The follower is against the cam for the shortest time when there is zero pressure from the Bourdon pressure spring. At maximum pressure, the follower is against the cam the greatest interval of time. The electrical pulse time increases proportionately as the pressure increases. This pulse is transmitted through telephone type wires to a receiver. The receiver contains an electromagnet which is activated by the pulse from the transmitter. The electromagnet armature acts on a differential gear assembly to position a pointer on the scale which is calibrated in units of pressure.

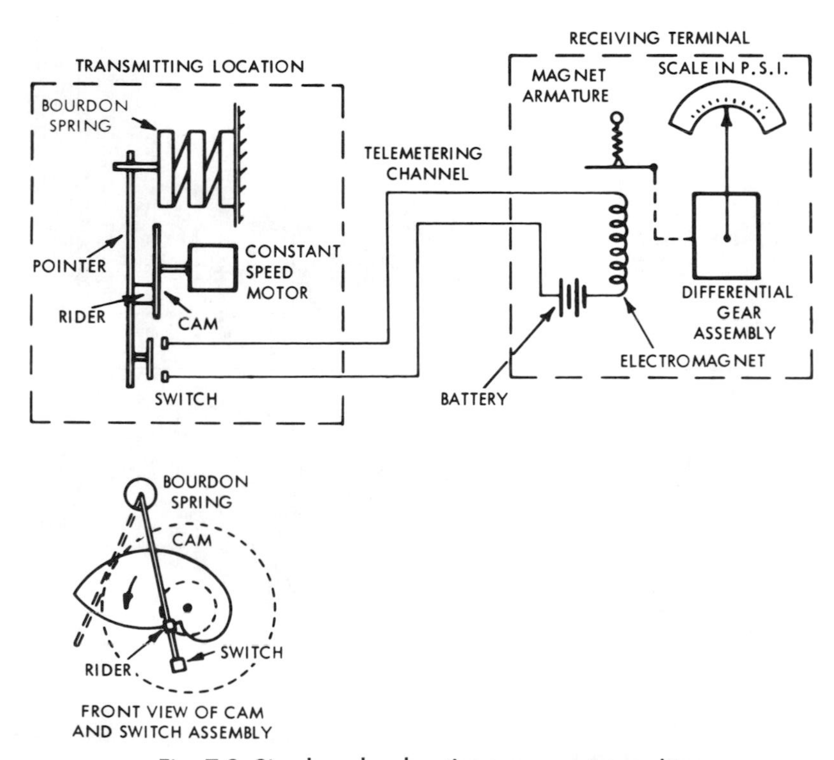

Fig. 7-8. Simple pulse duration pressure transmitter.

Frequency (or tone) telemetering. In a frequency or tone telemetering system, the measurement is transmitted as a function of electrical frequency. The transmitter uses any of the standard current or voltage signals from transducers as input. The signals are converted to frequencies that are proportional to the value of the signals. 9 to 15 Hz and 18 to 30 Hz are typical frequency ranges. (Hz stands for hertz.) The frequency signals are transmitted by telephone type wire to a receiving element. At the receiving element the frequency signals are reconverted to current signals. The current signal is the input for a recorder or an indicator.

The transmitted signal in tone telemetering is a pure, audible tone. The

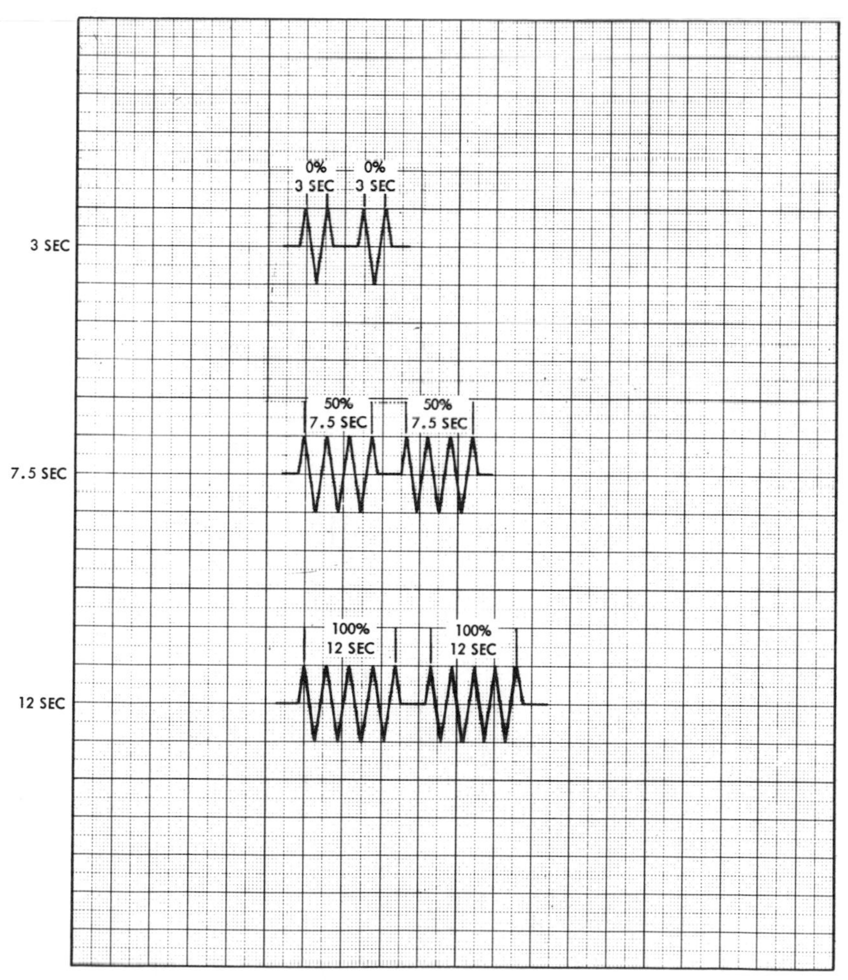

Fig. 7-9A. Transmitted signal in tone telemetering shown graphically.

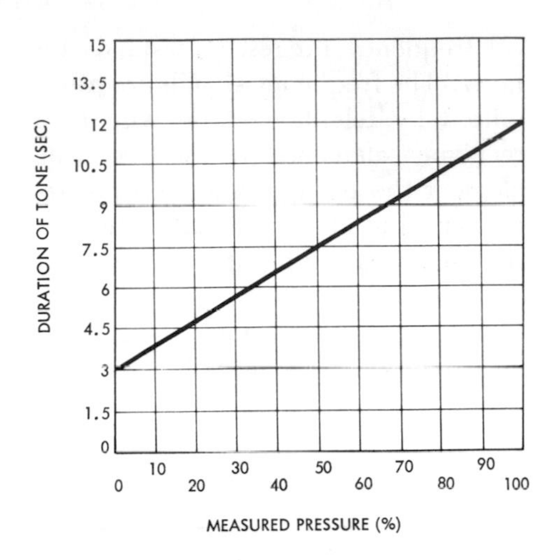

Fig. 7-9B. The relationship of the measured pressure and the duration of the tone.

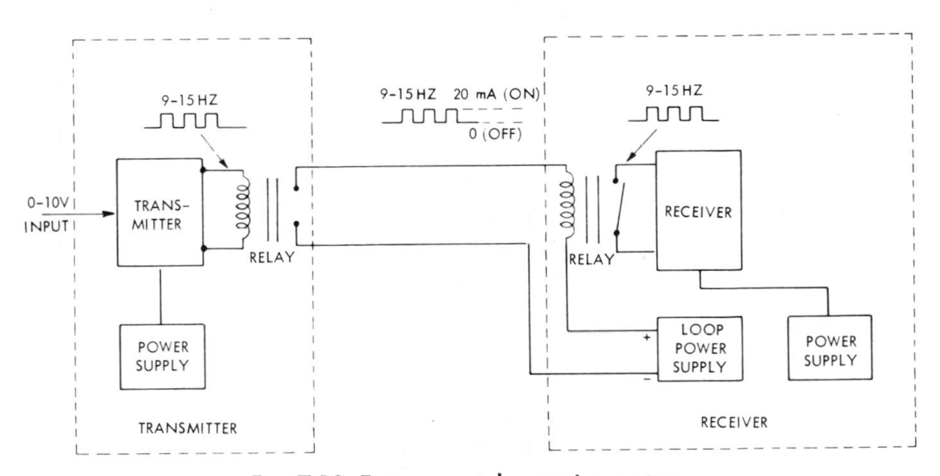

Fig. 7-10. Frequency telemetering system.

duration of the tone is proportional to the value that is being measured. See Figs. 7-9A and 7-9B. The tone signals can be transmitted on telephone type wire. They can also be transmitted by microwave or coaxial cables, similar to the cables used for closed circuit television. Fig. 7-10 shows a typical frequency telemetering system used for long distances.

Words to Know

telemetering   Ohm's law   psi

Review Questions

1. What is a transmitting transducer?
2. Name the three required components of a pneumatic transmission system.
3. Why is a restriction placed in a pneumatic transmitter?
4. What is the function of the sensing and the amplifier section of a typical pneumatic transmitter.
5. List the electrical characteristics which are used for transmitting such measurements as temperature, pressure, level, and flow.
6. Is a resistance temperature bulb a transducer?
7. What are the limitations of the electrical resistance type transmission system?
8. What is *telemetering?*
9. What are the two types of telemetering?
10. What method of transmission is used to transmit the measured variable over each of the following distances: a) 300 feet, b) 300 yards, c) 300 miles?

Chapter

8 Control

One of the principal objectives of process instrumentation is the automatic control of the measured variables. The measured variable is the temperature, pressure, flow, or level, etc. of the process material itself, or any ingredient or other material used during the process. The measured variable differs from application to application. Figs. 8-1A and 8-1B show examples of automatic control systems in simple diagram form.

Control Elements

An automatic control system basically consists of four elements: the *primary element*, the *measuring element*, the *controlling element*, and the *final element*.

Primary Element. A primary element is a device that detects or senses changes in the value of the controlled or measured variable. Examples of primary elements are thermocouples, orifice plates, floats, etc.

Measuring Element. The measuring element is the device (or apparatus) that receives the output of the primary element. It measures the amount the controlled (or measured) variable has deviated from the set point (the predetermined range required by the process). The measuring element can be nonindicating, or include an indicator or recorder. The signal is transmitted to the controller, as shown in Fig. 8-1B.

Controlling Element. The controlling element (or controller) uses changes in the value of the measured variable to alter the mechanical, pneumatic, or electrical source of power. (Hydraulic power can also be used.) The controlling element can actuate the source of power, increase or decrease its output, or turn it off, depending on its setting.

Final Element. The final element is a device that varies the energy supplied to the process. This energy is the manipulated variable so that the value of the measured or controlled variable is maintained within the desired range. See Fig. 8-1B.

These four elements combined form

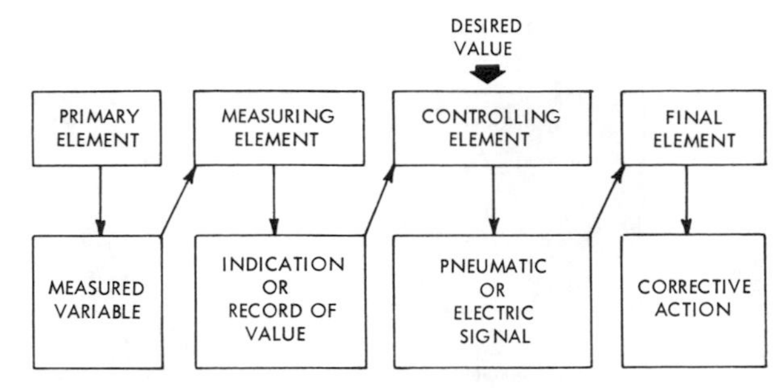

Fig. 8-1A. Simple diagram of an automatic control system.

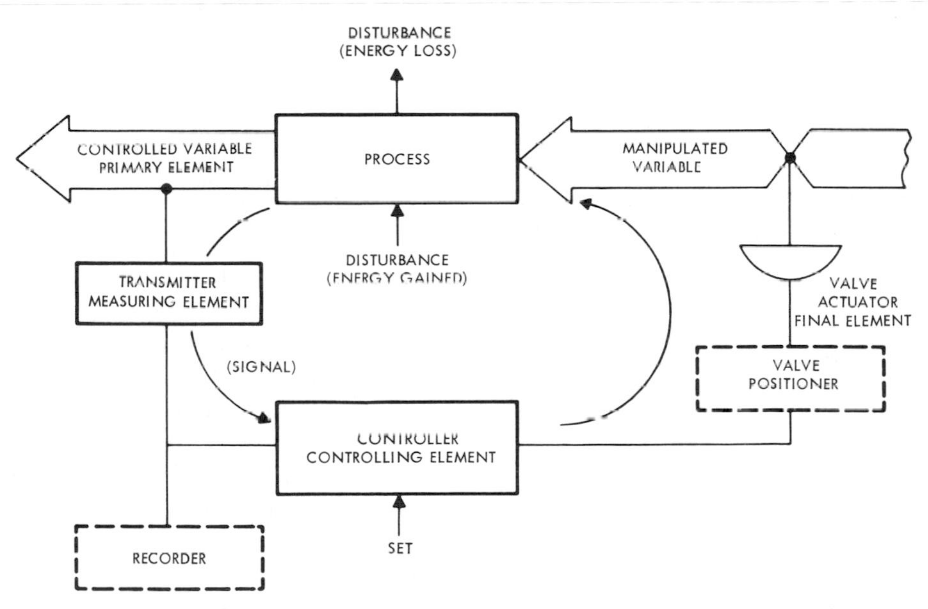

Fig. 8-1B. Diagram of an automatic control system showing the sequence of the various control elements. (The Foxboro Co.)

a self-operating controlling unit. No outside power source is required. Fig. 8-2 illustrates three such units.

Many self-operating controlling units are nonindicating (blind). When installed, they are regulated to the desired set point using a separate measuring device. This measuring device is also needed when the set point for the measured variable is changed. The power source in such devices is principally mechanical.

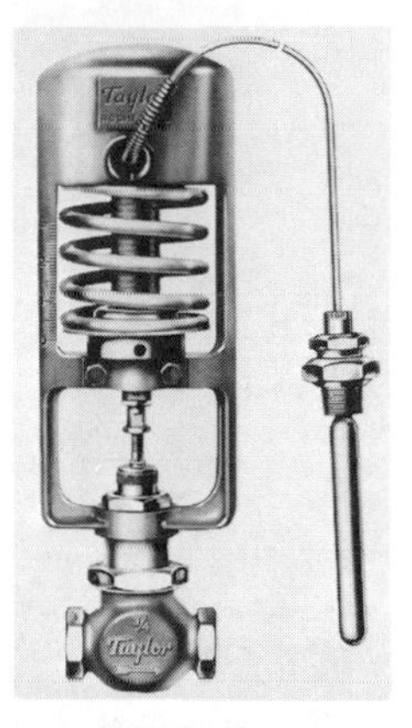

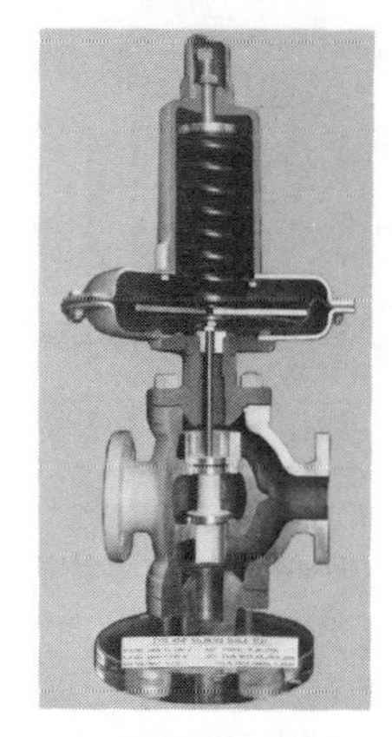

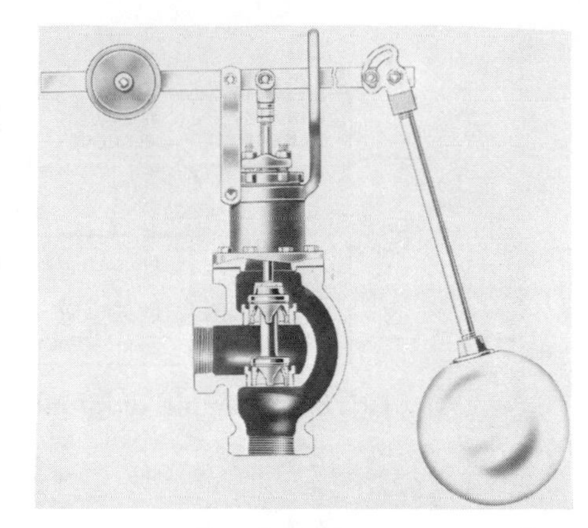

Fig. 8-2. Self-operating controlling units. Temperature controller, left, pressure controller, center, and level controller, right. (Taylor Instrument; A. W. Cash Co.; Fisher Governor)

Control Actions

The two general categories of automatic control are *open-loop control* and *closed-loop control*. In open-loop control there is no feedback of information to the controller. The process is generally regulated by time then the cycle stops. An example of open-loop control is an automatic car wash. The washing, buffing, rinsing, and drying steps in the cycle do not provide feedback to the controller regarding the effectiveness of the cleaning and drying process. An inspector must decide whether or not the process was satisfactory. Open-loop control when used in industrial applications is called feedforward control.

Closed-loop control provides feedback to the controller so that corrective action, if necessary, can be actuated.

The most common types of automatic controller actions available are:

1. On-Off action
2. Proportional action
3. Proportional action with reset
4. Proportional action with reset and derivative

The names of the actions describe the response of the final element to changes in the measured or controlled variable above or below the set point.

On-Off Action. In some industrial processes, acceptable control action is provided by a simple On-Off controller. As shown in Fig. 8-3, the final element only has two positions—open or closed. When the value of the measured or controlled variable is at or above the

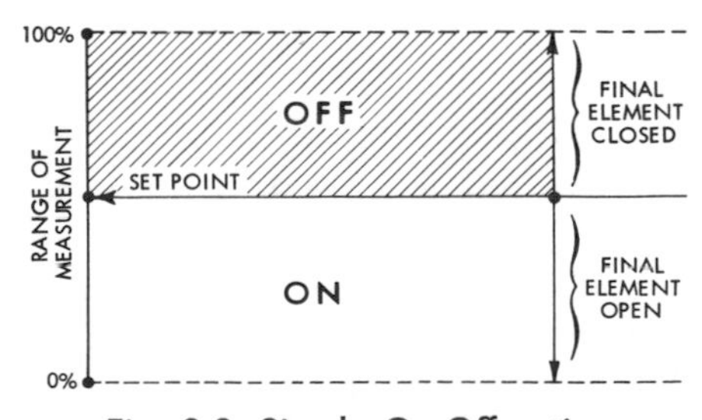

Fig. 8-3. Simple On-Off action.

set point, the final element is closed. When the range is below the set point, the final element is open. A common example of On-Off control action is a thermostat. If the thermostat is set for 68°F, the heating unit is off when the temperature is at or above 68°F. If the temperature drops below 68°F, the heating unit is actuated.

Proportional Action. Proportional action provides the control valve with variant positions between *On* and *Off*. The position of the final element is not simply open or closed, but varies, depending on how much the value of the measured variable is above or below this set point. See Fig. 8-4. The amount of energy to the process varies accordingly. Some processes require closer tolerances. The range of values of the measured variable which cause the final element to move from full open to full closed can be limited. However, the final element is not moved to a new position unless the value of the measured variable changes.

Proportional Action with Reset. As shown in Fig. 8-5, proportional action with reset enables the final element to assume intermediate positions. In addition, it can shift the relationship between the final element and the value of the measured or controlled variable. The controller is thus able to allow for changes in process characteristics, which proportional action alone cannot provide. The controller continues the corrective positioning of the final element until the measured variable returns to the desired value.

Proportional Action with Reset and Derivative. Derivative, or rate action, adds even more flexibility to the movement of the final element. Response is slow in some processes because a long period of time is needed to detect and correct changes in the measured variable. Derivative provides corrective positioning of the final element related to the rate at which the value of the measured variable is changing. See Fig. 8-6.

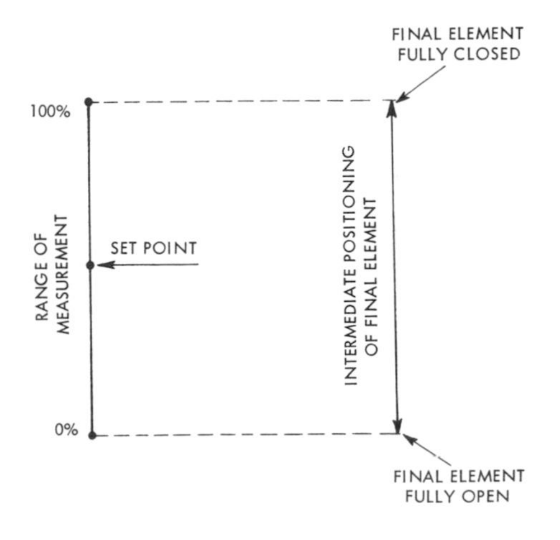

Fig. 8-4. Proportional action.

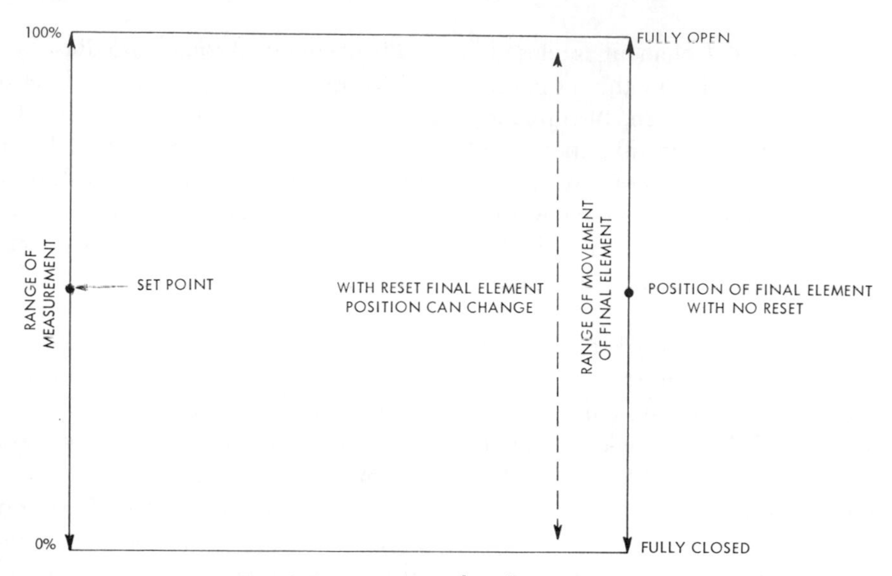

Fig. 8-5. Proportional with reset action.

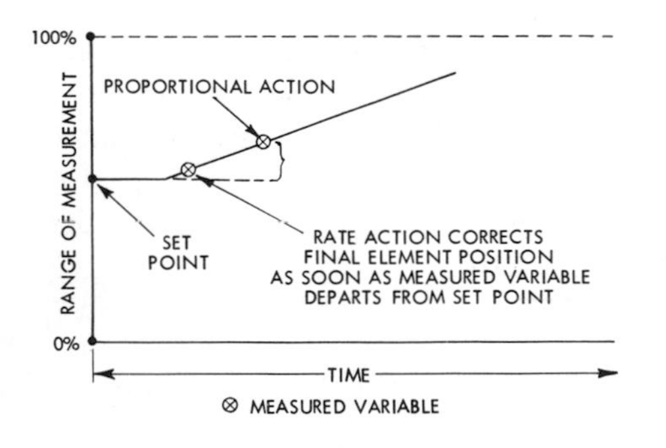

Fig. 8-6. Proportional with reset and derivative action.

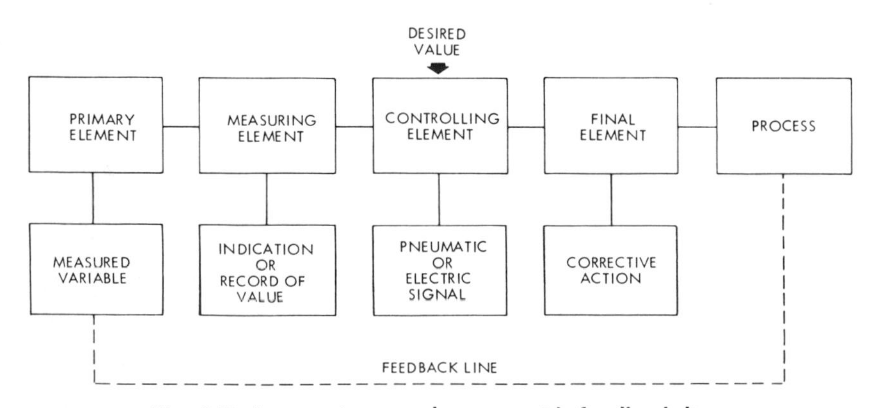

Fig. 8-7. Automatic control system with feedback line.

Reset and derivative are not independent control actions. They are refinements of proportional action. Feedback indicates to the controlling unit that corrective action has taken place and that this action has had the desired effect on the value of the measured variable. See Fig. 8-7. As indicated above, a control system which provides feedback is called a *closed-loop* control system.

The following is a discussion of the control actions as used in pneumatic control systems and then as applying to electrical control systems.

Pneumatic Control Systems

In a pneumatic control system, compressed air is supplied to the controlling element. As the value of the measured variable changes, the pneumatic output of the controlling element changes with it. A flapper-nozzle mechanism provides the means of controlling the pneumatic output. A typical pneumatic controller is shown in Fig. 8-8.

On-Off Action. Fig. 8-9 is an illustration of a typical pneumatic On-Off controller. When the measuring element detects the measured variable has deviated from the set point, the flapper-nozzle relationship changes. When the flapper is moved away from the nozzle, a minimum signal is sent to the air control relay. The air control relay in turn sends a maximum signal to the final element, causing it to move to one extreme position. See Fig. 8-10. When the flapper is moved against the nozzle, a minimum signal is sent to the final element. The final element moves to its other extreme position. See Fig. 8-11. There are no intermediate signals. The feedback in this system is provided by the process itself.

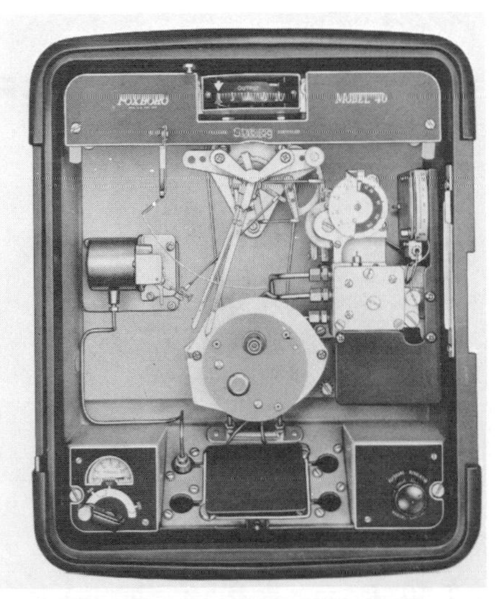

Fig. 8-8. Exterior and interior views of a typical pneumatic controller. (The Foxboro Co.)

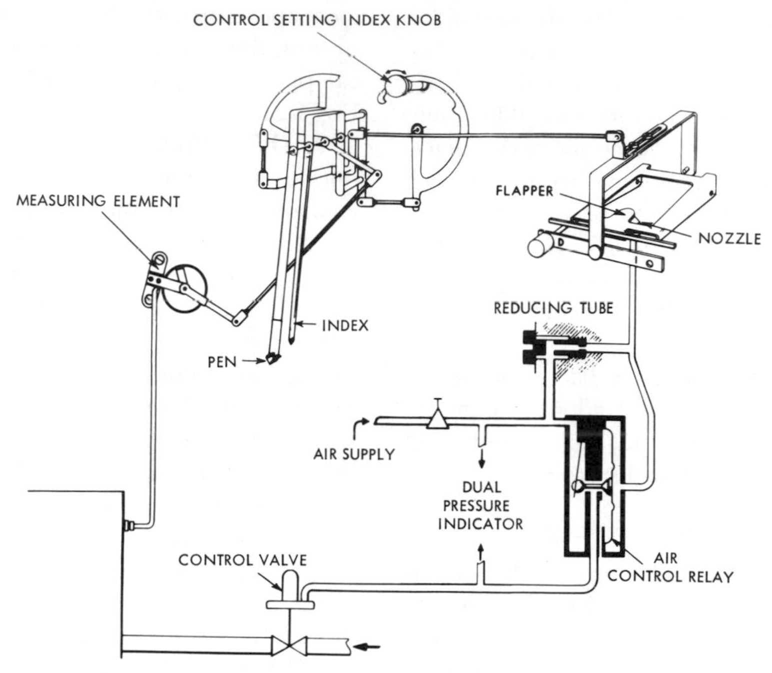

Fig. 8-9. Pneumatic On-Off controller. (The Foxboro Co.)

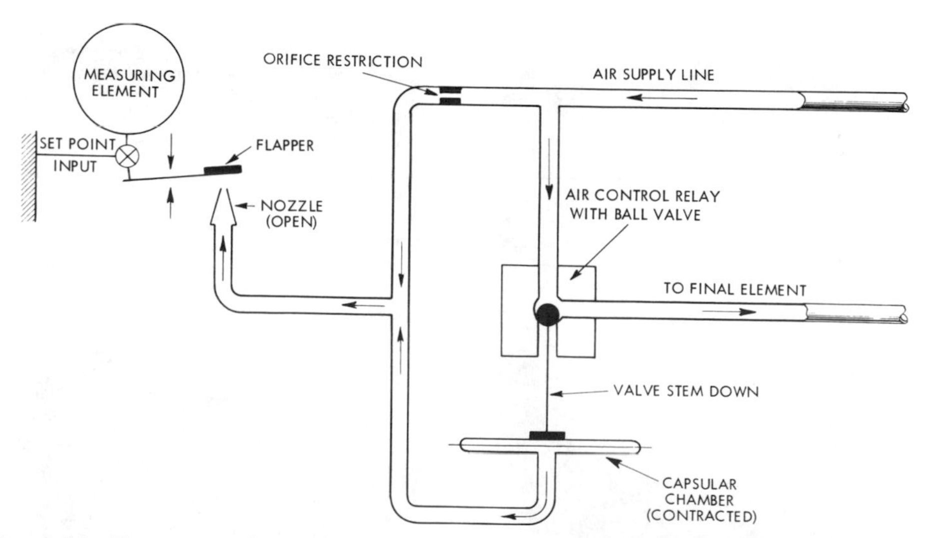

Fig. 8-10. Flapper-nozzle mechanism in pneumatic controller is open, positioning air relay valve so that maximum signal is sent to final element.

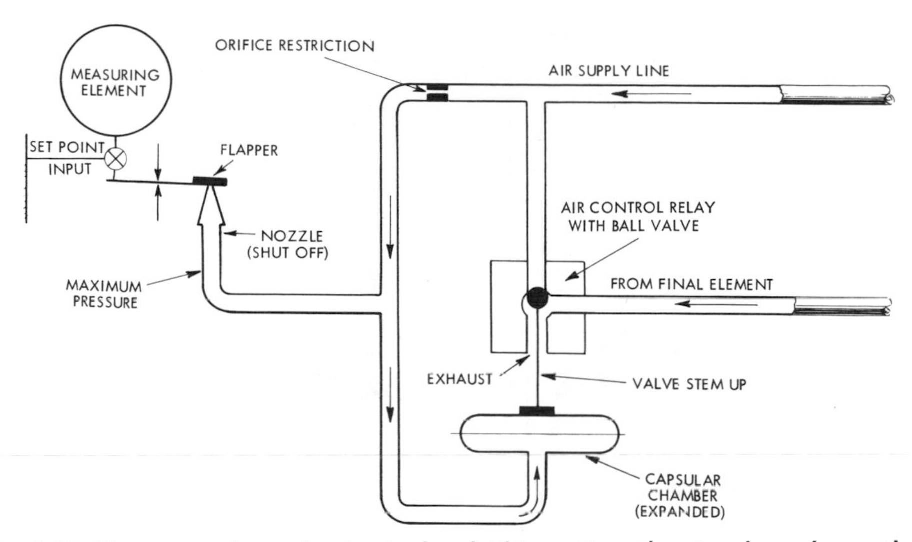

Fig. 8-11. Flapper-nozzle mechanism is closed. This positions the air relay valve so that a minimum signal is sent to the final element.

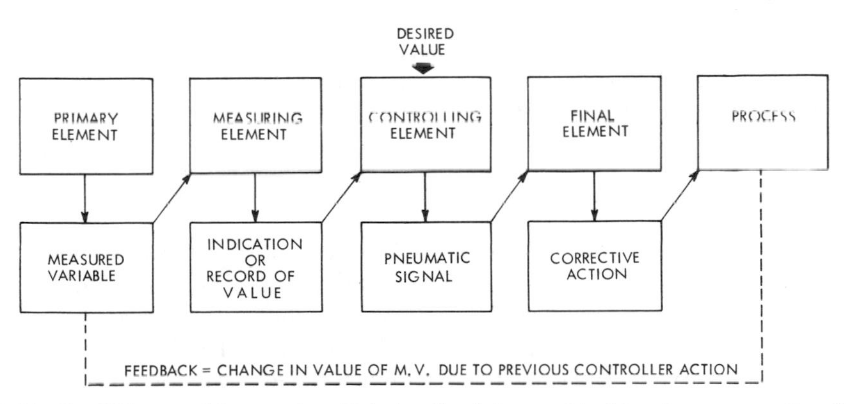

Fig. 8-12. On-Off control system in which feedback is provided by the process. Feedback equals change in value of measured variable.

See the simple diagram in Fig. 8-12. When the final element moves to either extreme position, it causes a maximum change in the energy to the process. This effects a change in the value of the measured or controlled variable. The change is sensed by the measuring element, and a new signal is sent to the controlling element, closing the loop.

Proportional Action. To achieve proportional action in a pneumatic

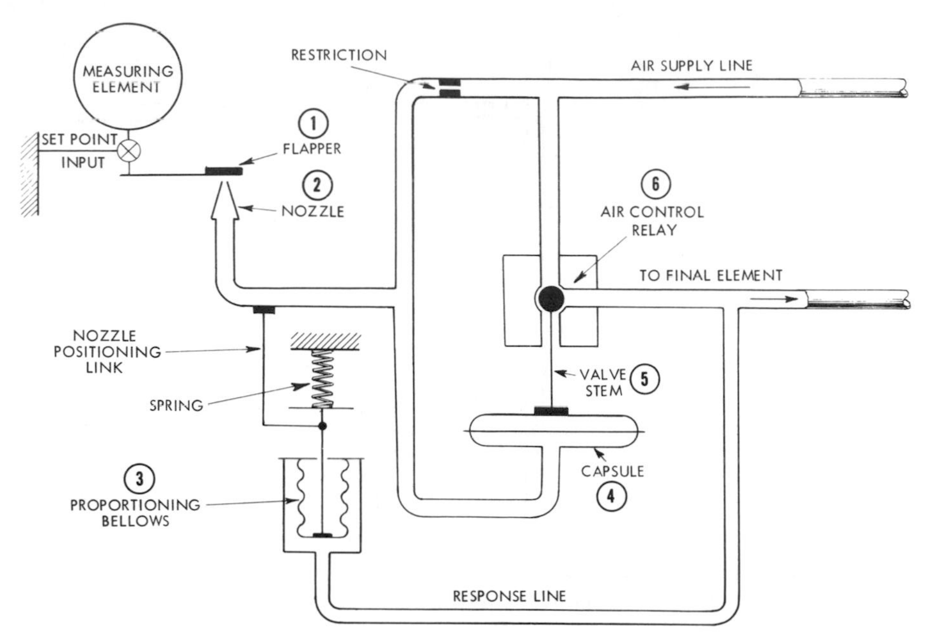

Fig. 8-13. Sequence of events in a pneumatic control system. The flapper (1) is positioned by the measuring element. The moveable nozzle (2) is positioned by the proportioning bellows (3), which expands or contracts as pressure in the response line changes. The capsule (4) expands on contracts as the pressure in the nozzle line changes. The valve stem (5) is positioned by the capsule. The air relay valve (6) regulates the pressure to the final element and the response line.

control system, the flapper-nozzle relationship is modified. This permits the use of more than two positions. A response line is added to the basic On-Off system, carrying the output signal from the control air relay to a mechanism that accurately positions the flapper or nozzle. If the measuring element positions the flapper, the additional mechanism positions the nozzle, and vice versa. In Fig. 8-13, the flapper is positioned by the measuring element, and the nozzle is positioned by the proportioning bellows and other parts of the additional mechanism.

The action of a pneumatic proportional action controller, using a bellows-actuated linkage to position the nozzle, is seen in Fig. 8-14. This illustration shows the direction in which the nozzle and bellows linkage move when the flapper moves in a downward direction. Fig. 8-15 shows the direction of motion for the opposite situation. The addition of the response line allows for many positions of the flapper and the nozzle. Each position of the measuring element (recording pen) requires a new flapper-nozzle position, and, therefore, a new controlled air pressure output signal.

Proportional Band. A change in the controlled air output is caused by deviation of the measured variable from

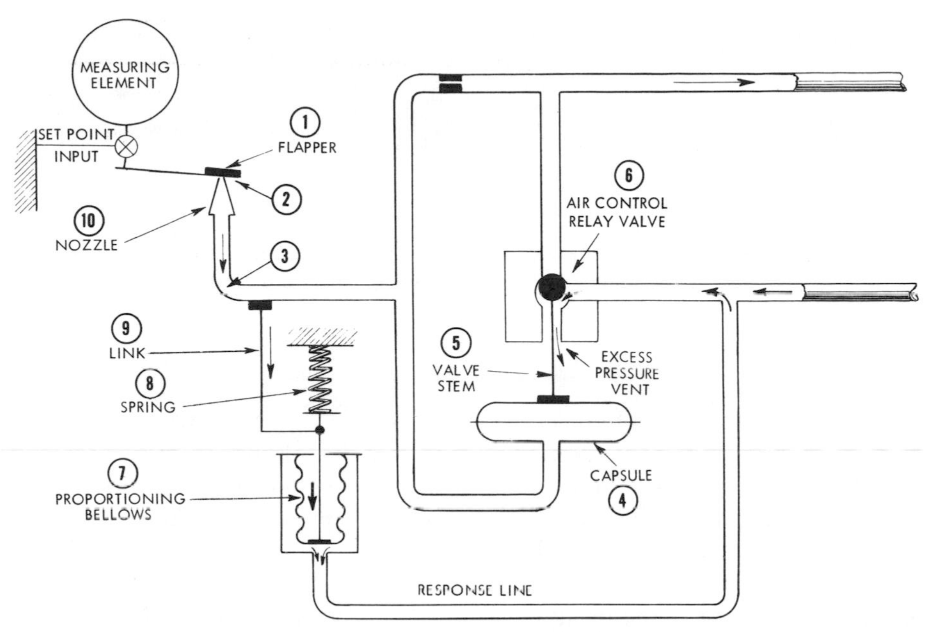

Fig. 8-14. The reduction of the air pressure to the final element and the response line causes the proportioning bellows (7) to expand. The spring (8) expands and moves the nozzle positioning link (9) downward so that the nozzle (10) moves downward.

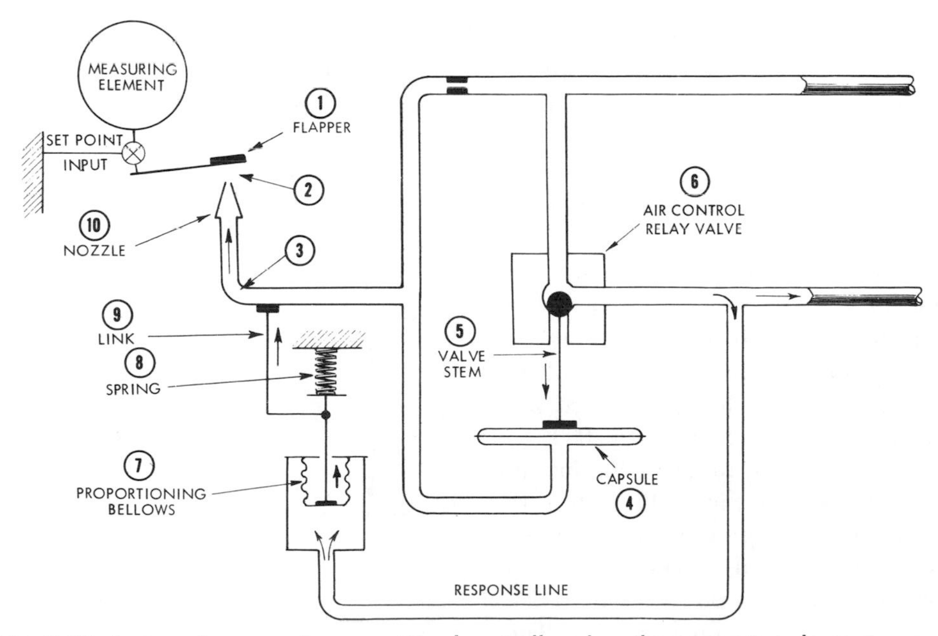

Fig. 8-15. Action of pneumatic proportional controller when the measuring element moves the flapper in an upward direction. The numbers show the sequence of events.

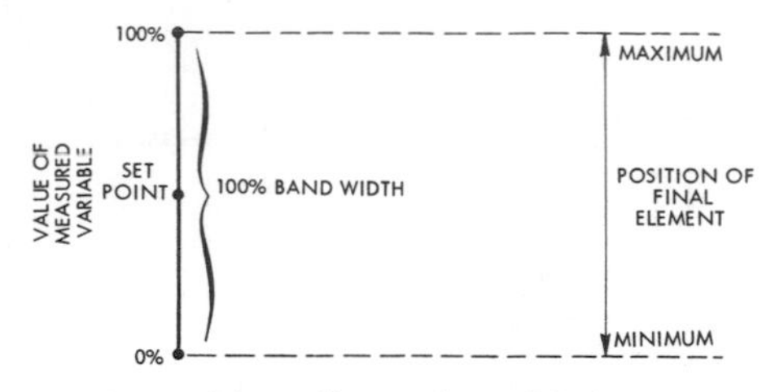

Fig. 8-16. The effect of a 100% proportional band on the final element.

Fig. 8-17. The effect of a 5% proportional band on the final element.

the set point. The amount of deviation which causes a maximum change in the controlled air output has to be adjusted. This adjustment establishes the proportional band of a controller which is the range of values of the measured variable the measuring element moves through as the controlled air output changes from minimum to maximum value. The proportional band is expressed as a percentage of the total range of the measured variable. The proportional band in the instrument referred to above is 100% because the measured variable must change from maximum to minimum value in order for the controlled air pressure to move from maximum to minimum position. See Fig. 8-16. The proportional band can be greater or smaller than 100%, depending on the requirements of the particular industrial application. Pneumatic controllers are available which have proportional band adjustment from 5% to 1000%.

The measuring range of the instrument shown in Fig. 8-17 is 0% to 100%. A 5% proportional band means the control signal moves from minimum to maximum position with a 5% change in the measured variable. The instrument can be set up so that the 5% range occurs between 20 and 25, 45 and 50, 73 and 78, etc. It depends on the location of the set point. For example, if the set point is at 50, the 5% range would be between 47½ and 52½.

A proportional band of 200% does not allow the controller to reach either maximum or minimum position. See Fig. 8-18. In this instrument, the measuring range is 0 to 100. The 200% proportional band requires the value of the measured variable to change a range of 200 to move from maximum to minimum position. The change

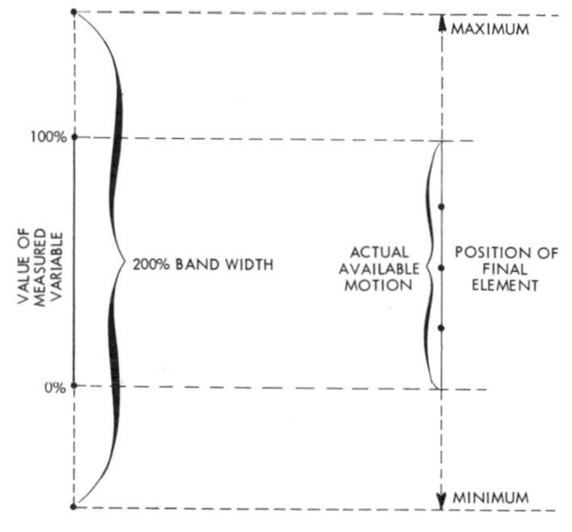

Fig. 8-18. The effect of a 200% proportional band on the final element.

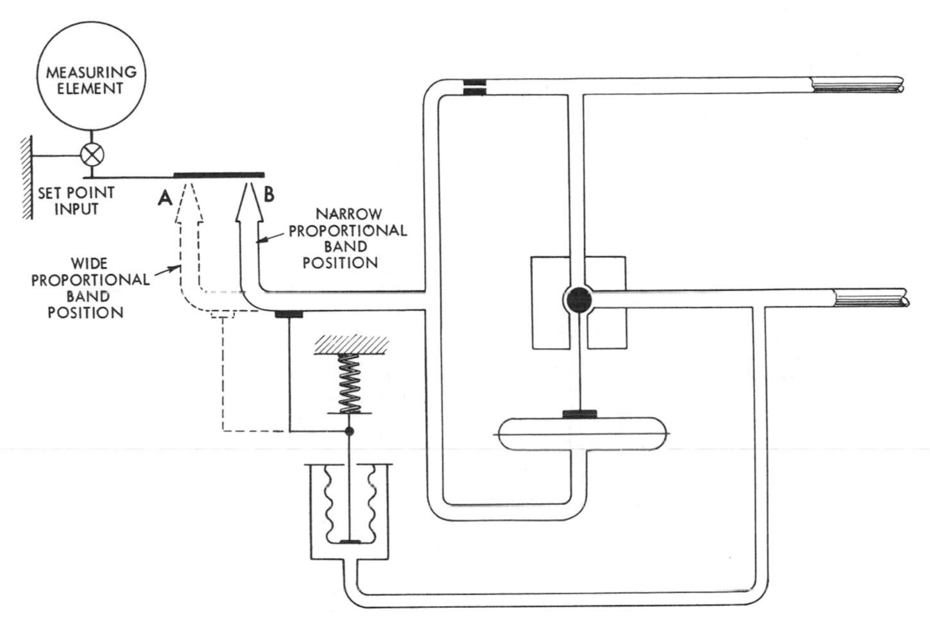

Fig. 8-19. Proportional band adjustment. When the nozzle is at point B, there is a narrow proportional band adjustment. When the nozzle is at point A, there is a wide proportional adjustment.

could occur between 0 to 200, −50 to +150, 50-250, etc. Since the range of the instrument is only 0 to 100, it is impossible for the value of the measured variable to move from maximum to minimum position. The effect of a wide proportional band is more precise control. A change of 100% in the range of the measured variable only provides a change of 50% in the controlled air pressure.

As shown in Fig. 8-19, the proportional band adjustment is made by regulating the motion of the nozzle for each unit of air pressure from the air control relay.

Automatic Reset. Automatic reset is added to the pneumatic controller by modifying the output signal from the air relay as it passes to the flapper-nozzle mechanism. Reset action requires the addition of an adjustable restriction, a capacity tank, and a reset bellows. See Fig. 8-20. The air signal from the air control relay, in addition to entering the proportional bellows, passes through the adjustable restriction and the tank to the reset bellows. The motion of the reset bellows opposes the motion of the proportional bellows.

The proportional and reset bellows are in balance only when the air control signal positions the controller so that the energy supplied to the process is correct and maintains the set point. See Fig. 8-21. When the measuring element continues to move the flapper because of the error between set point and control point, the proportional and reset bellows are not in bal-

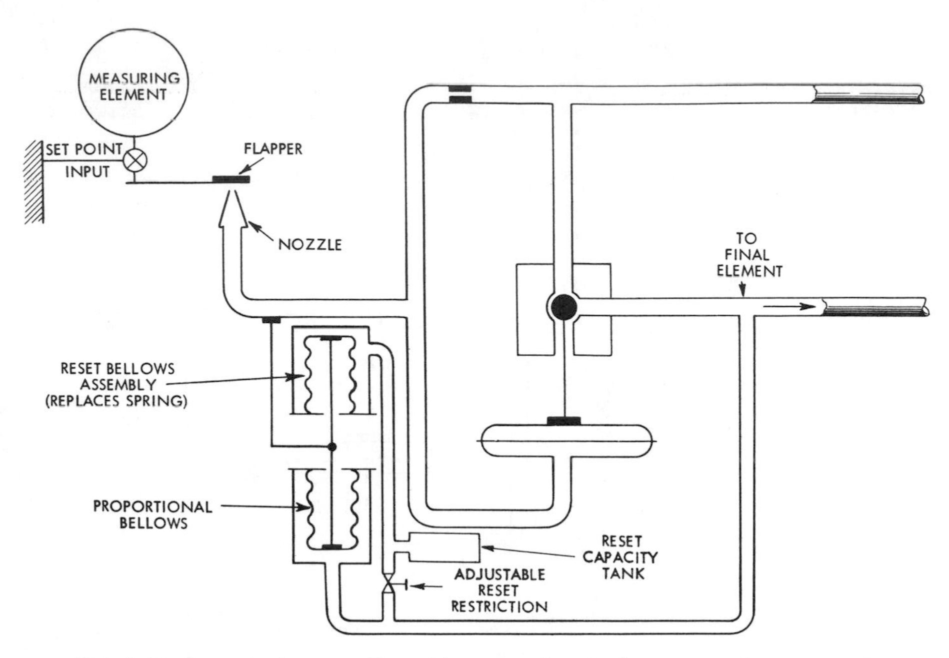

Fig. 8-20. Pneumatic controller with proportional plus automatic reset action.

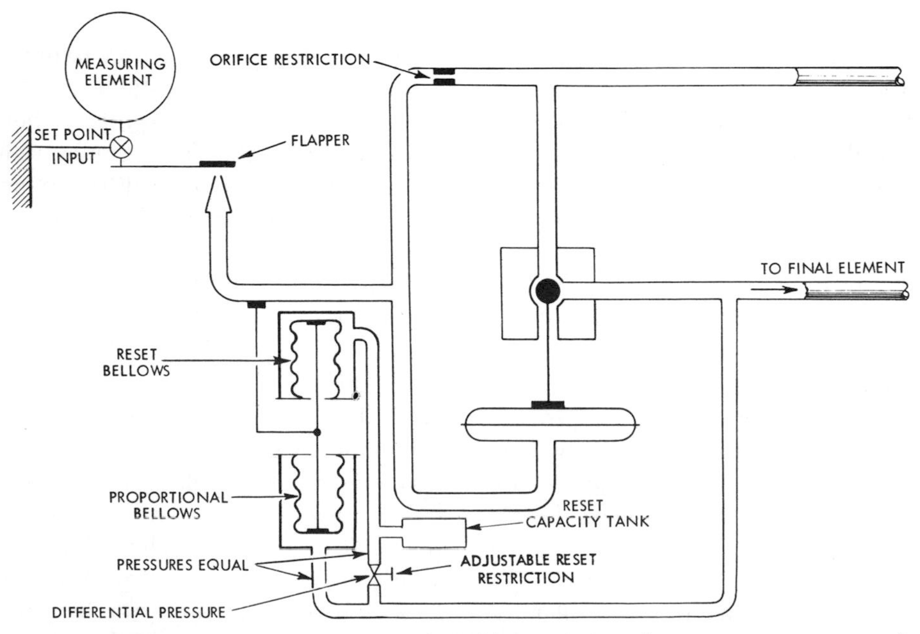

Fig. 8-21. Differential pressure across adjustable restriction determines time required to balance pressures between proportional bellows and reset bellows.

ance. As long as the bellows are not in balance, the air pressure signal to the controller continues to change. This permits the final element to provide energy to the process as required. The actual adjustment is measured in the number of times per minute the proportional and reset bellows are allowed to balance when there is an error. This is called *reset time*. Such action is necessary to take care of changes in the energy requirements of a process. The changes are referred to as load changes.

Derivative Action. Derivative action, sometimes called rate action, is added to a pneumatic controller by further modifying the output signal from the air relay.

Fig. 8-22 is an illustration of a controller having proportional action, with reset and derivative (rate action). This unit has adjustable reset resistance and an adjustable derivative resistance. There is also an additional capacity tank. The restriction and the capacity tank delay the corrective motion of the proportional bellows (outer and inner bellows). The amount of delay varies with the rate at which the difference between set point and control point is increasing or decreasing. The actual adjustment is in terms of *rate time*. This action is required for controlling processes in which response to energy changes is very slow. Such processes possess lags. The lags can be due to slow measurement or

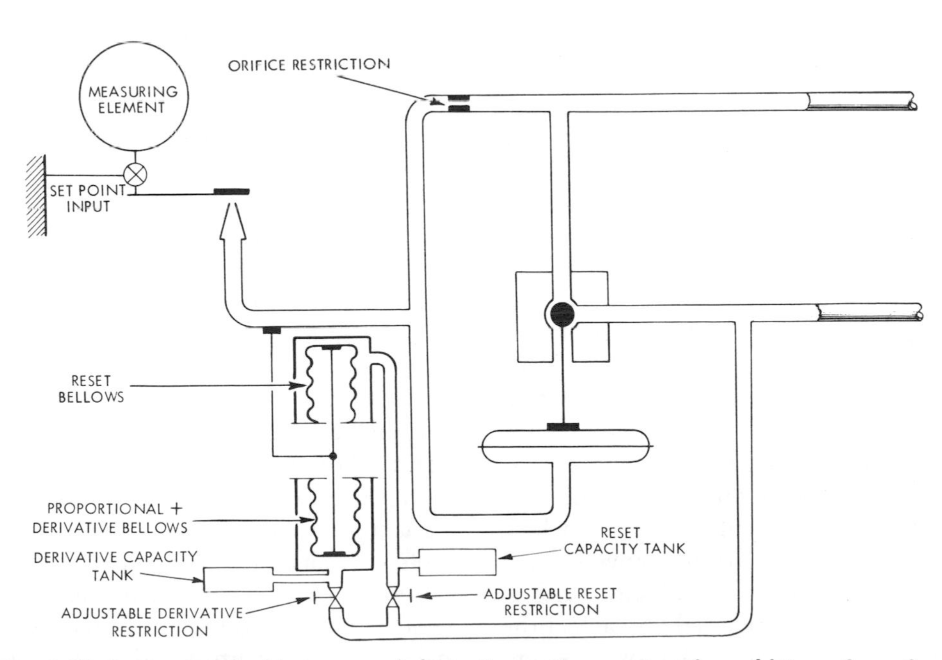

Fig. 8-22. Proportional with reset and derivative action requires the addition of another adjustable restriction and a capacity tank which delays change of pressure acting on the proportional with derivative bellows.

slow control. Since four controller actions are available, the proper application for each type will be discussed.

On-Off Control. On-Off control is most suitable in a process that quickly reacts to an energy change, when the energy change itself is not too rapid. There should be little or no load change in the process.

Proportional Control. Proportional control is most suitable in a process that reacts almost immediately after an energy change, when the energy change is quite slow. There can be load changes, but they should not be large or rapid or of long duration.

Proportional Action with Reset Control. Proportional action with reset is best suited for applications in which the process reacts with moderate speed to an energy change. The energy change itself can be fast or slow. There can be load changes, but they should not be rapid. The load changes can be greater and of longer duration than those which can be handled by proportional action alone.

Proportional Action with Reset Control and Derivative Action. A control unit having proportional action with reset control and derivative action is generally designed to handle any industrial process regardless of its characteristics. The adjustment of proportional band, reset time, and derivative (rate time) are interdependent. A change made in one adjustment does not benefit the control unit as a whole unless changes in the other adjustments are also made. Ideally, there is a setting recommended for each of the control actions for every process condition. Although there are mathematical methods by which approximation of the proper settings can be determined, a study of the control unit is recommended. An understanding of the effects of each control action adjustment provides better results. An in-depth understanding of the four controller actions is essential to the study of control.

Electrical Control Systems

Electrical controllers are available that provide all the control actions mentioned above.

On-Off Action. On-Off action can be achieved by any device that opens or closes an electrical circuit when the measured variable departs from the set point. Typical examples are thermostats, float switches, and pressure switches.

Proportional Action. Proportional action in an electrical controller requires an arrangement which provides an electrical output that changes as the measured variable departs from the set point. Many such arrangements are available. Fig. 8-23 is a block diagram of a typical electrical controller with proportional action. Two slidewires are provided for this controller. One slidewire is used for measurement and the other for control. The unbalance detector senses the difference between the value of the measured variable and the desired value (set point). This difference is transmitted to the amplifier, which transmits this signal to the control motor. The control motor responds by driving to the new position required. At the same time the control motor positions the final element and the control slidewire. The control slidewire transmits a signal to the unbalance detector, which nullifies

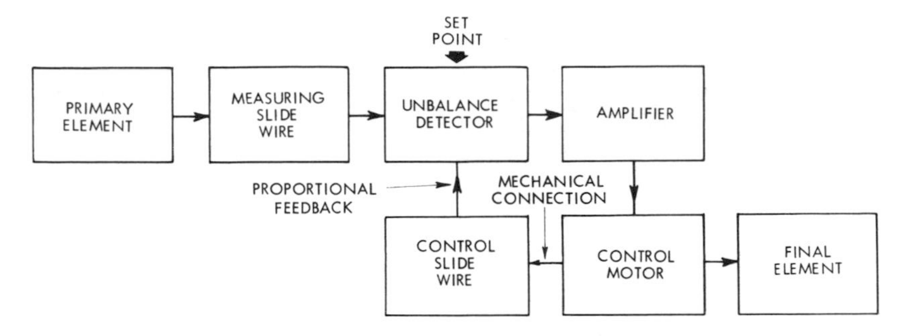

Fig. 8-23. Diagram of a typical electrical controller.

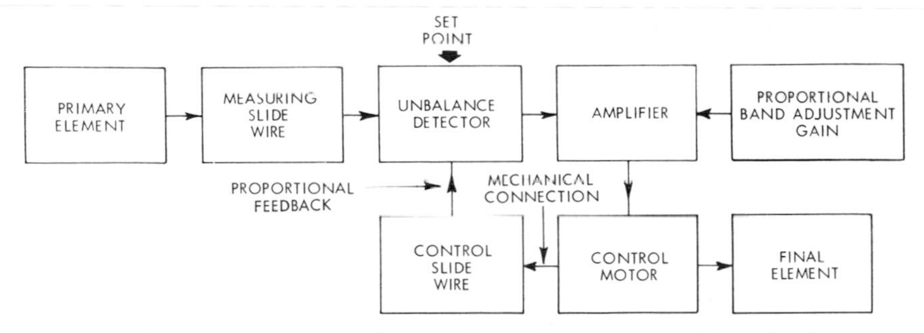

Fig. 8-24. Electrical proportional controller with proportional band adjustment.

the original unbalance. The unbalance detector does not produce an output signal. Therefore, the control motor does not receive a signal, and its movement stops, leaving the control slidewire and final element at their new positions. The position of the final element changes with each departure of the measured variable from the desired value.

To achieve proportional band adjustment in this controller, a method must be provided for permitting variation in the magnitude of the signal from the unbalance detector for each unit of departure of the measured variable from the set point. See Fig. 8-24. In an electric controller, this is frequently called a *gain adjustment*. With a narrow proportional band, the adjustment is made so that a small departure from set point creates considerable unbalance. The control motor must drive considerably to restore balance. With a wide proportional band the adjustment is made so that a large departure from set point creates little unbalance. This causes the control motor to drive only a small amount to restore balance.

Automatic Reset. Automatic reset can be added to an electric controller by modifying the feedback signal from the control slidewire, as shown in Fig.

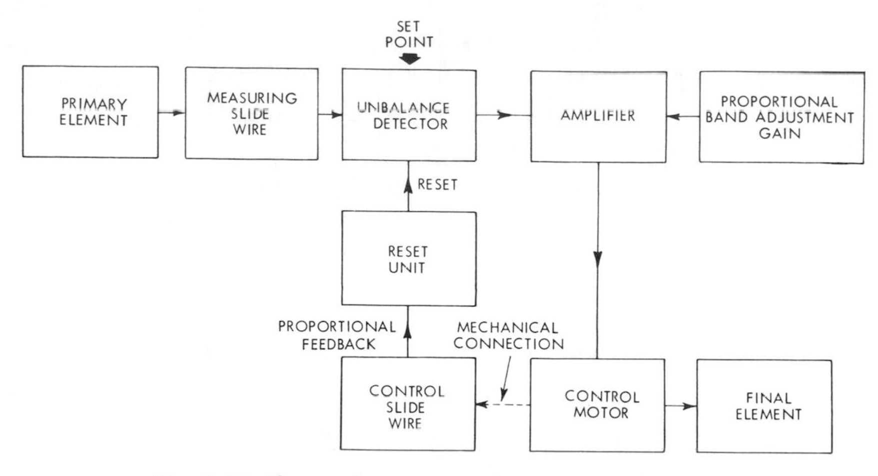

Fig. 8-25. Electrical proportional controller with reset.

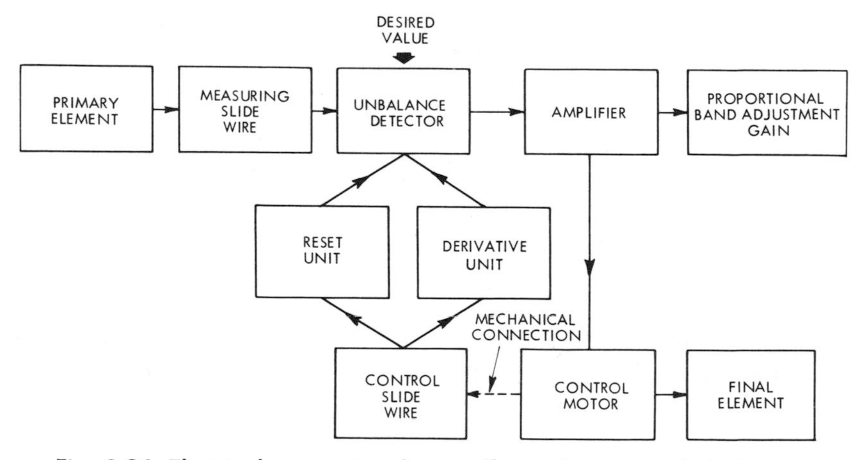

Fig. 8-26. Electrical proportional controller with reset and derivative.

8-25. The reset unit allows a shift in the relationship of the control slidewire position and the measuring slidewire position, as long as an unbalance between measured value and set point exists. If the position of the control motor with proportional controller action does not restore balance, the automatic reset unit continues to send a signal to the unbalance detector until a position is reached that restores balance to the system.

Derivative Response. Derivative (rate) response requires further modification of the proportional feedback signal from the control slidewire. See Fig. 8-26. The derivative unit receives the proportional feedback signal from the control slidewire and introduces a delay in the signal to the unbalance detector. The delay varies with the rate at which the unbalance occurs. The more rapid the unbalance, the longer the delay, and vice versa. As a result, the unbalance signal to the control motor is greater than if the feedback were to restore balance immediately. This allows the control motor to move a greater amount than with proportional action alone.

The above description of an electric controller refers to a system in which the final element can assume many positions. There are electric control systems in which the final element can be either *On* or *Off*, but throttling control is still required. A heating process that uses electric heating elements is a typical example. The heaters can only be turned On or Off. There is no intermediate position. In such a system, throttling is accompanied by proportioning the *time-On time-Off* cycles. See Fig. 8-27. Thus, for each particular

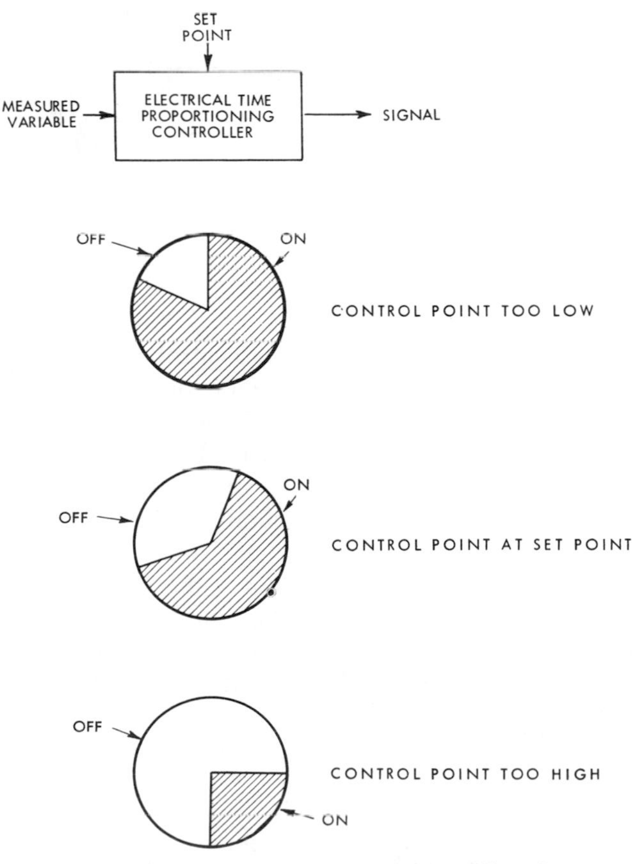

Fig. 8-27. Diagram of time-On and time-Off cycles.

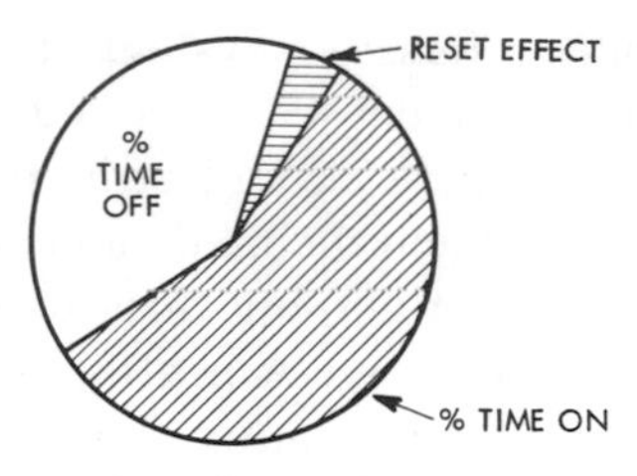

Fig. 8-28. The effect of automatic reset on the time-On, time-Off ratio.

set point there is a particular percentage of time *On*, percentage of time *Off* ratio. As the measured variable departs from the set point, the ratio changes. This arrangement is equivalent to proportional action.

Automatic Reset can be added to such a system. See Fig. 8-28. This allows the ratio to be changed as long as the measured variable is not at the set point.

Derivative response is generally not offered in such systems, hence the installation requires that lags be kept to a minimum.

Though schematically such a system resembles the control motor type system, the electrical circuitry and components are quite different. Time proportioning cannot be performed with the same equipment as position proportioning.

Final Elements

Final elements are those devices which actually regulate the level of energy of a process. These include valves, dampers, louvers, pumps, heating elements, etc.

All of these elements require operators that provide precise regulation. The operator must be compatible with

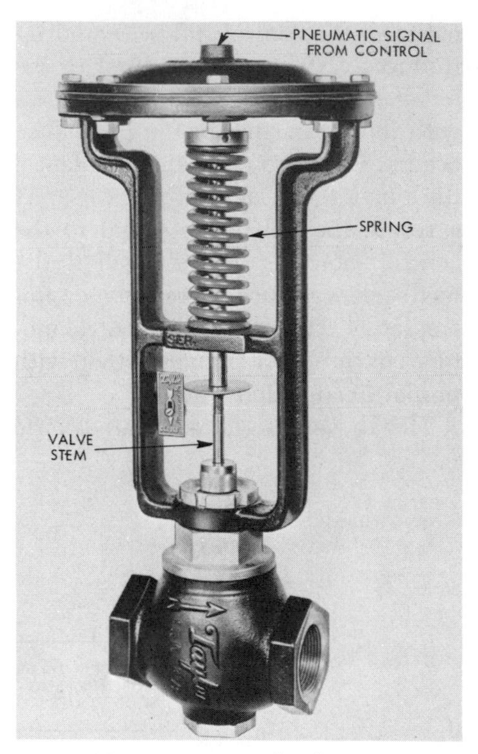

Fig. 8-29. Pneumatic diaphragm motor valve. (Taylor Instrument)

the control system used, whether it is pneumatic, electric, or hydraulic.

Typical of the pneumatic final elements is the diaphragm operated valve, shown in Fig. 8-29. The diaphragm operator receives the pneumatic signal from the controller and moves the valve stem up or down as the signal varies. The force on the diaphragm is balanced by the force of the spring. The body contains the throttling mechanism, which consists of a plug attached to the valve stem, and the valve seat, which is built into the body.

Some valves have an air-to-close diaphragm operator and a globe body.

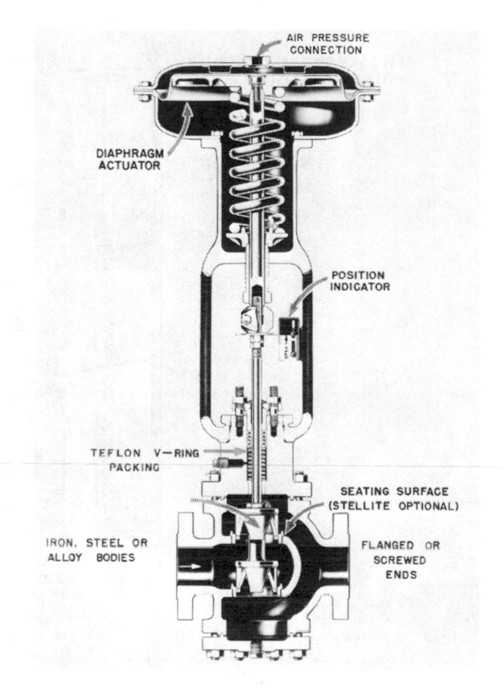

Fig. 8-30. Air-to-close operator. (Bailey Meter Co.)

See Fig. 8-30. Air-to-open diaphragm operators are also available as are many other types of bodies.

In addition to diaphragm operators, there are also pneumatic piston and cylinder types.

The positioning of dampers and louvers can also be accomplished using such pneumatic operators. Many other variations in addition to those mentioned here are available.

Electrical positioners can be operated by a motor or a solenoid. Solenoids permit only two positions, whereas motors can provide continuous positioning. The solenoids and motors can be ac or dc operated.

Electrical and pneumatic operations can also be combined. For instance, a solenoid-operated valve can be used to regulate the flow of air to a pneumatic final element. Or a pneumatic operator can be used to control the position of a rheostat which regulates the movement of an electric motor-operated final element.

Electro-hydraulic actuators are also available. In these actuators an electric motor drives a pump, using a hydraulic system. The pump in turn forces oil into an actuating cylinder. The cylinder moves the positioning shaft. Fig. 8-31 is an example of such an operator.

Pneumatic valve operators are sometimes affected by mechanical problems which make it extremely difficult to achieve accurate positioning of the valve stem. As a means of overcoming such problems, auxiliary valve positioners are generally used. The auxiliary valve positioners are combination balance and power amplifiers. The pneumatic signal from the controller enters the valve positioner which

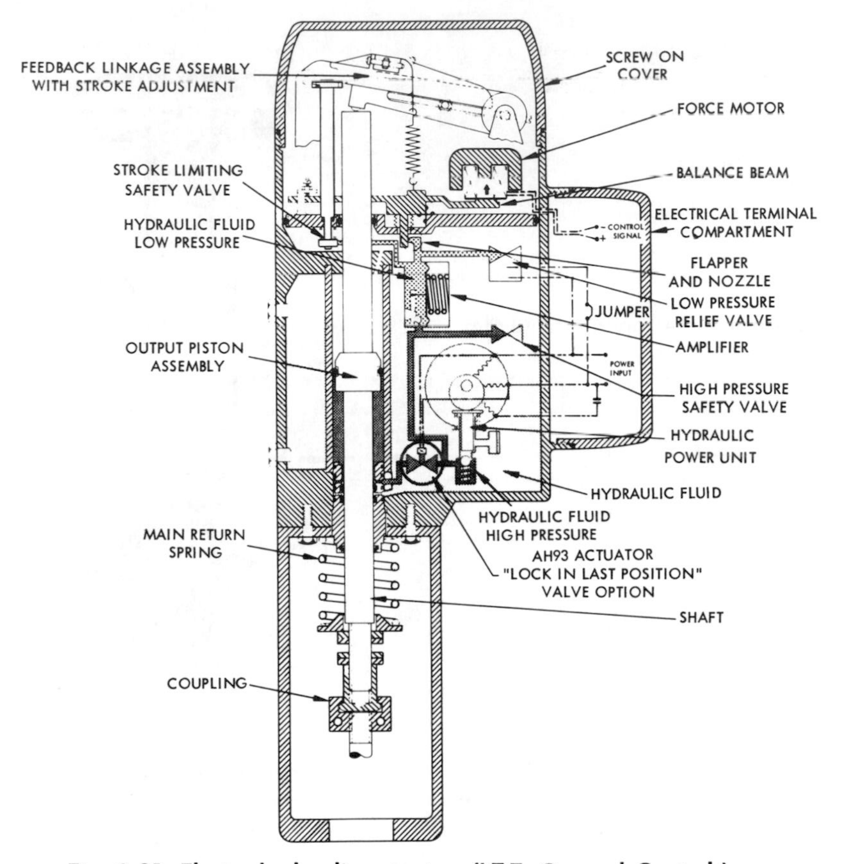

Fig. 8-31. Electro-hydraulic actuator. (I.T.T. General Controls)

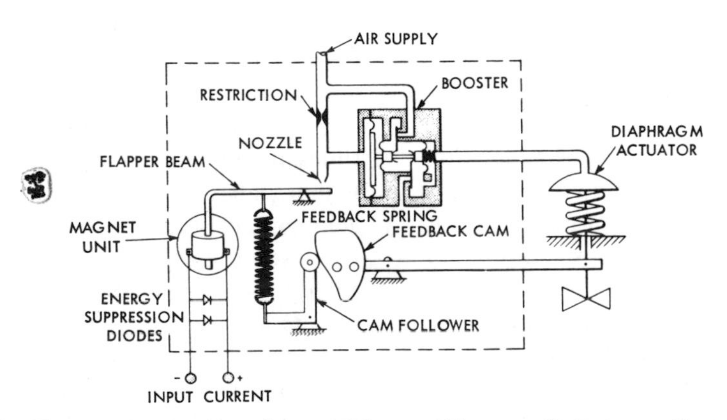

Fig. 8-32. Electro-pneumatic valve positioner. (Honeywell Process Control Div./Fort Washington, Pa.)

passes it to the valve operator. The positioner receives a mechanical signal by means of a link attached to the stem. The signal indicates the stem position. If the stem does not stop at the correct position determined by the controller, the valve positioner senses this and makes the necessary correction. A valve positioner can also be used to allow total valve motion even though there is only a limited pneumatic signal from the controller. Two valves, with positioners, can be operated from one controller. A portion of the controller signal operates one valve, and the remaining portion operates the other. This is sometimes required when the control of two separate fluids is needed to produce the desired value of the measured variable. Steam and cold water are occasionally controlled by a single temperature controller in this manner. Fig. 8-32 shows an electro-pneumatic valve positioner.

Words to Know

measured variable
primary element
measuring element
controlling element
final element
open-loop control
closed-loop control
feedback
proportional action
reset
reset time
derivative (rate) action
rate time
proportional band
flapper-nozzle control
load control

Review Questions

1. Describe the four elements essential to a *control system*.
2. What is a control system called when it includes all four of these elements?
3. What is the difference between open-loop and closed-loop control systems?
4. What is *feedback* as applied to a controller?
5. What are the four common automatic control actions? Describe the differences among them.
6. Describe proportional band adjustment.
7. In the pneumatic flapper-nozzle controller described in this chapter, automatic reset action takes place only when the proportional and reset bellows are not in balance. What causes this unbalance?
8. Derivative (rate) action, when applied to the pneumatic controller, delays the corrective motion of the proportional bellows. What effect does this have on the action of the final element?
9. Define *load change* and *lag*.
10. A process reacts very quickly to an energy change, but there is a

possibility of large and prolonged load changes. What control action is most suitable for this application?

11. What is the meaning of the term *gain control* in relation to an electrical controller?
12. How is automatic reset control achieved in an electrical control system which employs a measurement slidewire and a control slidewire?
13. How is the proportional band adjustment made on an electric temperature controller which varies the *time-On, time-Off* ratio of electric heaters?
14. The spring force on a diaphragm control valve is adjustable, which changes the air pressure for each valve position. Is this a reset adjustment?
15. What is the principal function of a pneumatic valve positioner?

part two

PART TWO discusses some of the same instruments that were examined in Part One, but in greater detail, and includes a chapter on electricity and electronics. Level measurement is not repeated since the principles involved are not sufficiently distinctive to warrant extensive discussion. For the most part, the devices used for level measurement are adaptations of those used for other measurements such as pressure or differential pressure. And although the transmission of process information is a valuable function of instrumentation, transmission is not repeated because the methods used are adaptations or complications of mechanical or electrical systems used in measurement and control.

This close-up shows the master control panel of a water pumping system used with various types of boilers. The memory relays permit sequential start-up and shutdown of various boilers in the system. Knobs start-up or turn off a motor, open or close a valve, etc., as required. The meters give readings of variables such as steam pressure, oil pressure, and vacuum. (Cleveland Electric Illuminating Co.)

Electricity

Chapter

9

Electrical energy is produced by chemical cells (batteries), light sensitive cells (photocells and solar batteries), temperature sensitive elements (thermocouples), pressure sensitive elements (crystals), and magnetic devices (rotating generators and alternators).

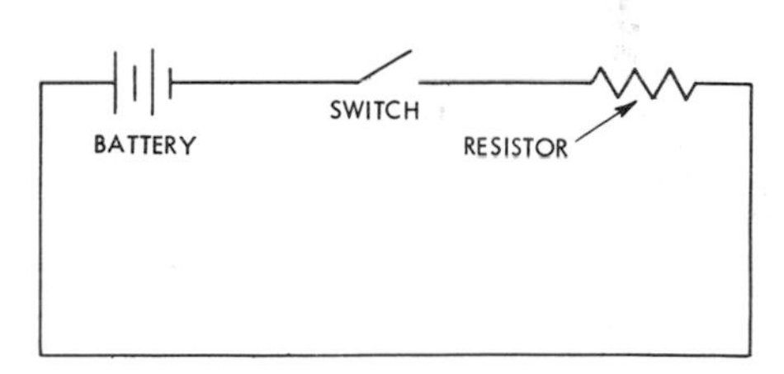

Fig. 9-1. A simple elecrical circuit.

Voltage is the measure of the potential energy difference between two points in an electrical circuit. The unit of voltage is the volt. It is sometimes called electromotive force or emf. Current is the rate of the flow of electrical energy passing between two points at different potential levels. The unit of current is the ampere. Resistance in an electrical circuit refers to the resistance to current flow. The unit of resistance is the ohm.

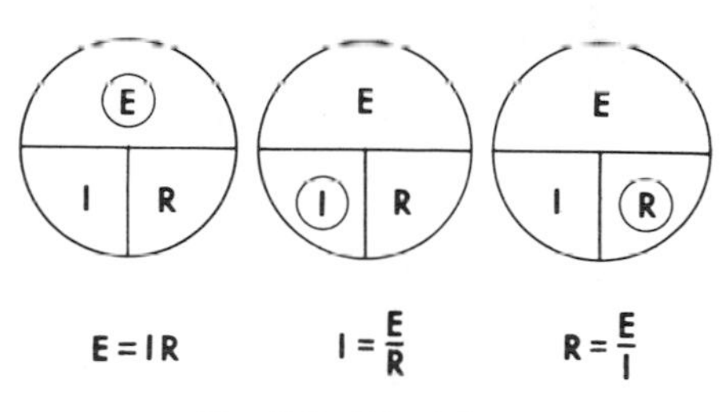

Fig. 9-2. Ohm's law.

Direct Current

The simplest direct current (dc) electrical circuit consists of a battery, a resistor, and a switch. See Fig. 9-1. When the voltage at the battery is equal to 1 volt, and the resistance is equal to 1 ohm, the current is equal to 1 ampere. This relationship follows Ohm's law, which can be expressed mathematically as E (voltage) = I (current) × R (resistance). Fig. 9-2 shows a simple way to remember Ohm's law.

Circuits in Series. There are frequently several resistances in an electrical circuit. When these resistances

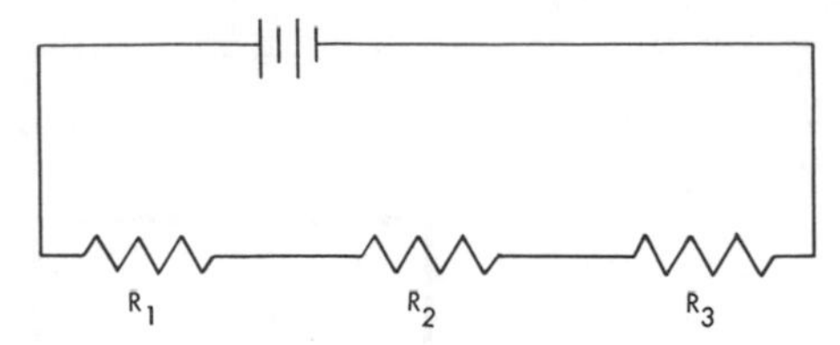

Fig. 9-3. Resistances joined in series.

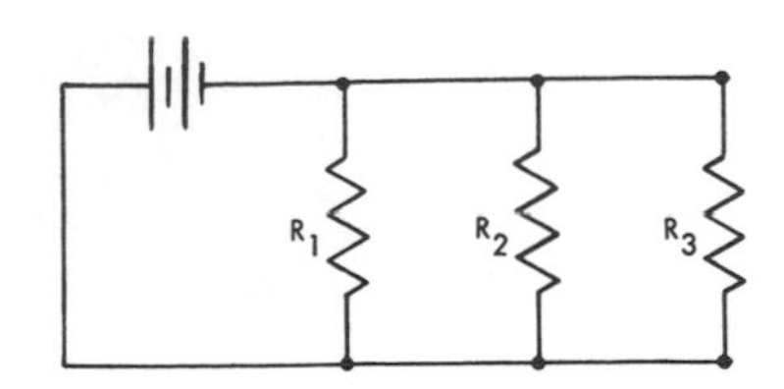

Fig. 9-4. Resistances in parallel.

are joined like the links of a chain, they are said to be in series. In this arrangement, as shown in Fig. 9-3, the resistances are additive. The total resistance is equal to the sum of the separate resistances. R total $= R_1 + R_2 + R_3$, and so on.

As the current passes through each resistance, there is a drop in voltage in accordance with Ohm's law: voltage loss through R_1 is equal to IR_1; voltage loss through R_2 is equal to IR_2. Therefore, the total voltage loss through several resistances in series equals IR_T. (T = total.)

The sum of the voltage drops in a circuit is equal to the sum of the applied voltages. This is called Kirchoff's law of voltage.

The characteristics of series circuits can be summarized as follows:

1. The total resistance is equal to the sum of the individual resistances.
2. The current through all the resistances is the same.
3. The sum of the voltage drops across the individual resistances is equal to the applied voltage.

Circuits in Parallel. When resistances in an electrical circuit are joined like the rungs of a ladder, they are said to be in parallel. In this arrangement, as shown in Fig. 9-4, the reciprocal of the total resistance equals the sum of the reciprocals of the individual resistances:

$$\frac{1}{R_T} = \frac{1}{R_1} + \frac{1}{R_2} + \frac{1}{R_3}\text{, and so on}$$

Since each individual resistance is connected to the applied voltage terminals, the voltage across each resistor is equal to the applied voltage. The current through each resistor varies with the value of the resistor. The battery current is equal to the sum of the currents through the parallel resistances. According to Kirchoff's law of current, the sum of the currents entering a junction point in an electrical circuit is equal to the sum of the currents leaving the junction point. See Fig. 9-5.

The characteristics of parallel re-

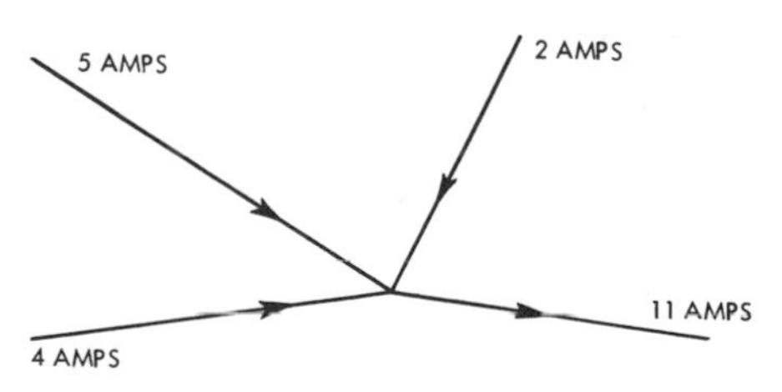

Fig. 9-5. A diagram of Kirchoff's law of current. The sum of 4 amps plus 5 amps plus 2 amps equals 11 amps.

sistance circuits can be summarized as follows:

1. The reciprocal of the total resistance is equal to the sum of the reciprocals of the individual resistances in the circuit.
2. The total current is equal to the sum of all the branch currents.
3. The ratio of the branch currents is the inverse of the resistance ratio:

$$\frac{I_1}{I_2} = \frac{R_2}{R_1} \text{ and } \frac{I_1}{I_3} = \frac{R_3}{R_1}, \text{ and so on.}$$

Networks. Resistors in a circuit can be arranged in combination series-parallel arrangements. When two or more components of an electrical network are in series, all the characteristics of series circuits must apply to these components. In the same way, when two or more components of an electrical network are in parallel, all the characteristics of parallel circuits must apply to these components. Electrical circuits can have several sources of voltage in addition to several series-parallel resistances. These are termed resistance networks.

The Wheatstone bridge, which is frequently used in instrumentation for measurement and control, is a special type of resistance network. To determine the voltages and currents, the Wheatstone bridge must be considered as three separate current loops. See Fig. 9-6.

$$I_T = I_1 + I_2 + I_3$$
I_1 = current through R_3
I_2 = current through R_2
I_3 = current through the galvanometer
$I_1 + I_3$ = current through R_1
$I_2 + I_3$ = current through R_4

When R_4 is the unknown resistance, the scale of the meter can be calibrated in units of resistance, since any change in R_4 causes a proportional change in I_3, which is the current through the meter. Because it is the current through the meter which causes its deflection, there is a position of the pointer for each value of R_4.

In the null-balance Wheatstone bridge, the meter is used to detect the condition when there is no current through it. In this type of bridge circuit, R_3 is an adjustable resistance, and R_4 is the resistance being measured. R_1 and R_2 are the fixed resistors. The scale is located at R_3 in such a position that the slider, in addition to establishing the value of R_3, also indicates the value of R_4.

Since there is no current through the galvanometer when the null-balance Wheatstone bridge is in balanced condition, $I_3 = 0$.

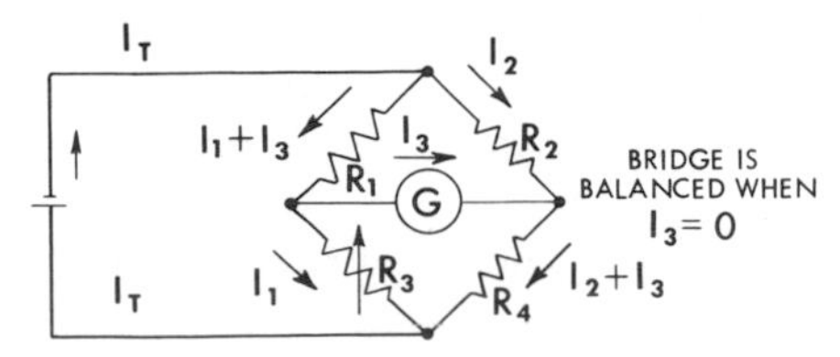

Fig. 9-6. The Wheatstone bridge type of circuit.

$$I_1R_1 = I_2R_2, \text{ or } I_1 = \frac{R_4I_2}{R_3}.$$

and

$$I_1R_3 = I_3R_4, \text{ or } I_1 = \frac{R_2I_2}{R_1},$$

Eliminating I_1,

$$\frac{R_2I_2}{R_1} = \frac{R_4I_2}{R_3}$$

and solving R_4,

$$R_4 = \frac{R_2R_3}{R_1}.$$

From this relationship it can be seen that for each value of R_4 there is a balancing value of R_3, making it possible to determine the value of R_4 by adjusting R_3.

Inductance. In addition to the components of an electrical circuit which resist the flow of current, there are those which resist a change in the current. This property of an electric component is called *inductance*. Fig. 9-7 is a simple circuit which demonstrates inductance.

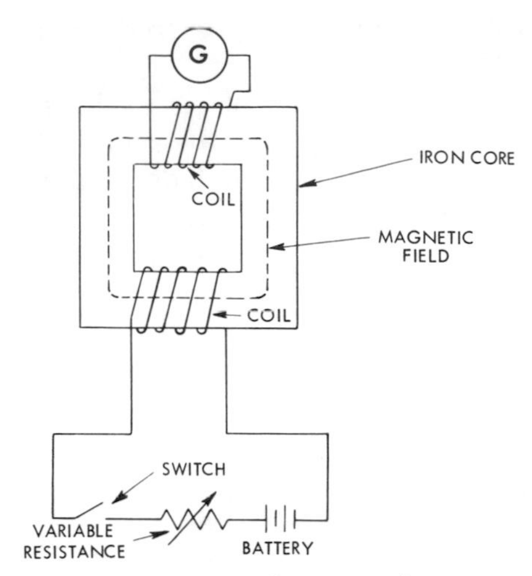

Fig. 9-7. A circuit showing inductance.

When the switch is closed, a current flows in the primary circuit. Magnetic lines of force are produced by the coil wrapped around the iron core. As the adjustable resistance is changed, the number of lines of force changes. This change causes a current to flow in the secondary circuit. The current in the secondary circuit flows in the direction which counteracts the effect of a change of current in the primary circuit. The voltage in the secondary circuit is called the induced *electromotive force* (emf). The magnitude of the voltage in the secondary circuit depends upon the rate of change of the primary circuit. This type of inductance, involving a primary and a secondary winding, is called *mutual inductance.*

The symbol for inductance is L, and the basic unit is the henry. When a current in the primary, which is changing at the rate of 1 ampere per second, induces an emf of 1 volt in the secondary, the circuit has a mutual inductance of 1 henry.

In addition to mutual inductance involving primary and secondary windings, there is also *self inductance,* which describes the property of a circuit to oppose any change in the current through it. See Fig. 9-8. Although even a straight wire has self-inductance, a coil is the most common inductor. The magnitude of the self-inductance is great enough to be of importance. When a current changing at the rate of 1 ampere per second induces a counter emf of 1 volt in a coil, the value of the inductance is 1 henry.

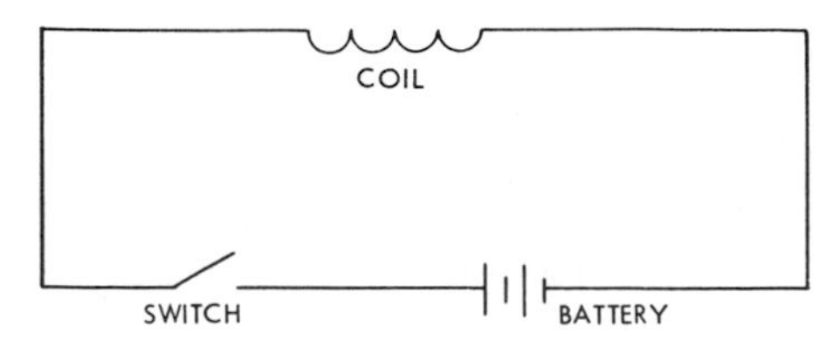

Fig. 9-8. A circuit showing self inductance. Self inductance produces an electromotive force in the circuit which tends to oppose the flow of the current.

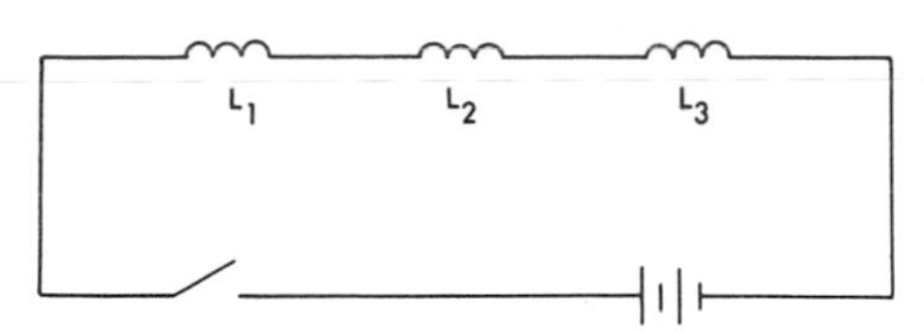

Fig. 9-9. Inductance in series.

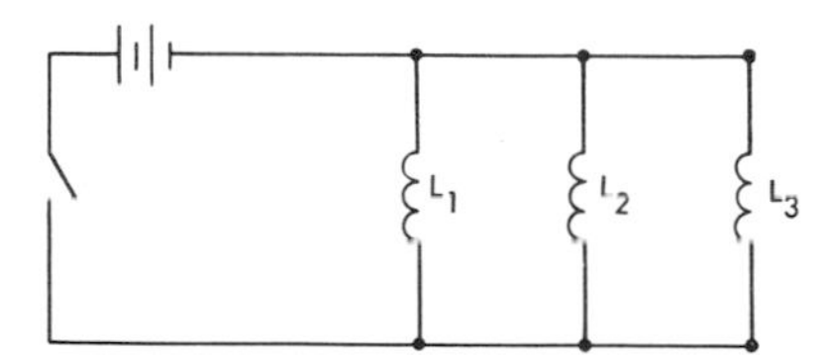

Fig. 9-10. Inductance in parallel.

Inductances may be joined in series or in parallel. When in series, as shown in Fig. 9-9, the total inductance equals the sum of the individual inductances: $L_T = L_1 + L_2 + L_3$, and so on.

When in parallel, as illustrated in Fig. 9-10, the reciprocal of the total inductance equals the sum of the reciprocals of the individual inductances:

$$\frac{1}{L_T} = \frac{1}{L_1} + \frac{1}{L_2} + \frac{1}{L_3}\text{, and so on.}$$

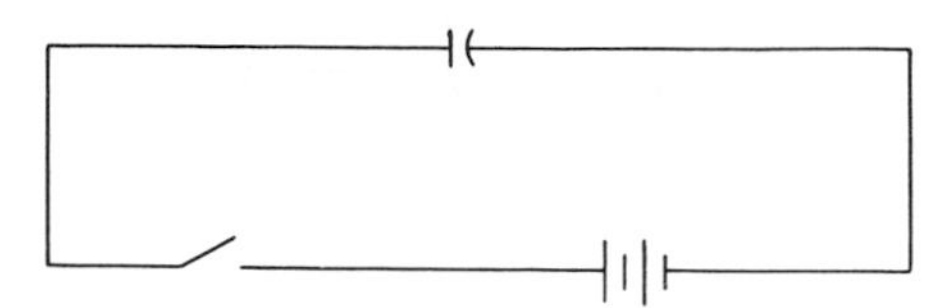

Fig. 9-11. Capacitor in circuit.

Capacitance. Inductance is the property of an electric circuit that resists a change in current. *Capacitance* is the property which resists a change in voltage. The circuit component which possesses capacitance is called a *capacitor.* A simple capacitor consists of two parallel metal plates separated by air or other nonconductive material called the *dielectric.* Fig. 9-11 illustrates a simple circuit containing a capacitor.

When the switch is closed, current begins to flow in the circuit. It continues to flow until the voltage across the plates of the capacitor equals the battery voltage. The current flow stops when capacitor voltage reaches this magnitude. The capacitor stores this voltage. The symbol for capacitance is the letter C. The unit of capacitance is the *farad.* A circuit has a capacitance of one farad when a charge of one coulomb is required to raise the circuit voltage by one volt:

$$C = \frac{Q}{V}.$$

A coulomb (Q) is a unit of quantity of electrical charge. It can be regarded as a certain number of charged particles (electrons). Another definition of the farad is: when a voltage, changing at the rate of one volt per second, causes a current of one ampere to flow into a capacitor, the capacitance is one farad. The farad is too large a unit for

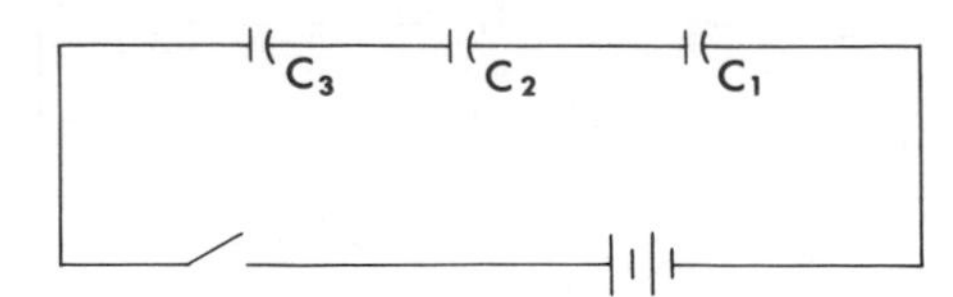

Fig. 9-12. Capacitors connected in series.

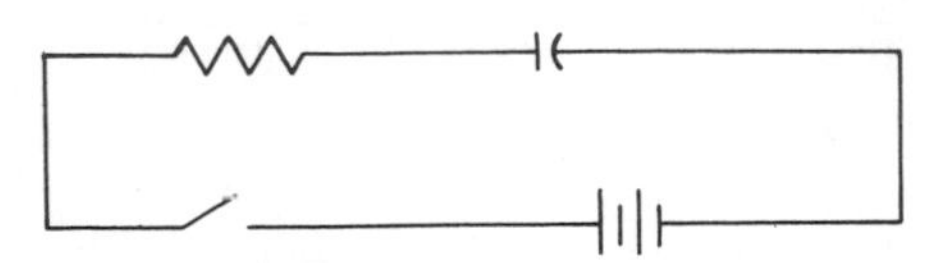

Fig. 9-14. The resistor protects the battery in the circuit.

most electrical circuitry, hence the microfarad is used. One microfarad equals one millionth of a farad. To express even smaller units of capacitance, the picofarad is used. One picofarad equals one millionth of a microfarad.

Capacitors can be connected in series or in parallel.

When the capacitors are connected in series, as shown in Fig. 9-12, the reciprocal of the total capacitance equals the sum of the reciprocals of the individual capacitances:

$$\frac{1}{C_T} = \frac{1}{C_1} + \frac{1}{C_2} + \frac{1}{C_3}, \text{ and so on.}$$

When the capacitors are connected in parallel, as shown in Fig. 9-13, the total capacitance equals the sum of the individual capacitances:

$$C_T = C_1 + C_2 + C_3, \text{ and so on.}$$

The build-up of voltage across a capacitor is called *charging,* and the dissipation of the voltage *discharging.* In actual circuits, a resistor is added to the circuit to limit the charging current, as shown in the resistor-capacitor network in Fig. 9-14.

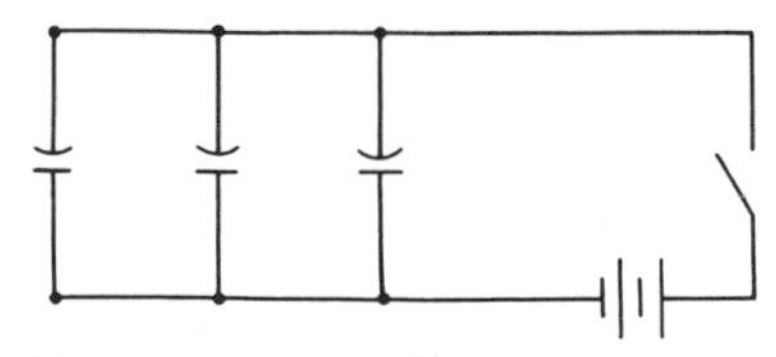

Fig. 9-13. Capacitors connected in parallel.

The effect of this current limiting resistor is to increase the time required to charge the capacitor. Fig. 9-15 is a graph of the rise of voltage across a capacitor being charged through a resistor. The curve starts the instant the switch is closed.

The time required for the capacitor voltage to reach 63.2% of its final value is termed the *time constant.* The time constant (t, in seconds) equals resistance (in ohms) times capacitance (in farads), or $t = RC$.

Fig. 9-16 is a graph of the discharge of the same capacitor in the same circuit.

Note that the capacitor discharge curve has the same shape as the charge curve – in fact is almost the mirror image of it. The time constant of discharge is the same as that for the charge to build up, or $t = RC$.

Resistance-capacitance (RC) networks have many applications in instrumentation. For example:

1. Reset action in an electrical controller which is accomplished in RC network, as shown in Fig. 9-17.
2. Derivative action in an electrical

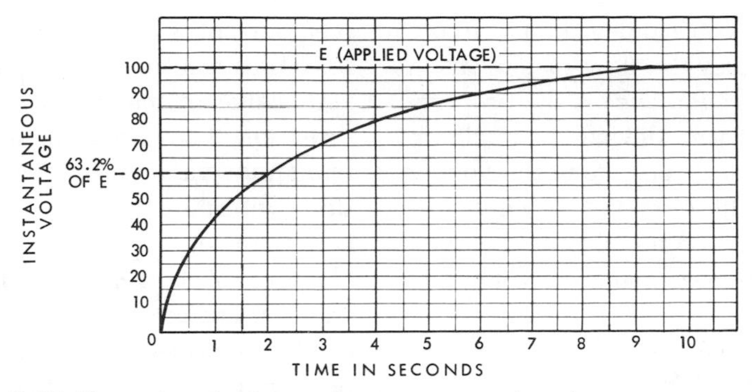

Fig. 9-15. The resistor increases the time required to charge the capacitor.

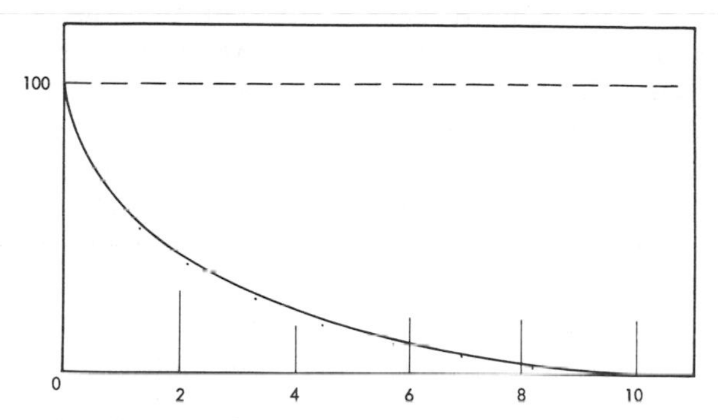

Fig. 9-16. Discharge time of the same capacitor.

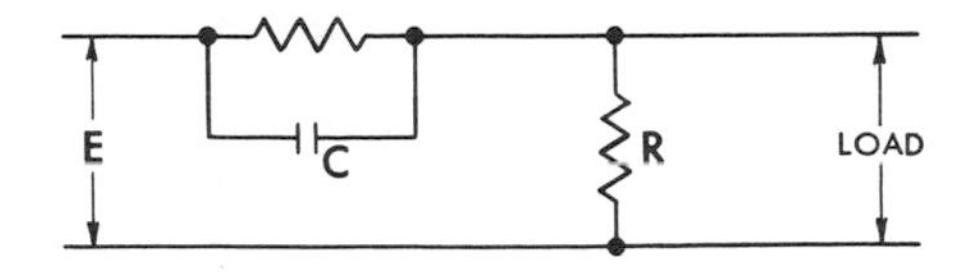

Fig. 9-17. Resistance-capacitor network used to obtain reset action.

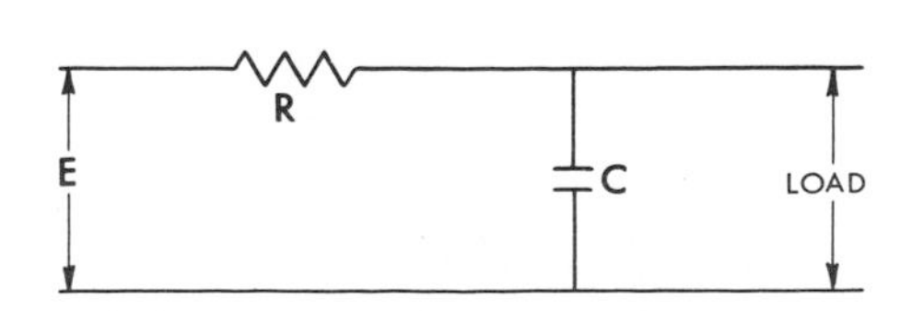

Fig. 9-18. Resistance-capacitor network used to obtain derivative action.

controller which is accomplished in an RC network as shown in Fig. 9-18.

3. Precise timers for short intervals made using these RC networks.

In conclusion, *resistance* is the property of an electrical device in a circuit, which opposes the flow of current in the circuit. *Inductance* is the property of an electrical device in a circuit which opposes a change in the flow of current in the circuit. *Capacitance* is the property of an electrical circuit which opposes a change in voltage in the circuit.

Alternating Current

All the preceding information pertains to direct current which can be defined as being current flow in one direction. Since the resistance of the circuit is of a fixed value, the voltage will also maintain a steady value.

In an alternating current (ac) circuit the voltage and current vary continuously between zero and maximum values. A comparison of ac and dc currents is illustrated in Fig. 9-19. In the diagram, the horizontal base line represents a zero value of current. The portion of the drawing above the line represents current values in a positive direction. The portion below the line represents negative current values. The dotted line, beginning at 10 amps in the left of the figure, represents a direct current of 10 amps. The curved lines illustrate an ac wave passing from zero value to maximum in a positive direction, back to zero, and then to maximum value in a negative direction, returning to zero value. Each complete set of positive and negative changes of current is called a full wave or *cycle*.

The number of complete cycles occurring in one second is referred to as the frequency of the circuit. The unit of frequency (f) is the *hertz* (Hz). One hertz equals one cycle per second. Alternating current, in the United States, is usually supplied at 60 hertz (cycles per second), although frequencies vary with individual countries.

In the diagram, the points of the ac wave are marked in angular degrees because of the angular rotation of the generator that produces alternating current. The amplitude of the curve at any point indicates the value of the current at that instant. The number of degrees in a complete cycle is always 360 even though its frequency can change. In the diagram, if the time required for one cycle is 1/60 of a second, the frequency is 60 hertz.

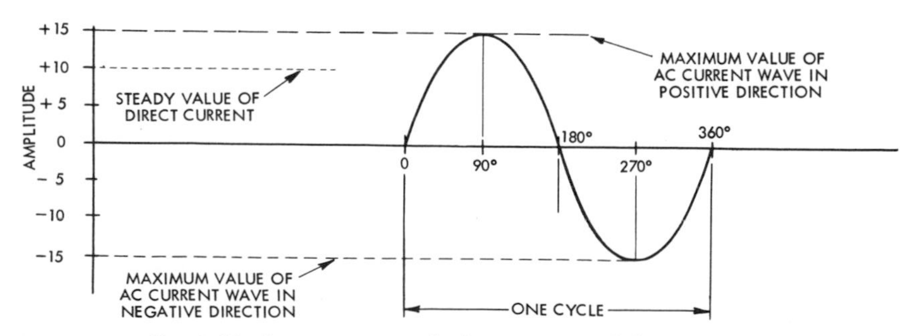

Fig. 9-19. A comparison of alternating and direct current.

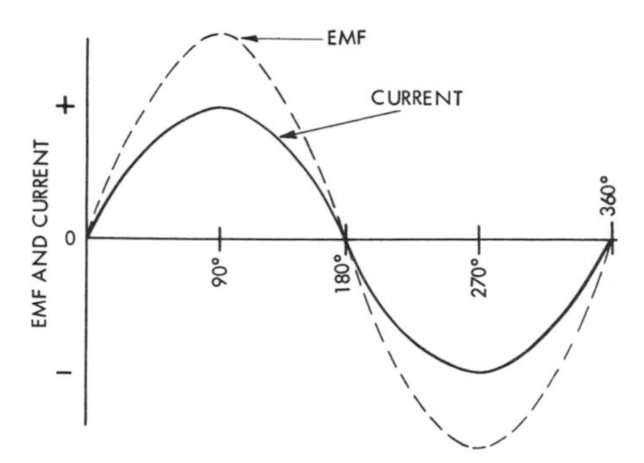

Fig. 9-20. Emf and current reach peak value at the same instant.

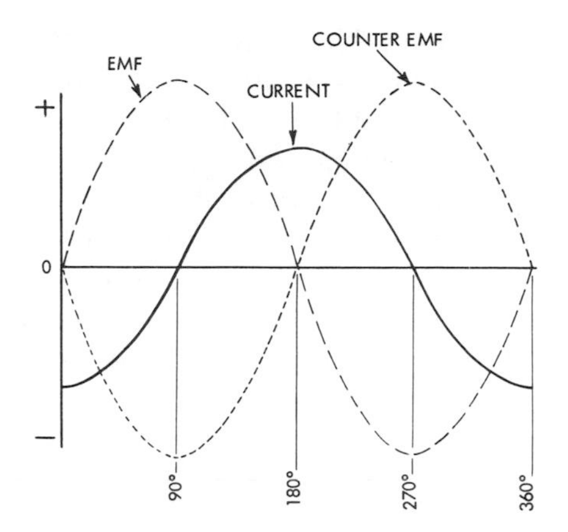

Fig. 9-21. Applied voltage and induced voltage in a single inductor circuit.

Fig. 9-20 shows the curves of the voltage and current in an alternating current resistance circuit. Note that both reach their peak value at the same instant.

Inductive Reactance. Consider now an inductor in an alternating current circuit. It should be remembered that an inductor opposes a change in current through it by creating a voltage which opposes the supply voltage. The voltage across an inductor must always equal the supply voltage. Since in an alternating current circuit the supply voltage is changing continuously, the opposing voltage also changes continuously. The value of the opposing voltage is proportional to the rate at which the current through the inductor is changing. Fig. 9-21 is a graph of the applied voltage and the induced voltage in a single inductor circuit. Note that when the applied voltage is at its maximum (positive) value, the opposing voltage is at its minimum (negative) value. When the opposing voltage is at its maximum value, the current must be changing at a maximum rate. When the opposing voltage is at the zero level, the current is not changing.

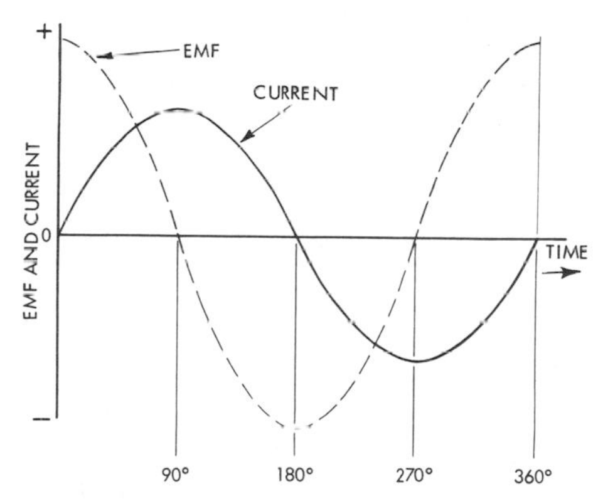

Fig. 9-22. Current lagging voltage by 90°.

It is now possible to draw a curve of the current on the graph. See Fig. 9-22. It will be seen that the current lags the applied and opposing voltage by 90° or 1/4 of a cycle. Since the curves of voltage and current both follow fixed patterns there is a constant voltage to current ratio for a particular inductance and applied voltage. Since, according to Ohm's law, the ratio E/I represents resistance, inductance in an ac circuit is similar to but not the same as resistance. A special term is applied to the resistive effect of inductance. In an alternating current circuit this term is *inductive reactance.* The unit of inductive reactance is the ohm and the symbol is X. A subscript $_L$ can be used to indicate inductance:

$$X_L = \frac{E_L}{I_L} \text{ or}$$

$$1 \text{ ohm inductive reactance} = \frac{1 \text{ volt across inductor}}{1 \text{ amp through inductor}}$$

The magnitude of inductive reactance is directly proportional to the frequency, but not to the magnitude of the applied voltage. The inductive reactance is expressed by the formula:

$$X_L = 2\pi fL$$

when

L = inductance in henrys
X_L = inductive reactance in ohms
f = frequency in hertz.

Changing only the magnitude of the applied voltage does not change the inductive reactance. This is true because the E/I ratio remains constant for a particular inductance.

Capacitive Reactance. Capacitance, it should be recalled, is the property of a circuit which resists a change in voltage. In a single capacitance alternating circuit the supply voltage is changing continuously, therefore the voltage across the capacitor must change continuously. This means that the capacitor must be continuously charging and discharging, thus apparently causing a current to flow through it, but actually discharging on each

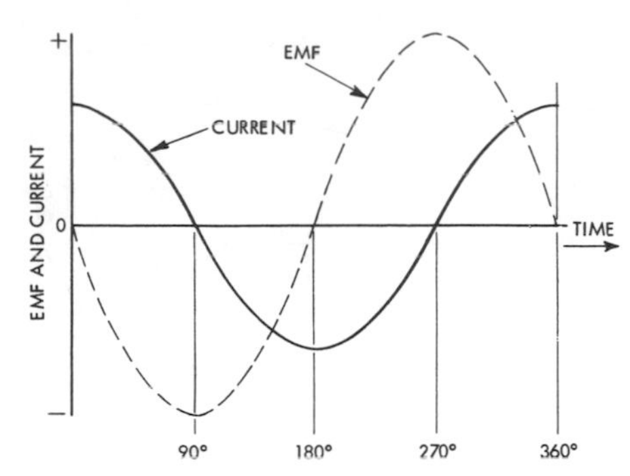

Fig. 9-23. Current leading voltage by 90°.

alternation or reversal of current. The current is proportional to the rate at which the voltage across the capacitor is changing. When the voltage is changing at its maximum rate, the current is at its maximum (positive or negative) value. When the voltage is not changing, the current is zero. From the graph Fig. 9-23, it can be seen that the current leads the voltage by 90° or ¼ cycle. The curves of voltage and current maintain a fixed pattern. Therefore, a constant ratio exists.

As has been pointed out under inductance, the ratio E/I represents resistance. Therefore, capacitance in an ac circuit behaves like resistance, except that it is not instantaneous. The term for this resistance to the flow of alternating current is *capacitive reactance.* The unit of capacitive reactance is the ohm, and the symbol is X_C:

$$X_C = \frac{E_C}{I_C} \text{ or}$$

$$1 \text{ ohm capacitive reactance} = \frac{1 \text{ volt across capacitor}}{1 \text{ amp through capacitor}}$$

The magnitude of capacitive reactance (X_C) is inversely proportional to frequency, expressed by the formula:

$$X_C = \frac{I}{2\pi fC}$$

when

X_C = capacitive reactance in ohms
C = capacitance in farads

(**Note:** Capacitive reaction acts the opposite of inductive capacitance.)

Changing the magnitude of the applied voltage does not change the capacitive reactance.

Impedance. The total resistance to current flow in an alternating current circuit is called *impedance.* The symbol for impedance is Z:

$$Z = \frac{E \text{ (applied voltage)}}{I \text{ (circuit current)}}.$$

The value of impedance is obtained by diagramming, using a right triangle with θ being the angle between the applied voltage and the resulting current.

For a circuit containing a resistance and an inductance in series the value of $Z = \sqrt{R^2 + X_L^2}$. See Fig. 9-24.

For a circuit containing a resistance and a capacitance in series the value of $Z = \sqrt{R^2 + X_C^2}$. See Fig. 9-25.

For a circuit containing a resistance, an inductance and a capacitance in series the value of Z =

$$\sqrt{R^2 + (X_L - X_C)^2}.$$

In this discussion of reactance and impedance, the mathematical relationships have been reduced to a minimum. The purpose of the discussion was to introduce the student to the terms and explain them as simply as possible.

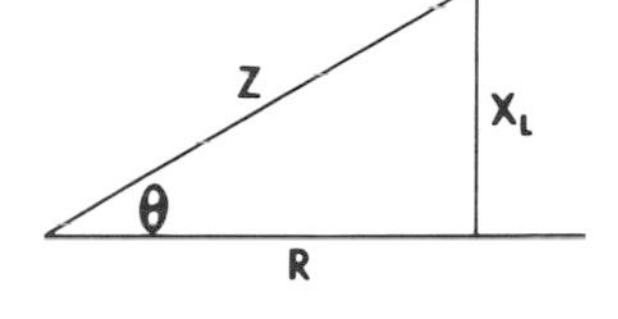

Fig. 9-24. Inductive resistance.

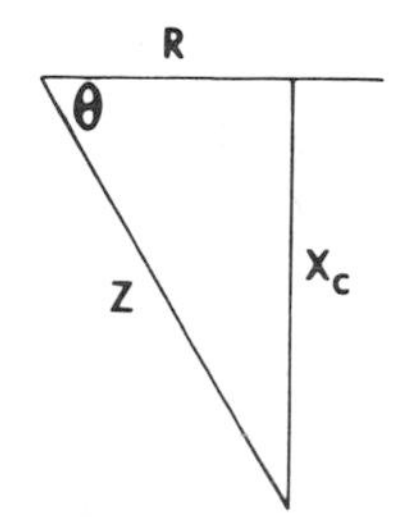

Fig. 9-25. Capacitive resistance.

Electronics

Electron Tube Fundamentals

Although the use of electron tubes (frequently referred to as *vacuum* tubes) is rapidly declining, they are still included for study in this text because their principles and function are similar to their replacements. Also, early instrumentation which is still functional includes electron tubes among its components.

The electron tube operates on the principle known as *thermonic emission,* which refers to the freeing of electrons electrically heated in a vacuum. The heat is supplied by a wire filament heated to incandescence. The higher the temperature, the more electrons are emitted. See Fig. 9-26. Another electrical conductor *(plate),* located in the tube and supplied with a positive electric charge from outside the tube, gathers the electrons which are negatively charged. This directional flow of electrons constitutes a direct electric current. The higher the positive charge on the plate, the higher the plate current. A simple two element vacuum tube is called a *diode.* See Fig. 9-27.

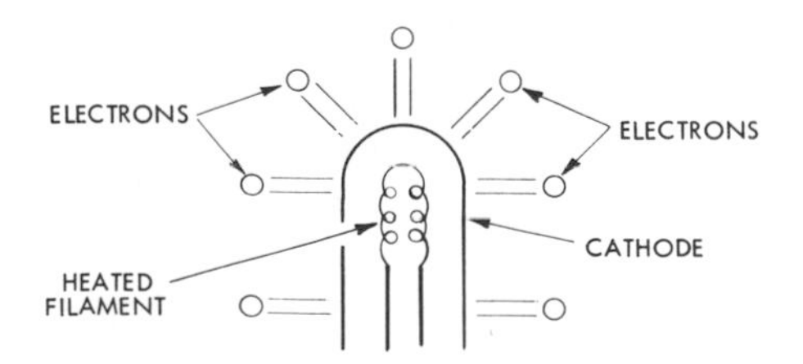

Fig. 9-26. Thermionic emission.

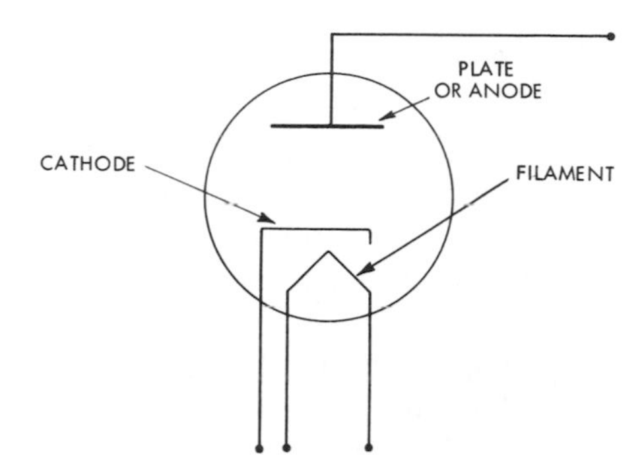

Fig. 9-27. Elements of a diode.

Rectification. Since the current flow in an electron tube is always in one direction (from negative to positive) a two element tube (diode) can be used to change alternating current to direct current. This is known as *rectification.* The plate is positive with respect to the *cathode* (the element which emits electrons) only during one half cycle of the alternating current. The resulting direct current occurs only during this half cycle, and hence is not continuous, nor is it of constant magnitude. This is known as *half-wave rectification.* See Fig. 9-28. It is possible to make the direct current continuous by adding a second plate which is positive with respect to the cathode during the other half cycle of the alternating current. This is termed *full-wave rectification.* See Fig. 9-29.

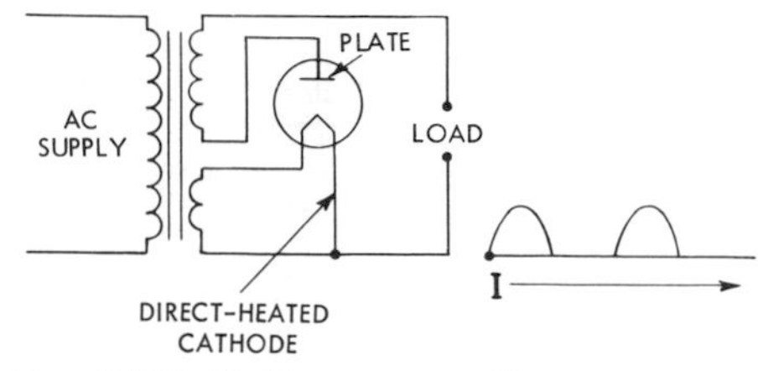

Fig. 9-28. Half-wave rectification.

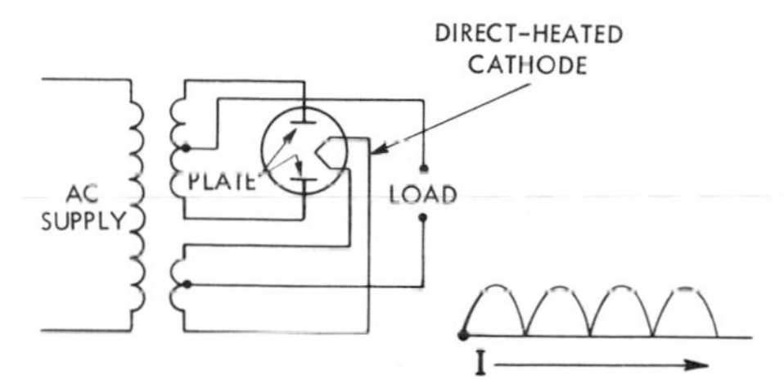

Fig. 9-29. Full-wave rectification.

Amplification. A third element called a *grid,* when inserted between the cathode and the plate, permits regulation of the number of electrons reaching the plate. Such a tube is called a *triode.* See Fig. 9-30. When the grid is positive with respect to the cathode, more electrons pass to the plate. When the grid is negative with respect to the cathode, few electrons reach the plate. A small change in the voltage of the grid has the same effect as a large change in the plate voltage. This is what is meant by *amplification,* and the ratio of grid voltage and plate voltage which produces equal results is called the amplification factor of the tube. An amplification factor of 20, as shown in Fig. 9-31, means that if the grid voltage is changed by 1 volt, the effect on the plate current is the same as if the plate voltage were changed by 20 volts.

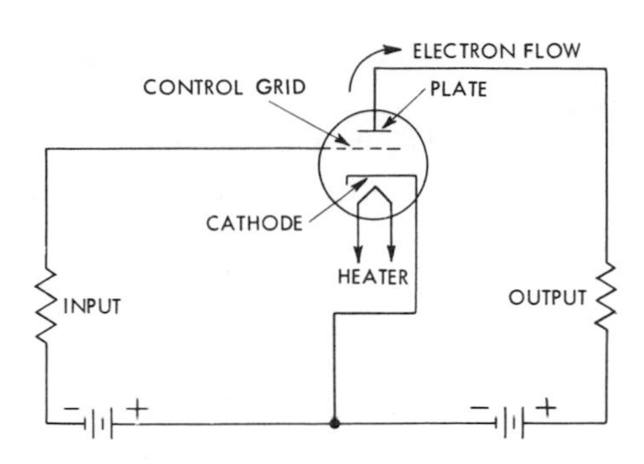

Fig. 9-30. Elements of triode.

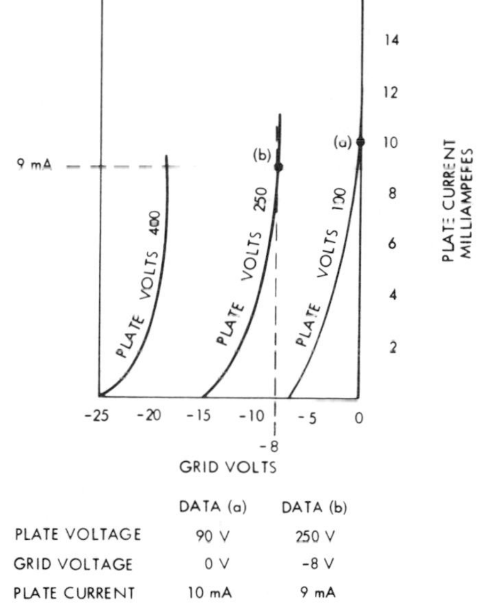

	DATA (a)	DATA (b)
PLATE VOLTAGE	90 V	250 V
GRID VOLTAGE	0 V	-8 V
PLATE CURRENT	10 mA	9 mA

Fig. 9-31. Characteristics of a tube with an amplification factor of 20.

The symbol for the amplification factor is the Greek letter μ (pronounced *moo*). The amplification factors of electron tubes range from about 3 to 100. When the amplification factor is less than 8 the tube is said to have a low μ. Medium μ tubes are those whose amplification factor is between

8 and 30. Above 30, the tube has a high μ.

The amplification characteristics of a tube are generally described in graphic form. The curves describe tube performance when the plate current is made to flow through resistances of various magnitudes.

Oscillation. The amplification can be modified by adding or subtracting some of the plate voltage and the grid voltage. This arrangement is called *feedback*. When a part of the plate voltage is added to the grid voltage, the feedback is positive. When it is subtracted, the feedback is negative. Positive feedback increases the amplification. When all the grid voltage is supplied by positive feedback and the magnitude of the feedback is great enough, the tube becomes an oscillator.

Negative feedback decreases the amplification. As negative feedback increases, tube performance according to the characteristic curve decreases. Negative feedback also tends to eliminate distortion of the plate current.

The three elements of the vacuum tube form three small capacitors. The capacitances are those between:

1. cathode and plate
2. cathode and grid
3. grid and plate

They are called interelectrode capacitances and can cause instability. The capacitance between the grid and plate, which is the most disturbing, can be reduced by the addition of a fourth element, a screen. The four element tube, as shown in Fig. 9-32, is called a *tetrode*. The screen is placed between the grid and the plate, and is positive with respect to the cathode.

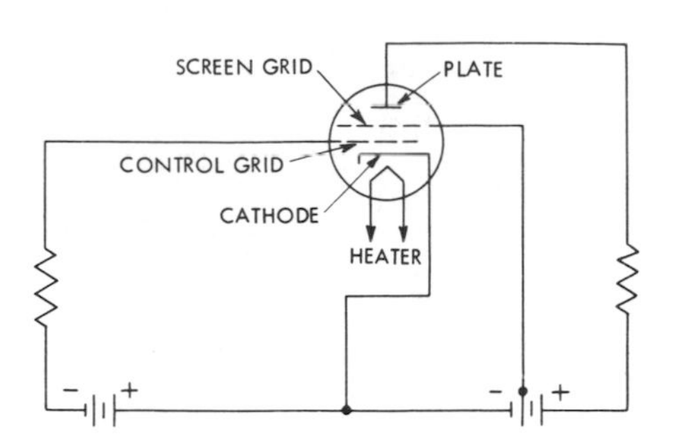

Fig. 9-32. Elements of a tetrode.

It attracts electrons from the cathodes, but most of these pass through the positive with respect to the cathode, screen to the plate. Thus, the screen reduces the capacitance between grid and plate. In addition, because the plate current is more dependent on the screen voltage than on the plate voltage, it is possible to obtain higher amplification with a tetrode.

Because of the closeness of the screen to the plate it is possible for the electrons which pass to the plate at very high speed to dislodge electrons already there and cause them to be attracted to the screen. This would lower the plate current. To eliminate this possibility a fifth element (a *suppressor*) is added between the screen and the plate. The five element tube, shown in Fig. 9-33, is called a *pentode*. The suppressor is generally connected to the cathode and is negative with respect to the plate. Because of its negative charge, the suppressor repels the dislodged electrons back to the plate. Thus, the pentode is more efficient than the tetrode or triode and permits greater amplification.

Amplification is the most common application of electron tube circuits in instrumentation. For instance, the volt-

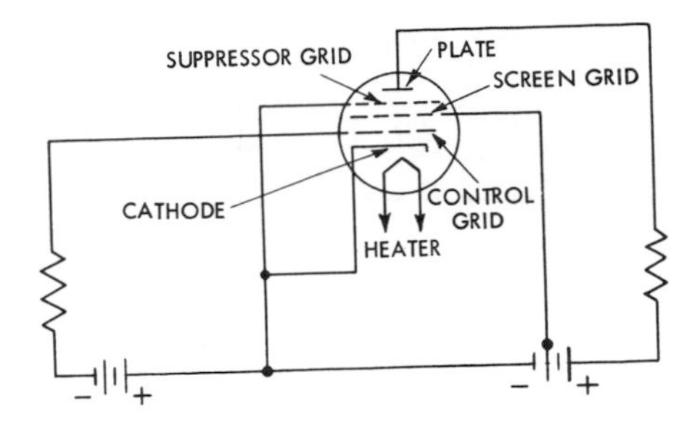

Fig. 9-33. Elements of a pentode.

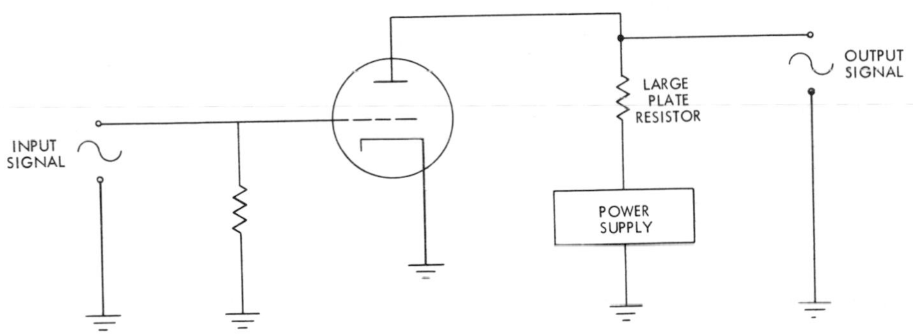

Fig. 9-34. A triode voltage amplifier.

age output of a thermocouple requires amplification, enabling it to operate a motor. As has been stated, the dc thermocouple voltage is converted to an ac voltage before amplification. A simple electronic amplifier, shown in Fig. 9-34, uses a single triode with appropriate resistors to establish the voltage levels on cathode grid and plate. The plate resistor is large, making the voltage variations across it consistently greater than the variations of the input voltage. The plate voltage establishes the output voltage.

The addition of a second tube, shown in Fig. 9-35, and associated resistors results in a total amplification equal to the product of the amplification of each section. The result is a two-stage amplifier. Note that a capacitor is connected across the resistance to the cathode of each tube.

These capacitors maintain a constant voltage across the resistors. There is also a capacitor between the output of the first stage and the second. This capacitor blocks the dc potential of the plate of the first tube from entering the second stage. Other methods for joining the two stages are also used. An inductor or transformer can be substituted for the capacitor. This is the arrangement on which all amplifiers are based. Although multi-state amplifier circuits are considerably more complicated, close examination reveals the presence of several fundamental circuits.

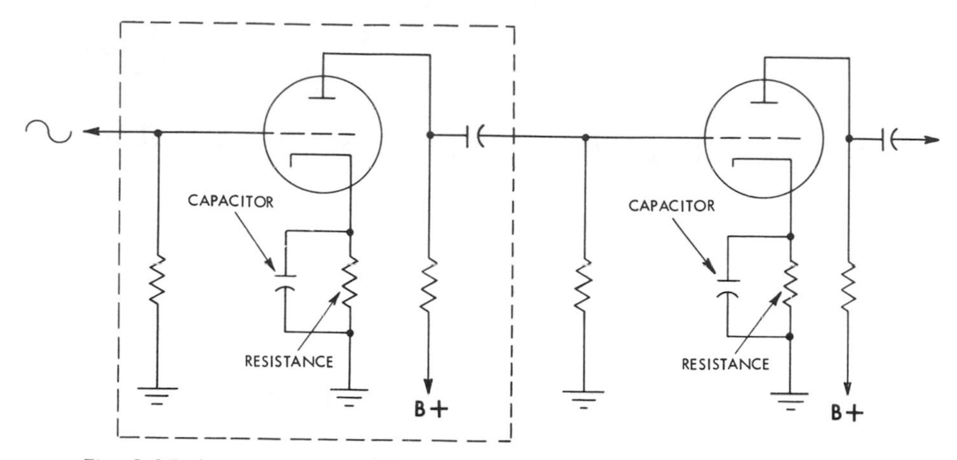

Fig. 9-35. A two-stage amplifier circuit. One stage is within the dotted lines.

Transistor Fundamentals

Transistors have replaced electron (vacuum) tubes. The small size of transistors, their low operating voltage, stability and long life are some of their advantages over electron tubes. Transistors are made of materials such as silicon and germanium. Such materials are called *semiconductors*. They have greater resistance to the flow of current than metallic conductors, but not as great a resistance as insulators.

Semiconductor crystals used in the manufacture of transistors can be either N-type or P-type. The N-type is one which, by the addition of a donor impurity, possesses a quantity of free electrons. It is therefore a material which depends upon the flow of free electrons for its conductivity.

The P-type is one, which, by the addition of an acceptor impurity, is left with a quantity of holes. These holes are electrical charge carriers similar to electrons but possessing instead a positive charge. Therefore, the P-type is also a material which depends upon the flow of positively charged holes for its conductivity.

There are five principal classifications of semiconductors:

1. diode
2. bipolar or junction transistor
3. silicon controlled rectifier (SCR)
4. unijunction transistor
5. unipolar or field effect transistor

Semiconductor Diodes. A semiconductor diode is usually a single crystal of semiconductor material which is artificially created. One half of the crystal is made N-type and the other made P-type by the addition of the appropriate impurities during processing. See Fig. 9-36.

If an external voltage source is applied to a diode so that the negative lead is attached to the N-type half and the positive lead attached to the P-type half, current flows through the diode. The negatively charged electrons are attracted to the positive P-

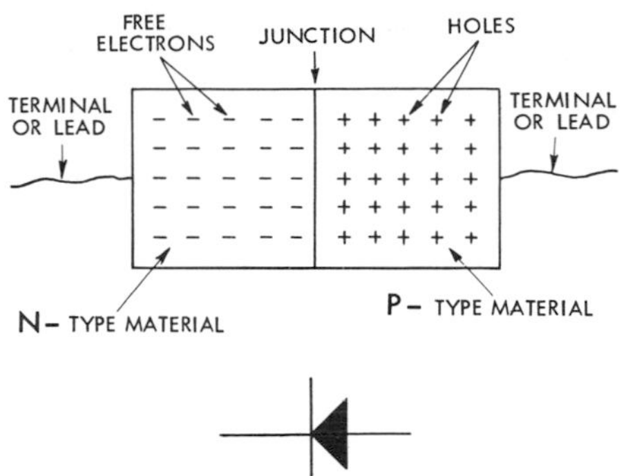

Fig. 9-36. A semiconductor diode and its schematic symbol. Electron flow is opposite to direction of arrow.

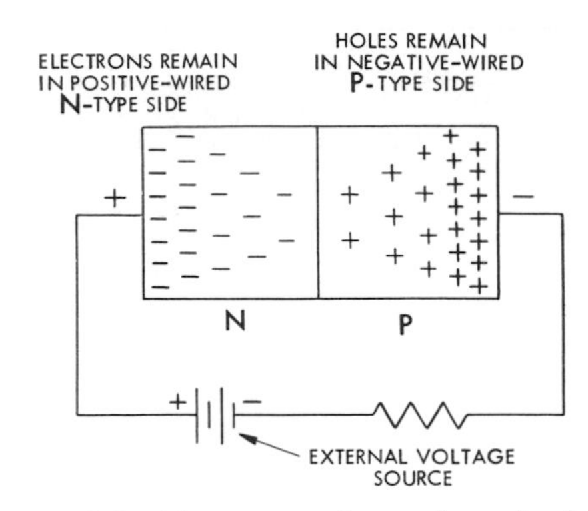

Fig. 9-38. No current flows through the diode when the externally applied voltage makes the N-type side positive and the P-type side negative.

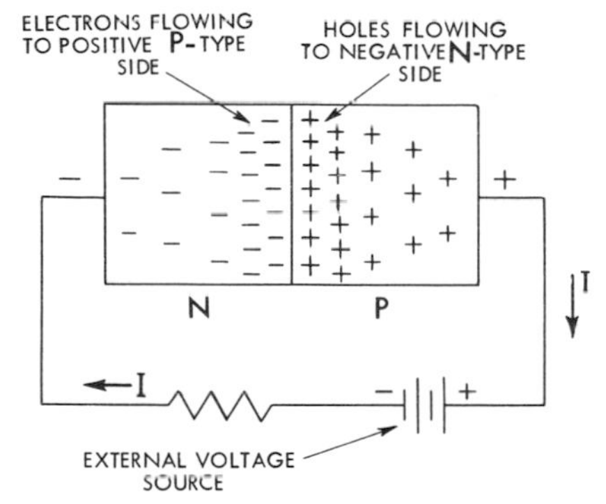

Fig. 9-37. Current flows through the diode and in the external circuit when the externally applied voltage makes the N-type side negative and the P-type side positive.

type side, and the positive holes are attracted to the negative N-type side. See Fig. 9-37. However, if the externally applied voltage is reversed, making the lead to the N-type side positive and the lead to the P-type side negative, no current flows through the diode. The electrons are repelled from the negative P-type side and the positive holes are repelled by the positive N-type side. See Fig. 9-38.

The practical value of the semiconductor diode lies in its ability to conduct electricity when the N side is wired negatively and the P side positively, and to stop conducting when the N side is wired positively and the P side negatively. The most common use of this selective conduction is the process known as *rectification*.

Rectification. The process known as rectification, as we have seen in the section on electron tubes, results in converting alternating current (ac) into direct current (dc). The simplest form of rectifier circuit is the half-wave circuit. See Fig. 9-39. In this circuit an ac voltage E_S is applied to the series combination of the load and the diode. During one half of the input cycle the P region of the diode is positive and the N region negative, and current flows through the diode giving an output voltage E_R across the load. During the other half of the input cycle, the N

Worst enimy of transisters
Heat

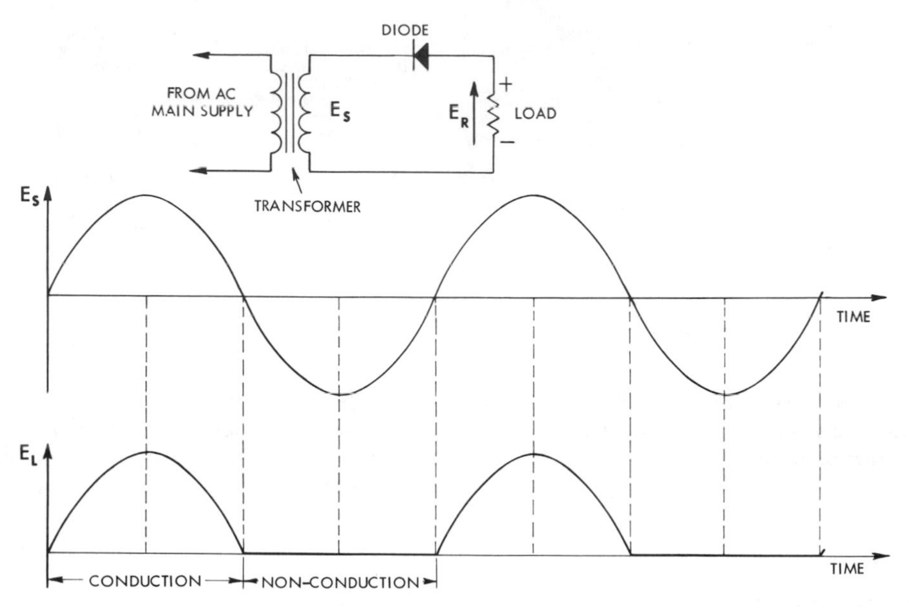

Fig. 9-39. Only one half of the wave form, which represents the ac current phase of the acceptable polarity, passes through the half-wave rectifier.

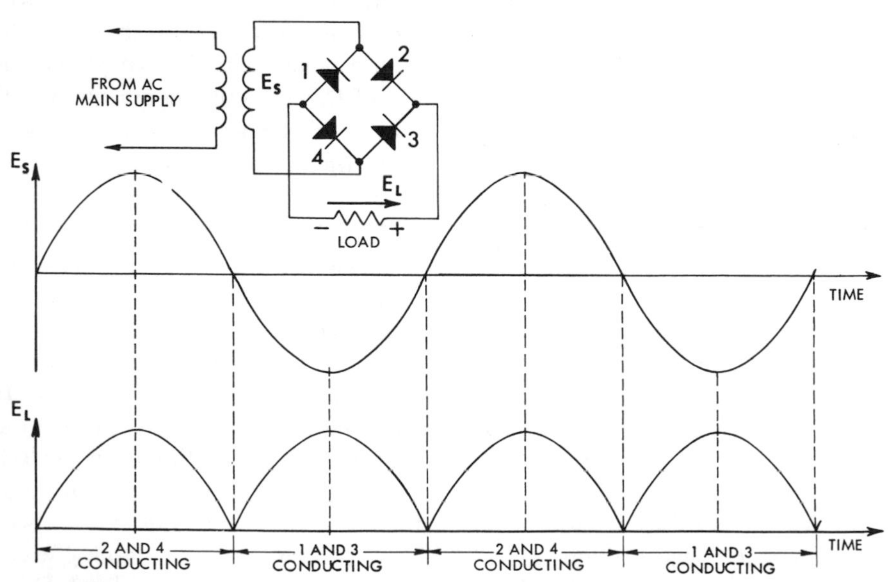

Fig. 9-40. With the circuit arrangement, all ac current is passed by the alternating pairs of diodes. The effect is full wave rectification.

region of the diode is positive and the P region is negative. The diode will not conduct, resulting in no output voltage.

A better, more common form of rectifier circuit uses four diodes in a bridge circuit. It is known as a full-wave bridge rectifier circuit. See Fig. 9-40. During the first half cycle when the top lead of the transformer is positive and the bottom lead negative, diodes 1 and 3 do not conduct. However, diodes 2 and 4 conduct, resulting in a voltage across the load. During the next half-cycle when the voltage from the transformer has reversed, diodes 2 and 4 do not conduct. Diodes 1 and 3 do conduct, again giving an output voltage. In both cases, the output voltage is of the same polarity.

The output voltage from the rectifier circuits is not steady dc. The dips in the output voltage can be filled in by placing a capacitor across the load. A capacitor so used is called a filter capacitor. Usually several capacitors are used in conjunction with resistors or inductors.

Junction Transistors. A bipolar transistor is a single crystal of silicon or germanium containing three dissimilar sections: N-type, P-type, and N-type, or P-type, N-type, and P-type. See Fig. 9-41. The three sections are referred to as the emitter, base, and collector. The base is always the center region. In both types there are two separate N-P junctions, hence the term *bipolar*.

NPN Transistors. Fig. 9-42 shows an NPN transistor connected to the external power source necessary to provide a current flow through the transistor. Disregarding the N-type collector for a moment, we see that the nega-

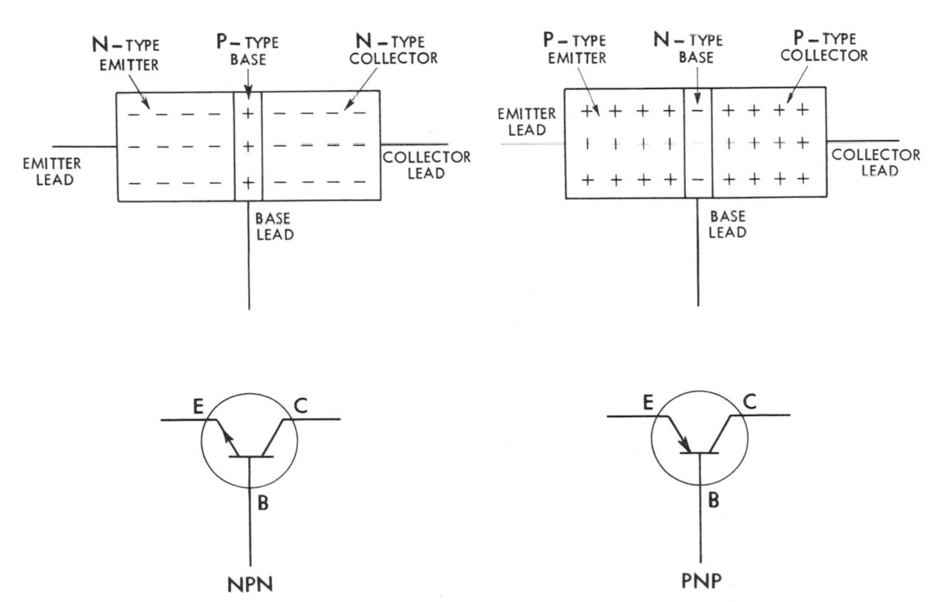

Fig. 9-41. The construction and the schematic symbols for NPN and PNP transistors.

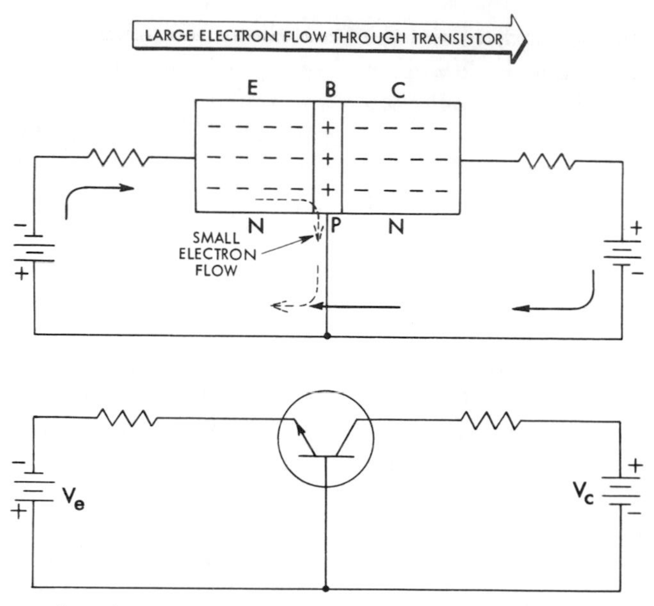

Fig. 9-42. A complete biasing circuit for an NPN transistor which is connected in a common base configuration.

tive lead of one battery is connected to the N-type emitter. The positive lead is connected to the P-type base. The negative terminal connection has the effect of repulsing the negatively charged free electrons in the N-type emitter to the emitter-base junction. At the same time, the positive terminal connection to the base repulses the positive holes to the emitter-base junction where the electrons and holes readily combine. The net effect is an electron flow across the junction, sustained as long as the battery is connected.

If the battery leads are reversed, the positive battery lead attracts the free electrons of the N-type emitter away from the junction. The negative battery lead similarly attracts the holes away from the junction. The net effect is to prevent electrons from flowing across the junction.

We see then that the emitter-base connection resembles the simple diode. When the battery leads are connected as shown in Fig. 9-42, current crosses the emitter-base junction. The emitter and base are said to be forward biased. When the battery leads are reversed, no current flow is possible. The emitter and base are then said to be reverse biased.

Checking the base-collector connections would seem to indicate that no current can flow through the base-collector junction. The base and collector are reverse biased. It would further seem that, since there can be no current flow from the base to the collector, there can be no current flow through the transistor. However, the transistor is constructed with an extremely thin base. Electrons readily cross the emitter-base junction. The extreme thinness of the base means that relatively few electrons crossing the emitter-base junction combine with holes of the P-type base material. The positive voltage at the N-type collector terminal attracts most of the electrons *through* the base region, across the base-collector junction, and to the collector. Enough electrons combine with holes to create a small emitter-base current. This completes the emitter-collector circuit, and establishes a current flow through the transistor.

PNP Transistors. PNP transistors are generally connected so that the N-type region is made negative and the P-type region positive for the emitter-to-base junction. The N-type region is wired positive and the P-type region negative for the base-collector junction.

One would again expect, from diode theory, that a current would flow from the emitter to the base, but that no current would flow from the base to the collector. This, however, is not correct. In the case of a PNP transistor, holes are attracted from the emitter to the base because the base has been negatively biased by the external voltage source. The collector is, in turn, negative with respect to the base. Since the base is thin, most of the holes attracted to it from the emitter flow through it into the collector. In a given transistor, the percentage of holes leaving the emitter that get into the collector is relatively constant for a wide range of applied bias voltages. This value is usually about 98%. The remaining 2% of the holes are conducted out of the base lead.

The NPN transistor shown in Fig. 9-42 operates primarily by establishing a large electron flow through the transistor from the emitter to the collector. The PNP transistor, however, depends for its operation on the flow of *holes* from emitter to collector. Although there is a flow of electrons through the transistor circuit in a direction opposite to that of the flow of holes within the transistor, the main current-carrying activity within the PNP transistor itself (not the entire circuit) is the flow of positively charged holes.

The Silicon-Controlled Rectifier (SCR). The silicon-controlled rectifier (SCR) is a type of diode with very low resistance when conducting and very high resistance in the Off state. Because of these characteristics SCR's are mainly used in high-power control and switching circuits, and they may be used to control either ac or dc current.

In operation, the SCR is similar to the thyratron tube and falls under the general solid-state category of thyristors. In construction, the SCR is a four-layer semiconductor. It is biased similar to a normal rectifier but will not conduct current until it is gated.

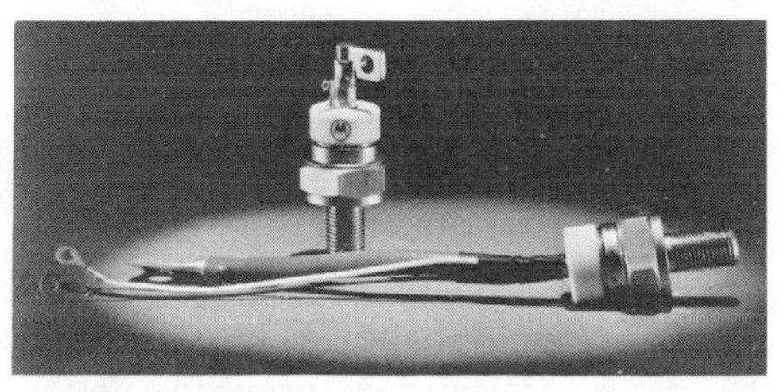

Fig. 9-43. A typical silicon-controlled rectifier (SCR). (Motorola Semiconductor Products, Inc.)

Fig. 9-43 is a photograph of a typical SCR in its case.

The Unijunction Transistor (UJT). The unijunction transistor is a one-junction solid-state device. It is best used as an oscillator. The UJT oscillator (relaxation type) is described by its waveform, which is sawtooth in appearance.

The UJT is used in application with a capacitor. When charging, the output waveform of the UJT looks like a normal sine wave. When the capacitor is charged to maximum, current stops flowing through the UJT and the capacitor then discharges rapidly through the UJT. The UJT and the capacitor act together, producing a slow rise and a very quick descent of the waveform, hence the name *relaxation oscillator.*

The UJT can also be used to trigger silicon-controlled rectifiers (SCR's) at distinct time intervals. This allows the SCR to turn on circuit functions such as motor controls.

The UJT has three elements, as shown in Fig. 9-44: an emitter lead and two base leads (1 and 2). It is formed from a large N-type silicon bar and a small P-type particle bonded to its side or top. When reverse voltage is applied across the emitter and base 1, the UJT transistor acts as a resistor. If forward voltage is applied to the emitter, the resistance between the emitter and base 1 decreases, therefore increasing current through the bases. More emitter voltage causes more base current, less emitter voltage causes less base current.

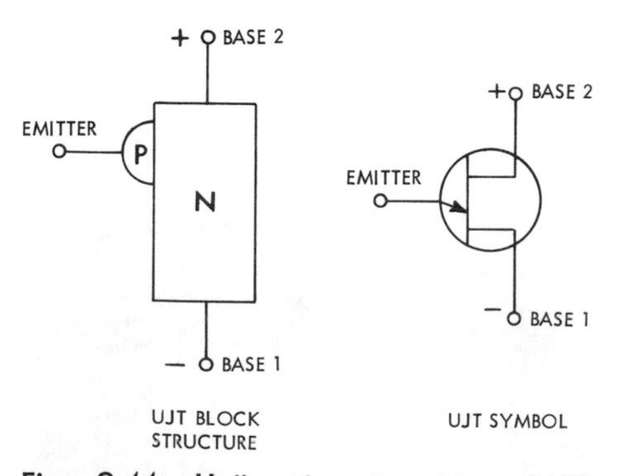

Fig. 9-44. Unijunction transistor (UJT), showing functional diagram of block-structure type at left, and schematic diagram at right.

Transistor Amplification. The function of the transistor, when used as an amplifier, is simply to use a small current flow between base and emitter to control a larger current flow between emitter and collector. The ability of a transistor to increase (or amplify) a signal current depends upon the fixed percentage of current which flows from the emitter to the collector. When a voltage is applied to the base in such a manner as to increase the forward biasing, or voltage, the emitter and collector currents are increased. The percentage of amplification is a fixed characteristic of the particular transistor.

For example, assume that a transistor has an emitter current of 1 milliampere. Using the 98% and 2% values mentioned above, the collector current would be 0.98 ma and the base current 0.02 ma. If a voltage of 4 millivolts is applied to the base, the base current and the collector current would both be increased by 10%. Thus by merely applying the voltage to the base, the transistor has amplified the current.

Junction transistors can be used as amplifiers in several different circuit

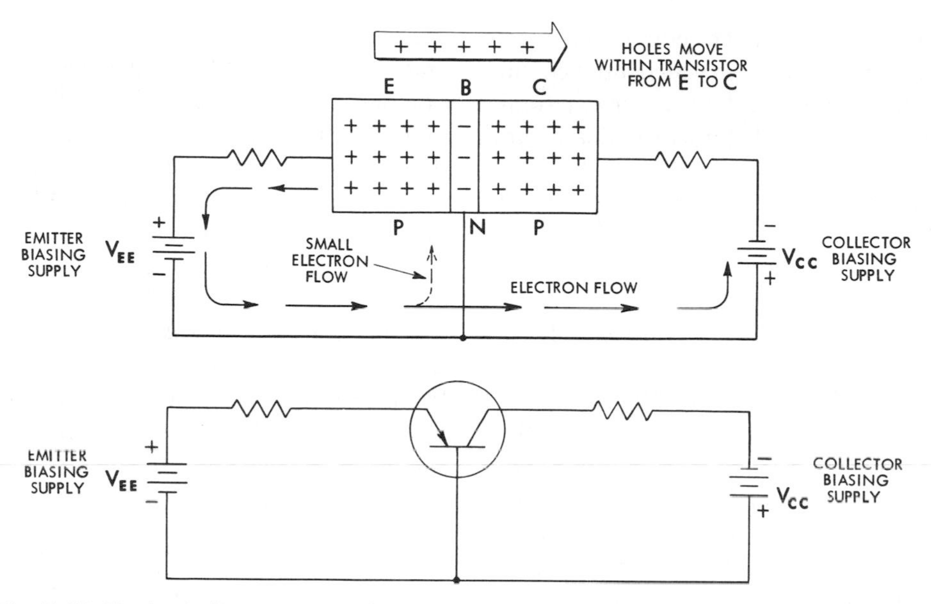

Fig. 9-45. The basic biasing circuit for a PNP transistor which is connected in a common base configuration.

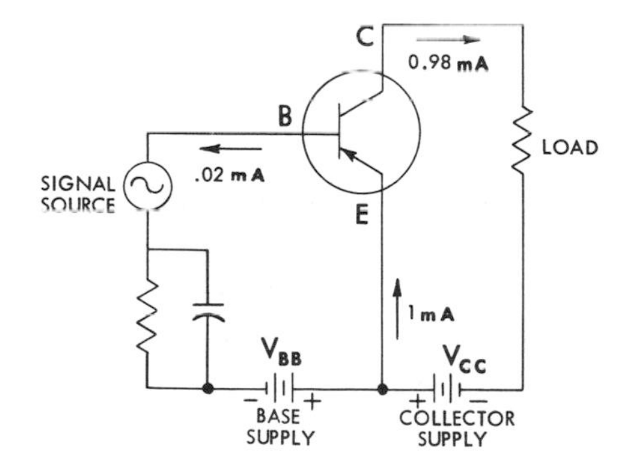

Fig. 9-46. A transistor used as an amplifier. It is connected in a common emitter configuration.

arrangements. One possibility is the common-base amplifier arrangement, in which the base is common to both the input and the output. See Fig. 9-45. This type is used infrequently. Its current gain factor is always less than one, which does not mean that this particular transistor arrangement is useless in amplifier design. While current output can be less than input, output voltage can be greatly increased.

Another amplifier circuit is the common-emitter circuit shown in Fig. 9-46, in which the emitter is common to the input and output. The common-emitter amplifier is the most widely used circuit configuration. Collector current (output) can be much higher than base current (input).

Feedback in Amplifiers. As with electron tubes, when some of the signal coming out of a transistor amplifier is fed back to the input, the process is called *feedback*. Negative feedback results when the output signal being fed back opposes the input signal. This results in a lower gain for the amplifier, but it makes the amplifier more linear and more stable against temperature and power supply voltage variations. If enough negative feedback is employed, the gain of the amplifier becomes almost independent of the transistor. This type of amplifier is usually referred to as an operational amplifier.

If the voltage being fed back into the input tends to increase the input signal, the result is positive feedback. Positive feedback increases the gain of an amplifier but reduces its temperature and voltage supply stability. If too much positive feedback is used, the output of the amplifier becomes independent of the input signal and oscillation results.

Transistor Oscillators. An oscillator is a circuit that is used to generate a varying voltage from a dc voltage source. There are two general types of oscillators: (1) Sinusoidal oscillators, and (2) Non-sinusoidal. Sinusoidal oscillators generate a sinusoidal (appears as a sine wave on an oscilloscope or when graphed) voltage. This is done by applying just enough positive feedback to start and maintain oscillation. Fig. 9-39 illustrates an ac sinusoidal pattern. Non-sinusoidal oscillators generate a voltage that has a square or sawtooth shape. This is done by applying many times the amount of positive feedback that would be required for oscillation. Oscillators are used to generate audio frequency voltages for telemetering and remote control applications.

Field Effect Transistors. Field effect, or unipolar transistors are made from a single crystal of silicon which is completely P-type except in two small regions on opposite sides of the crystal which are N-type. The two N-type regions are usually electrically connected and called the *gate*. One end of the main body is called the *source*, and the other end the *drain*. See Fig. 9-47.

A dc voltage source, in series with a load resistor, is placed between the source and drain leads. See Fig. 9-48. The controlling signal or input voltage is placed in series with the source and gate leads. The polarity of the dc source is such that the N-type gate is made positive with respect to the P-type body of the transistor. This causes a reverse bias on the gate-body junction and no current flows from the gate or control circuit.

The operation of this transistor is based upon a phenomenon called *conductivity modulation*. As the signal voltage applied to the gate lead increases, the reverse bias on the gate-body junction increases, causing the holes in the body in the region of the junction to move away. This causes the effective cross-sectional area of the body to decrease and the current flowing from the source to drain to decrease. Since this current is large com-

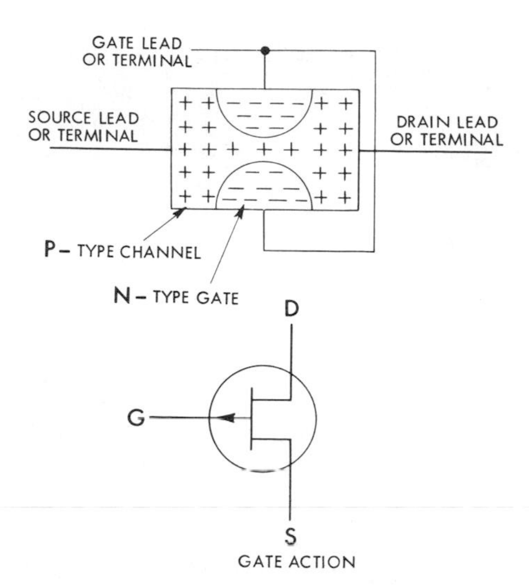

Fig. 9-47. The gate action and schematic symbol for a field effect (unipolar) transistor.

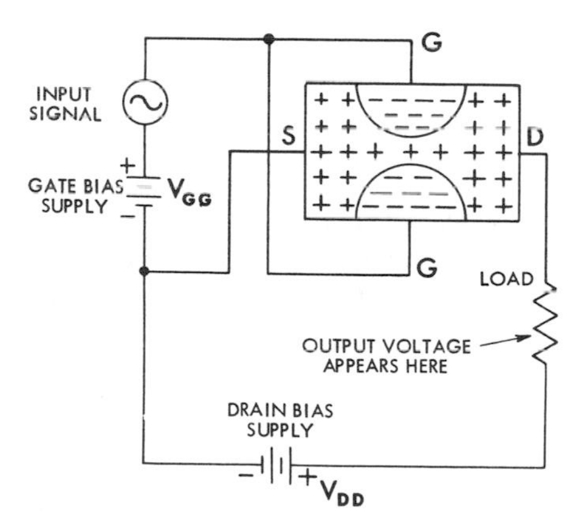

Fig. 9-48. A field effect transistor wired for use an an amplifier.

pared to the nearly zero current in the gate circuit, the amplification of the device is extremely high.

Almost any circuit that can be built around electron tubes can be duplicated with transistors. Transistors offer many advantages over tubes, including small size, stability, remarkable linearity, and longevity. Transistors are used in most control and reset applications.

Micro-Miniaturization

The result of many years of research and testing has placed semiconductor manufacturers in a position to build just about what they want. They have defined the state of the art as microminiature and have accomplished astounding feats in the manufacture of these miraculous devices. In the Fig. 9-49, a photograph of some of these *chips* are shown. These are actual microcircuit components. The center chip is superimposed over a ten cent piece (dime), to show the actual size. Fig.

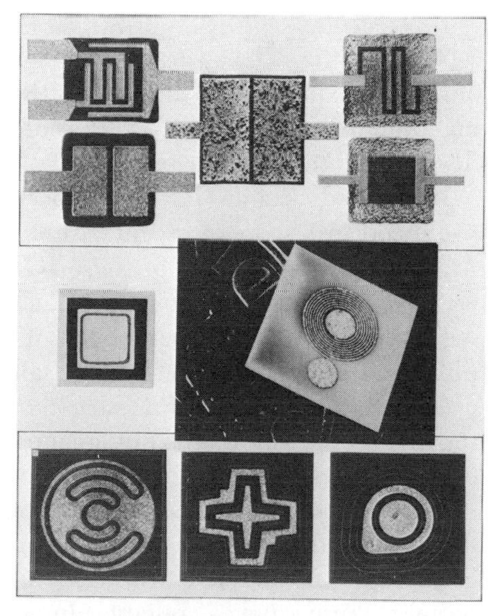

Fig. 9-49. Microcircuit components. Entire circuits are only a fraction of the size of a dime. (Motorola Semiconductor Products, Inc.)

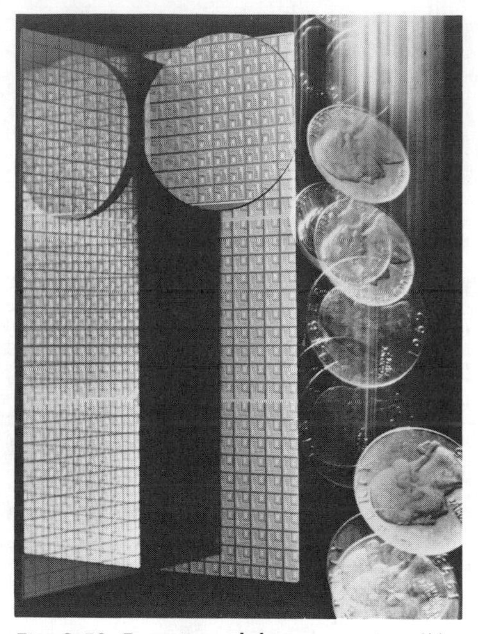

Fig. 9-50. Transistor slab construction. (Motorola Semiconductor Products, Inc.)

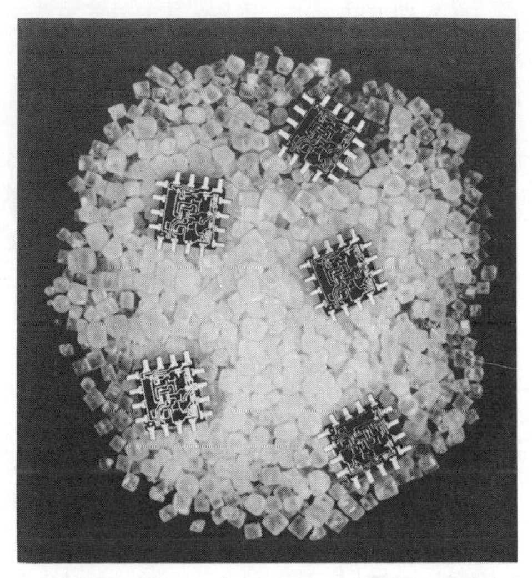

Fig. 9-51. Magnified photo of beam lead operation amplifiers built as integrated circuits. They are shown in comparison to grains (actually cube-shaped) of table salt. (Motorola Semiconductor Products, Inc.)

9-50 shows the slab construction which aids in mass production and allows the manufacturer to produce solid state devices at a much lower price.

Integrated circuits (IC's) have been produced by a score of manufacturers. As an illustration of one of these IC's we have taken an operational amplifier with beam leads. The operational amplifier in Fig. 9-51 is sitting in table salt used in testing. (The cubic shape of the salt crystals shows the high magnification used in making the photograph.) The structures extending from the IC are the beam leads, made of gold. They are bonded to the IC internally, providing a reliable interconnection. The operational amplifier shown consists of 13 transistors, 2 diodes, and 15 resistors of various sizes. This is a typical IC. There are many types used for many purposes.

A book at least the size of the present text would be required to cover the designs, manufacturing processes, and applications of these tiny circuits which more and more are replacing solid-state circuits made of *discreet* (individual) devices. IC's appear to indicate the direction of electronics in the near and forseeable future.

Operational Amplifiers. Just as transistors made possible a considerable reduction in the size of electronic equipment, integrated circuits have improved on this. Integrated circuits are actually chips containing numerous transistors, diodes, and resistors. To appreciate their size, they are shown in Figs. 9-49, 9-50, and 9-51 along side

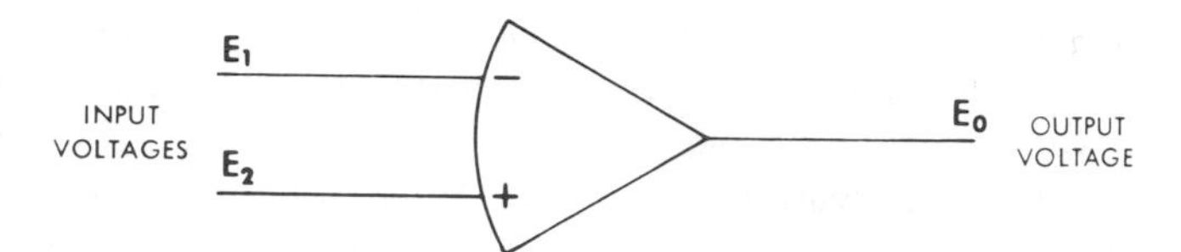

Fig. 9-52. An operational amplifier.

familiar objects and materials. Integrated circuits, multifunction circuits in solid-state form, are used in operational amplifiers which are found extensively in instrumentation because of their computing capability. A typical operational amplifier is shown schematically in Fig. 9-52. Operational amplifiers are connected to resistors and capacitors for many specific functions such as signal conversion, amplification, impedance matching, addition, subtraction, multiplication, division, and differentiation and integration.

Words to Know

voltage (E)
coulomb
direct current (dc)
alternating current (ac)
Ohm's law
current (I)
resistance (R)
Kirchoff's law
Wheatstone bridge
inductance
electromotive force (emf)
capacitance (C)
capacitor
dielectric
farad (F)
hertz (Hz)
henry (H)
impedance (Z)
electron (vacuum) tube
rectification
amplification
oscillation
feedback
cathode
tetrode
pentode
semiconductor
diode
junction (bipolar) transistor
silicon controlled rectifier (SCR)
unijunction transistor (UJT)
field effect (unipolar) transistor (FET)

Review Questions

1. Define *voltage, current,* and *resistance.*
2. List three characteristics of parallel resistance circuits.
3. Describe the two types of inductance.
4. Electrical resistance refers to the resistance to current flow. What does capacitance resist?
5. What is the time constant of a capacitor? How is it determined?
6. Batteries, photocells and thermocouples produce voltages for direct current systems. What device

is used to produce voltage for alternating current systems?

7. What are the positive or negative half cycles of alternating current called?
8. In an alternating current circuit, with a particular inductance the ratio of voltage to current is constant when the voltage is constant. What other electrical characteristic is determined by the voltage/current ratio?
9. Why does the current through a capacitor in an alternating current circuit lead the voltage?
10. What is the term used to describe the sum of inductive reactance and capacitive reactance? Express this relationship by using the proper symbols.
11. Describe the three principal functions of the *electron tube.*
12. What are the advantages of a transistor over a triode electron tube?

Temperature

Chapter

10

Heat is a form of energy. When heat is transferred to a substance, the energy of the molecules of that substance is increased. The amount of heat required to raise the temperature of one pound of water one degree Fahrenheit is called a British Thermal Unit (BTU). The amount of heat required to raise the temperature of one pound of any substance one degree Fahrenheit is called the *thermal capacity* of that substance.

The ratio of the thermal capacity of a substance to the thermal capacity of water is called the *specific heat* of that substance. Specific heats of some solids and liquids are given in Table 1. Specific heat has no unit of measurement, but is the same numerically as the thermal capacity.

The amount of heat, Q, required to raise the temperature of a mass, M, of a substance having a thermal capacity (specific heat) of C from temperature T_1 to T_2 is expressed;

$$Q = M \times C (T_1 - T_2)$$

If M and C are known, temperature measurements determine the change in the amount of heat possessed by a body in going from one temperature to another.

Solids expand as heat is applied to them. For example, a metal rod heated

TABLE I SPECIFIC HEAT OF VARIOUS MATERIAL

MATERIAL	SPECIFIC HEAT (C)
ALCOHOL	0.59
COPPER	0.093
GLASS	0.14
MERCURY	0.033
PLATINUM	0.032

TABLE 2 COEFFICIENT OF EXPANSION FOR VARIOUS MATERIALS

MATERIAL	$\alpha/F°$	$\beta/F°$
ALCOHOL		0.00061
WATER		0.000115
MERCURY		0.0001
GLASS	0.000005	0.000015
COPPER	0.000009	0.000039
PLATINUM	0.000005	0.000016
INVAR	0.0000008	0.0000027

uniformly will increase in length uniformly. The increase per degree of temperature rise is called the *coefficient of linear expansion*. The coefficient (α is the symbol used) differs for each material. Table 2 shows the coefficient of linear expansion for various materials. For example, a one inch copper strip expands to 1.000009 inches long when heated 1°F. This is expressed by the following formula:

$$Lt_2 = Lt_1 [1 + \alpha (T_2 - T_1)]$$

(Lt_2 is the *length* at temperature T_2; Lt_1 is the *length* at temperature T_2.)

The expansion of a solid due to heat affects not only its length, but all its dimensions. The thermal expansion of the volume of a substance can be expressed:

$$Vt_2 = Vt_1 [1 + \beta (T_2 - T_1)]$$

(Vt_2 is the *volume* at temperature T_2; Vt_1 is the *volume* at temperature T_1.)

β is the *coefficient of volumetric expansion* and is approximately equal to 3α. Liquids also expand, when heated, according to the same equation. Coefficients of volumetric expansion of various materials are also listed in Table 2.

Mercury-in-Glass Thermometers

The volumetric expansion of mercury is over six times greater than that of glass. See Table 2. The mercury-in-glass thermometer is based on this difference in volumetric expension. Basically this thermometer consists of a fine bore glass tube, joined to a reservoir and sealed at the top. A measured quantity of mercury is enclosed in the reservoir. See Fig. 10-1.

When the thermometer is exposed to heat, the mercury expands more than the glass, and is forced to rise up the tube. The mercury rises a certain amount for each degree of temperature. Reference marks are made on the thermometer, using a regulated bath which can establish and maintain temperatures very closely. The spaces between the reference marks are evenly subdivided. A larger number of reference marks enables the thermometer to be read more accurately.

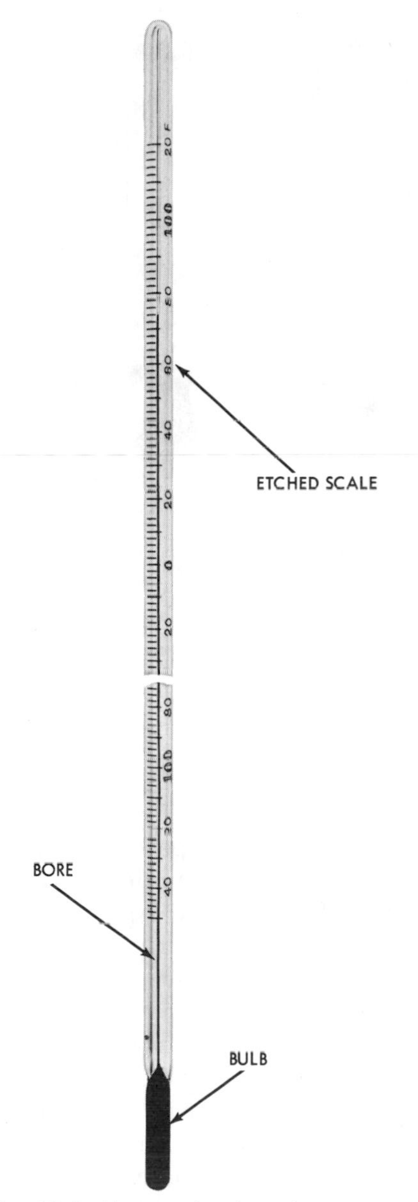

Fig. 10-1. Mercury-in-glass thermometer.

Some mercury-in-glass thermometers are calibrated to be completely immersed; other are calibrated for partial immersion. The thermometer should be immersed as recommended by the manufacturer to obtain the most accurate readings.

The most common type of mercury-in-glass thermometer used for process measurements is the industrial thermometer. See Fig. 10-2. In this type the glass tube is not marked or scaled. Instead the graduations are engraved on metal plates. Both the tube and the scales are enclosed in a metal case. The lower portion of the glass tube extends out of the bottom of the case into a metal bulb chamber. The thermometer is fitted with an external pipe thread which enables it to be screwed into a pipe.

As shown in Fig. 10-2, the industrial thermometer is available in vertical, horizontal, or oblique angle shapes. The chamber contains a liquid with excellent heat transfer characteristics which improves thermal conductivity between the metal bulb chamber and the glass thermometer. The chamber can be fitted into a secondary chamber called a *separable socket*. This prevents damage to the thermometer and permits replacement without draining the process pipe. However, the separable socket slows down the response of the thermometer to temperature changes.

Fig. 10-3 diagrams the response characteristics of various industrial thermometers. The dots on the curves indicate the time required for the thermometers to reach 63.2% of their final value. This is referred to as the *time constant* of the thermometer. Since it is more advantageous to know the time required to reach 90, 95, or 99% of final value, manufacturers frequently state the time required to

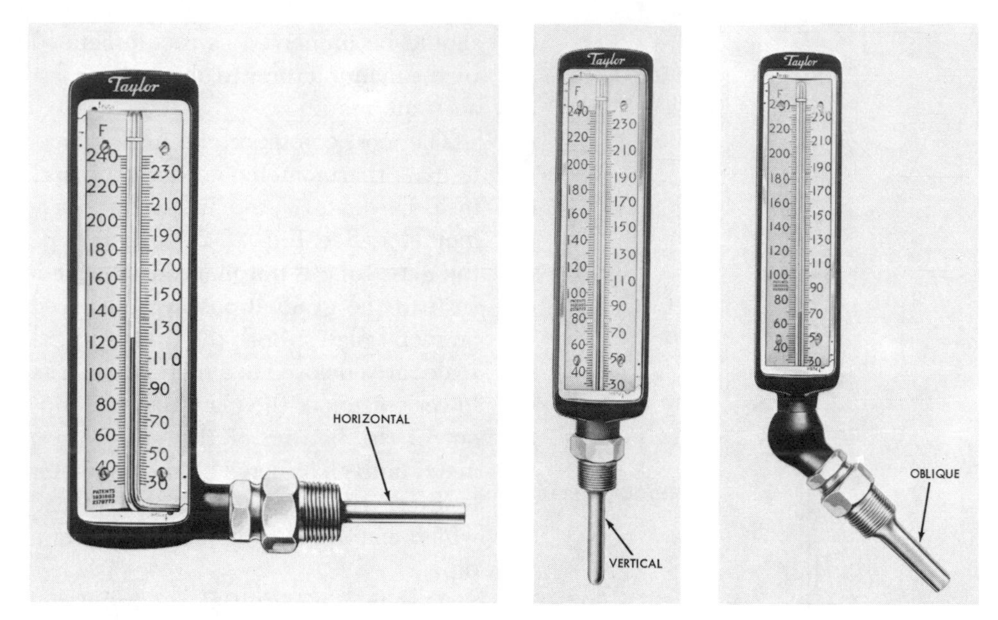

Fig. 10-2. Industrial thermometers are available in various styles. The threaded bases allow direct mounting on pipes or containers. (Taylor Instrument)

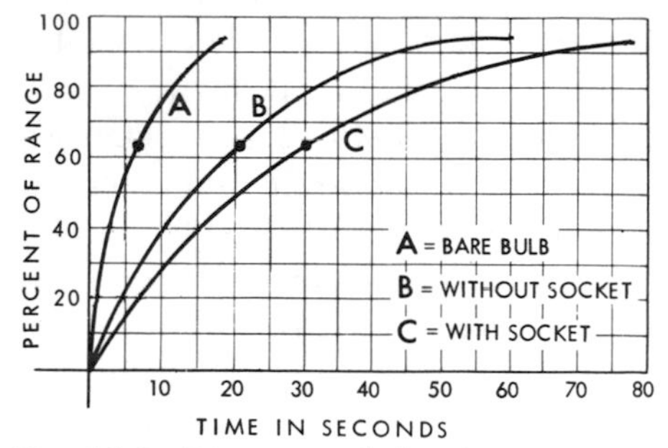

Fig. 10-3. Diagram of the time response of various thermometers.

reach these higher percentages when they list the response speed of their thermometers.

Bimetallic Thermometers

Alloys whose coefficients of thermal expansion can be closely controlled make the bimetallic thermometer a very dependable temperature measuring device. For example, Invar, a nickel alloy, scarcely expands when exposed to heat. The alloys are welded together and rolled to specifications, thus providing the bimetallic material used in these thermometers. Alloys with widely differing rates of thermal expansion are used in applications which only require a small temperature range. If the application requires a large temperature range, alloys with more similar expansion rates are used. *Flexivity* is the term used to describe the thermal activity of a bimetal. The movement of the bimetallic strip is proportional to its flexivity.

Fig. 10-4 shows a bimetallic element which employs a multiple helical ar-

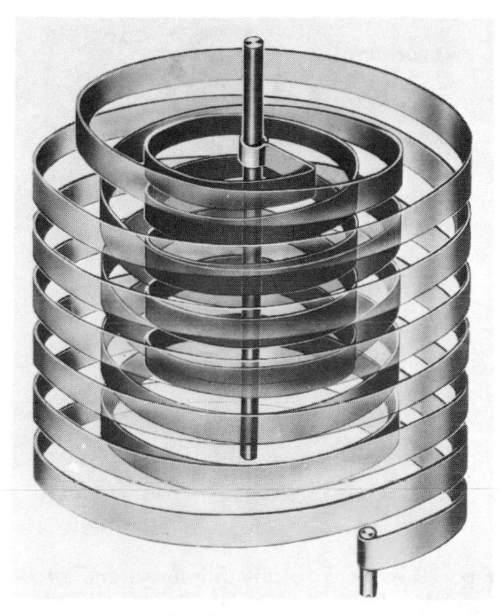

Fig. 10-4. Helical element for bimetallic thermometer. (Weston Instruments, Inc.)

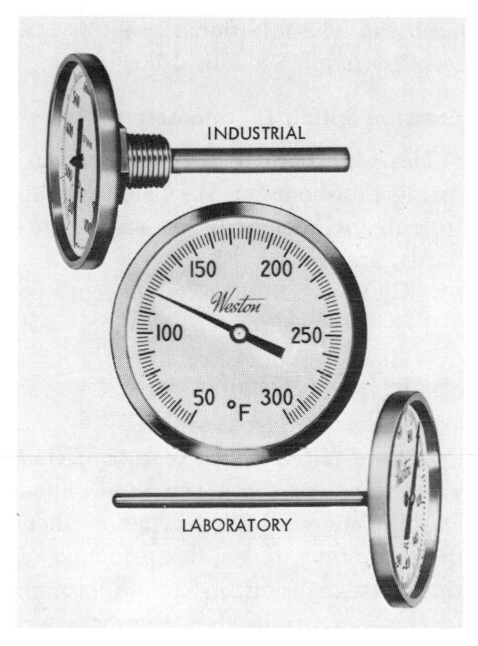

Fig. 10-5. Bimetallic thermometers are available for precision laboratory work as well as for a variety of industrial applications. (Weston Instruments, Inc.)

rangement, or coils within coils. This construction allows for a long bimetallic element to be used inside a small space. The long bimetallic element has greater expansion when it is exposed to heat, and hence greater accuracy and response. One end of the bimetal is fixed to the bottom of the stem, and the opposite end is fixed to the shaft attached to the pointer.

Bimetallic thermometers for laboratory applications have slender stems which allow for greater accuracy and speed of response. Those designed for industrial usage generally have thicker stems and heavier construction. See Fig. 10-5. Special thermometers are also available which have alarm contacts and secondary pointers for indicating maximum and minimum tem-

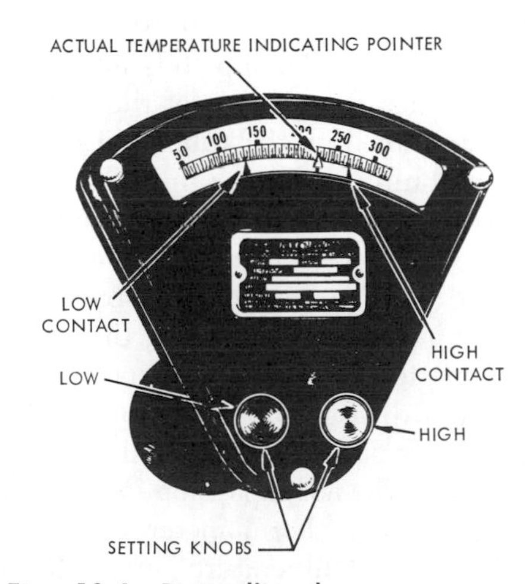

Fig. 10-6. Bimetallic thermometers are made to suit a wide range of applications. They can feature secondary pointers and alarm contacts. (Bristol Division of Acco)

perature. The additional pointers are reset by hand. See Fig. 10-6.

Pressure Spring Thermometers

The four basic classes of pressure spring thermometers, as listed by the Instrument Society of America Standards, are:

Class 1. Liquid-filled (except mercury)
Class 2. Vapor pressure
Class 3. Gas-filled
Class 4. Mercury-filled

Liquid-filled and Mercury-filled Thermometers. Both the liquid-filled and the mercury-filled types of thermometer operate on the principle of thermal expansion, in similar fashion to the mercury-in-glass thermometer. When the bulb is immersed in a heated substance, the liquid expands. This causes the pressure spring to unwind. See Fig. 10-7. The indicating, recording, or controlling mechanisms are attached to the pressure spring and are actuated by its movements. The liquid or mercury is put into the system under pressure and completely fills it.

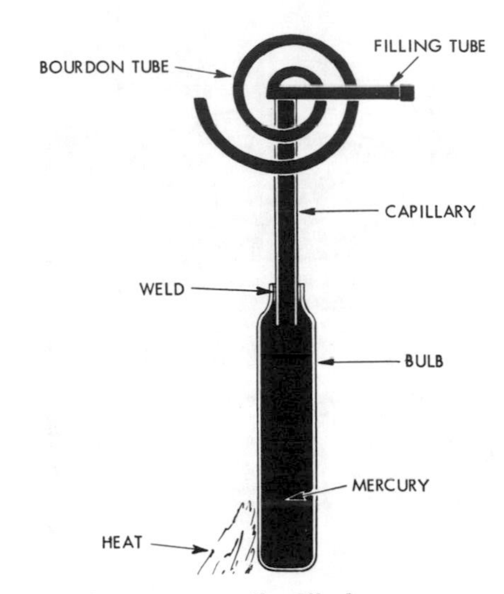

Fig. 10-7. In a totally filled system, pressure develops when the bulb is heated because there is no room for the fluid to expand.

The measuring range of the system is determined by the volume of liquid in the bulb. The wider the range, the greater the required volume. The bulb expands with temperature, but since this expansion is small compared to the expansion of the liquid, the effect is negligible. If the coefficients of volumetric expansion of the bulb and the liquid are similar, the overall effect is to reduce the net expansion for a given range. If the thermometer has been calibrated (temperature graduations applied), allowances can be made for slight variations in the coefficients.

The thermometer is designed to detect changes of temperature at the bulb. There is, however, the possibility of error caused by temperature variations along the tubing or at the pressure spring. The bulb of the liquid-filled system contains most of the liquid. In long systems, however, the volume of the tubing can allow ambient temperature changes. Several methods have been devised to eliminate such error. Systems employing these modifications are known as *compensated systems*. Only mercury and liquid-filled systems require this compensation.

As shown in Fig. 10-8, fully compensated systems employ a second pressure spring and special tubing. This tubing is not connected to the bulb but is terminated just short of it. The motion of the compensating spring opposes any errant motion of the primary

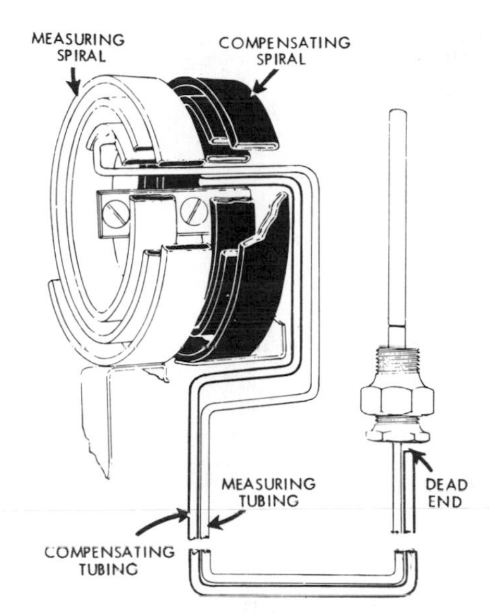

Fig. 10-8. A fully compensated liquid-filled thermometer.

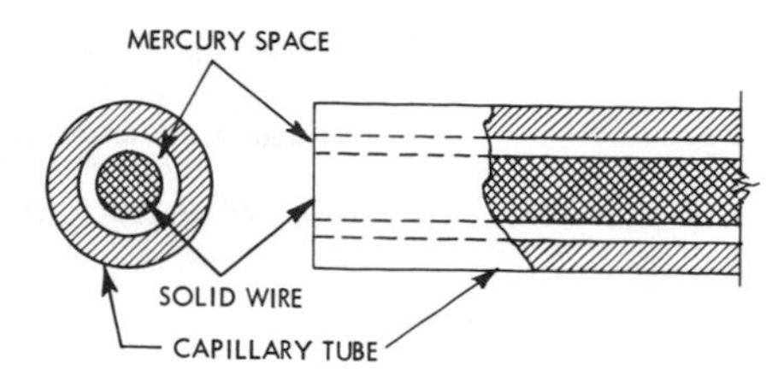

Fig. 10-9. Self-compensating capillary tubing.

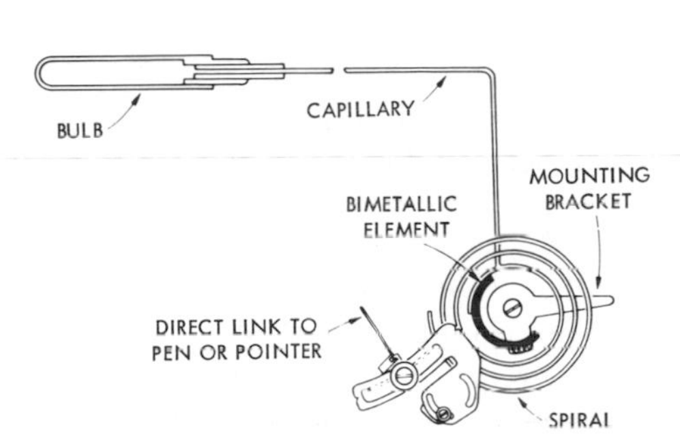

Fig. 10-10. A compensation system for applications under 50 feet. (Honeywell Process Control Div./Fort Washington, Pa.)

pressure spring by means of linkage. When temperature along the primary tubing varies and causes the primary spring to expand, the compensating spring also expands, thus nullifying the motion to the indicating mechanism. Only the expansion of the fluid in the bulb can affect the indicating mechanism.

A special type of tubing which contains a solid wire provides a method of self-compensation. Whenever the tubing expands because of temperature variations, the wire also expands, keeping constant the effective volume of the system. See Fig. 10-9.

A bimetallic strip can be used to compensate for ambient temperature change at the case. The strip is tied in with the movement of the spiral integrally (as shown in Fig. 10-10) so that it causes the spiral to rotate in the opposite direction of the rotation caused by the thermal system. Since opposing forces are exerted, the pen or pointer does not move with ambient temperature changes. Compensation at the case is all that is needed when tubing length does not exceed 50 feet.

Another type of compensation, generally referred to as *self-compensated,* is used for applications requiring tubing over 50 feet to a maximum of a 125 feet in length. The capillary and the spiral are compensated for at the case and along the entire length of the tubing. A close tolerance invar wire, having negligible expansion because of temperature changes, is used in the

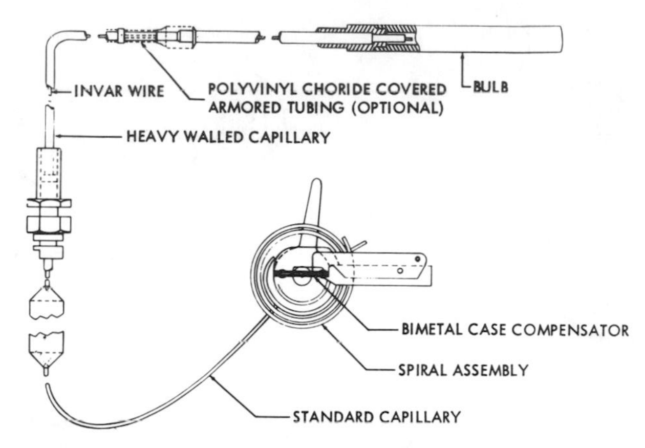

Fig. 10-11. A compensation system for applications ranging from 50 to 125 feet. (Honeywell Process Control Div./Fort Washington, Pa.)

capillary. The space between the wire and the tubing is filled with mercury. The mercury and the tubing are properly regulated by the invar wire and change equally, which maintains a constant internal pressure, regardless of ambient temperature changes along the capillary. A bimetallic strip on the spiral compensates for ambient temperature changes at the case. See Fig. 10-11.

Gas-filled Pressure Spring Thermometers. The gas-filled thermometer (See Figs. 10-12A and 10-12B.) depends on the increase in pressure of a confined gas (kept at constant volume) due to a temperature increase. The relationship between temperature and pressure in this type of system partially expresses Charles' law:

$$\frac{P_1}{P_2} = \frac{T_1}{T_2}$$

(In this formula
P_1 = initial pressure
P_2 = increased pressure
T_1 = initial temperature
T_2 = increased temperature
Temperature and pressure are in absolute values.)

The system is filled under high pressure. The increase in pressure for each

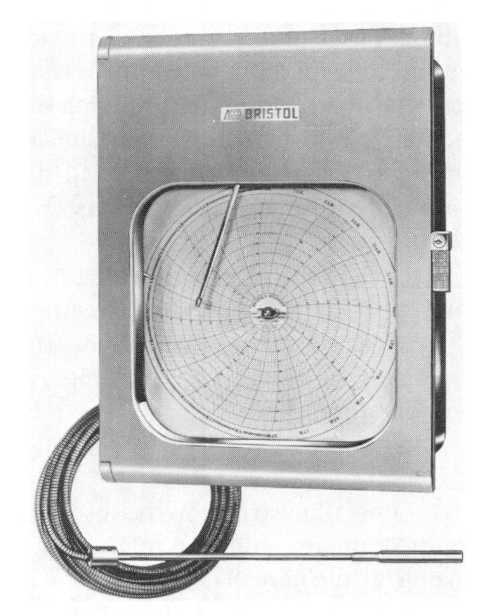

Fig. 10-12A. A pressure-spring thermometer/recorder used for long distance (remote) readings. (Bristol Division of Acco)

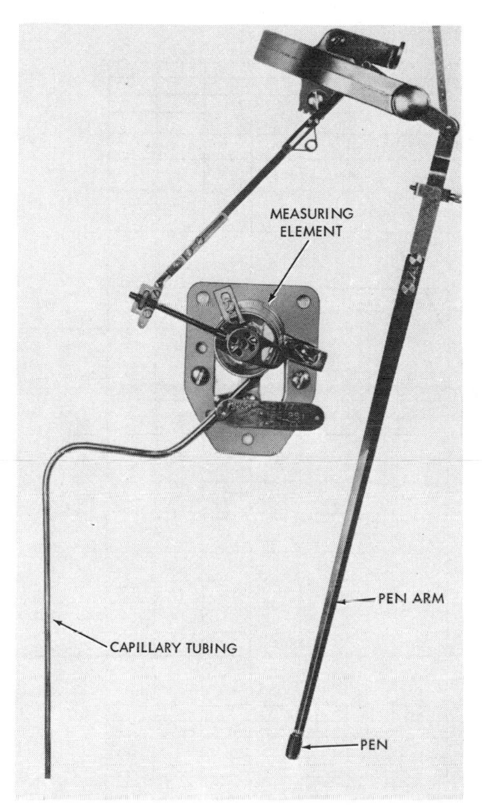

Fig. 10-12B. The measuring element for a pressure-spring thermometer/recorder. (Bristol Division of Acco)

degree of temperature rise (the Kelvin scale is used) is therefore greater than if the filling pressure were low. Nitrogen is the gas most often used for such systems because it is chemically inert and possesses a favorable coefficient of thermal expansion.

Except for the size of the bulb, the gas-filled system is identical to the liquid-filled types. The gas-filled bulb must be larger and its volume must be considerably greater than that of the rest of the system. Special bulbs consisting of a length which has a small diameter tubing can be used to measure the average temperature along the bulb.

Vapor Pressure Thermometers. Unlike the liquid-filled and gas-filled systems which depend upon volumetric expansion for their operation, vapor pressure thermometers depend upon the vapor pressure of a liquid which only partially fills the system. The liquid in this type of system can expand, but as it is heated its vapor pressure increases. See Fig. 10-13. Water in a pressure cooker responds in a similar manner. The pressure increases as the water is heated and changes to steam (water vapor).

Vapor pressure does not increase linearly (unit increase in pressure for each unit of temperature rise). At lower temperatures, the increase of vapor pressure for each unit of temperature change is small. The change of

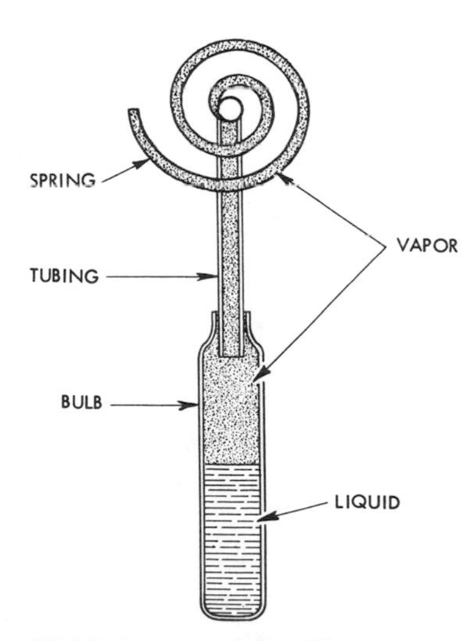

Fig. 10-13. In a partially filled system, the liquid expands when the bulb is heated and vapor pressure increases.

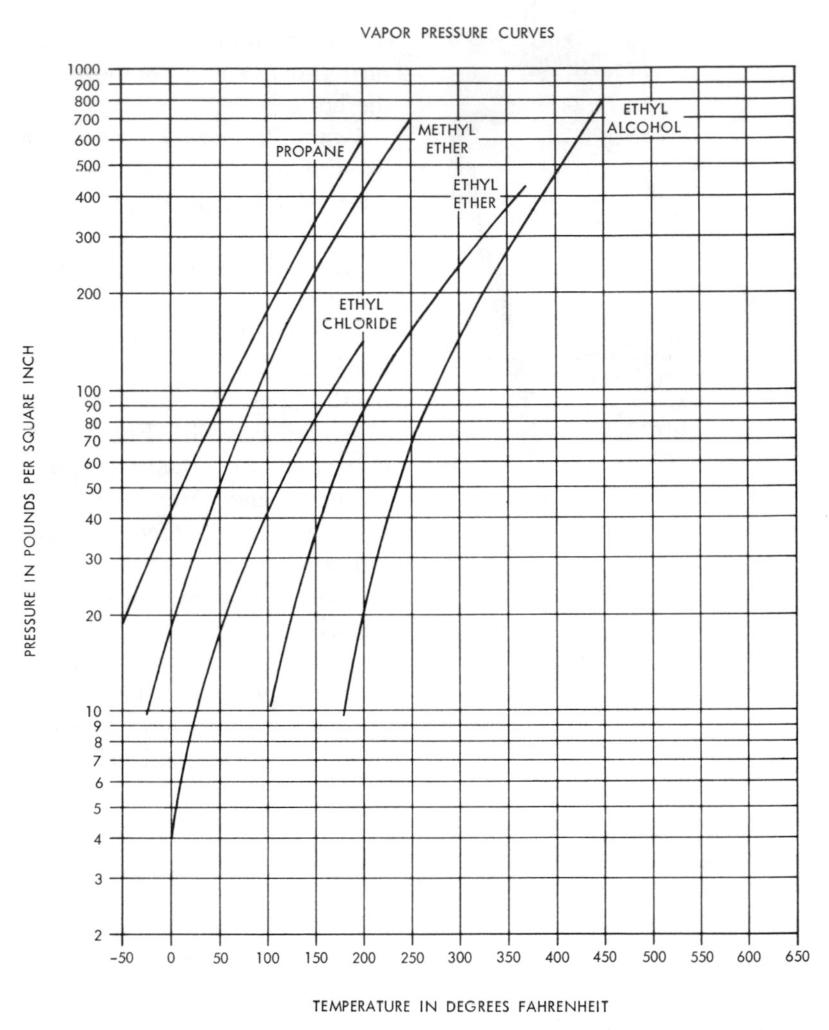

Fig. 10-14. Rate of vapor pressure increase for various chemicals.

vapor pressure at higher temperatures is much greater. The vapor pressure is said to change logarithmically. See Fig. 10-14.

Another characteristic of the partially-filled vapor pressure thermometer is the interchange of vapor and liquid when the temperature of the sensitive bulb is altered from a value lower than the temperature of the pressure spring to a higher value. If the temperature of the bulb is lower than that of the pressure spring, the liquid remains in the bulb, and the vapor occupies the tubing and spring. See Fig. 10-15. If the bulb temperature is higher than that of the system, the vapor is in the bulb, and the liquid oc-

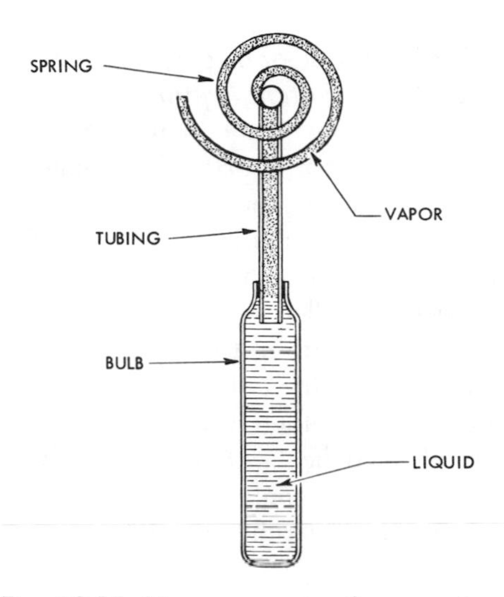

Fig. 10-15. Vapor pressure thermometer. Liquid remains in the bulb when bulb temperature is lower than the rest of the system. Vapor is in the tubing and spring.

cupies the pressure spring. See Fig. 10-16. The simple partially-filled vapor pressure thermometer should not be used if the temperature of the bulb does not *always* remain higher or lower than the temperature of the remaining system.

Manufacturers have developed a method of filling the vapor pressure thermometer which eliminates the above problem. This method employs a dual filled system in which two dissimilar liquids (a vaporizing and a non-vaporizing type) are used. The liquid which vaporizes is, of course, temperature sensitive. It is called the *actuating liquid*. The vapor pressure is create's acts on the second, or non-vaporizing liquid. This second liquid is called the *transmitting liquid*, since it transmits the vapor pressure created by the actuating liquid to the pressure spring in much the same fashion as any hydraulic fluid. See Fig. 10-17. There is no

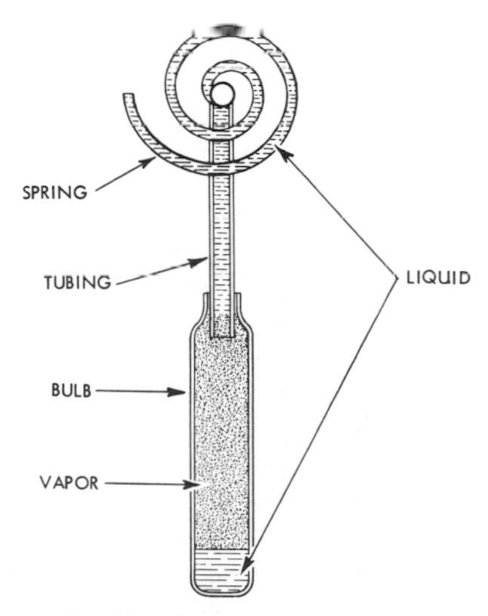

Fig. 10-16. When bulb temperature is higher than that of the rest of the system, the vapor is in the bulb, and the liquid is in the tubing and spring.

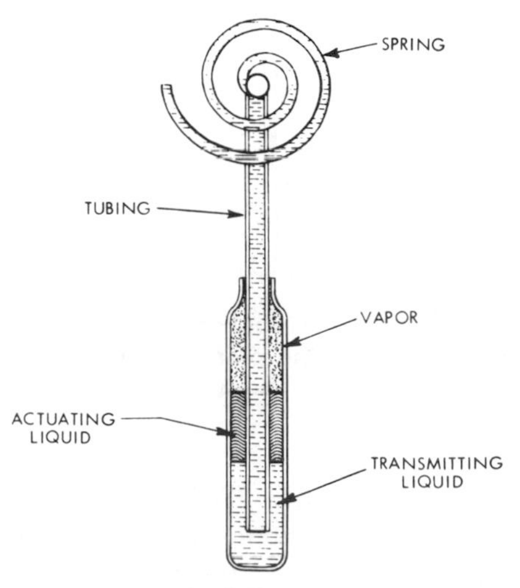

Fig. 10-17. A dual filled pressure system.

transfer of vapor and liquid in this dual filled system. The actuating liquid and the vapor created by it remain in the bulb at all times.

Bulb location. The difference in height between the bulb and the pressure spring can also introduce error, especially in the partially-filled vapor pressure system. Since this system is not filled under pressure as are the totally-filled systems, any column of fluid can create a pressure which causes an erroneous reading. In advance of purchase, the manufacturer should be advised if the bulb in the required application is to be located above the pressure spring, and at what specific height. The system should then be installed precisely as specified by the manufacturer. See Fig. 10-18.

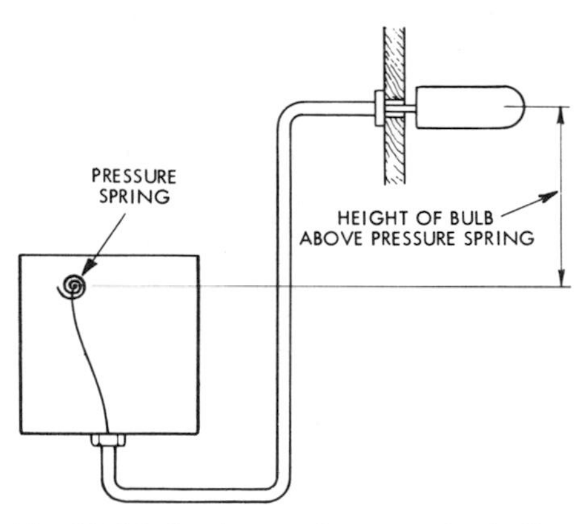

Fig. 10-18. In this application, the bulb is installed above the pressure spring.

Because totally filled (liquid mercury and gas) systems are filled under high pressure, the pressure caused by the height of the bulb has a negligible effect, and little or no error is involved.

Response to temperature change. A second important consideration in studying pressure spring thermometers is response to temperature change. How the system is installed can influence the speed of response. The bulb must be installed so that it only senses the temperature of the process or material into which it is immersed. It should be shielded from the possibility of reflected heat. At no point should the bulb be in contact with cold metal, which will lower the temperature reading.

Of the various types of pressure spring thermometers, the gas-filled systems have the fastest response, followed by the vapor pressure systems and the liquid-filled systems. See Fig. 10-19. The response of all systems is faster if the process to be measured is a liquid rather than a gas. See Fig. 10-20. Response is also affected by the speed or flow of the measured material past the bulb. A fast process flow will

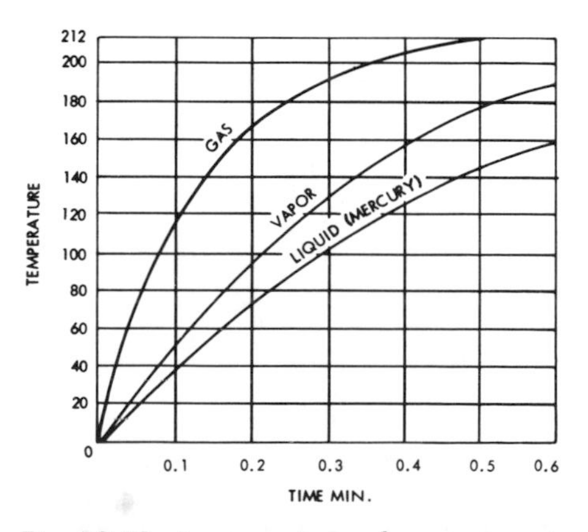

Fig. 10-19. A comparison of response rate for various type of thermometers.

provide a fast thermometer response.

Bulb size is another factor to be considered. A large bulb area will effect a quick response. A special, large area bulb for measuring air or gas temperatures is shown in Fig. 10-21. When the

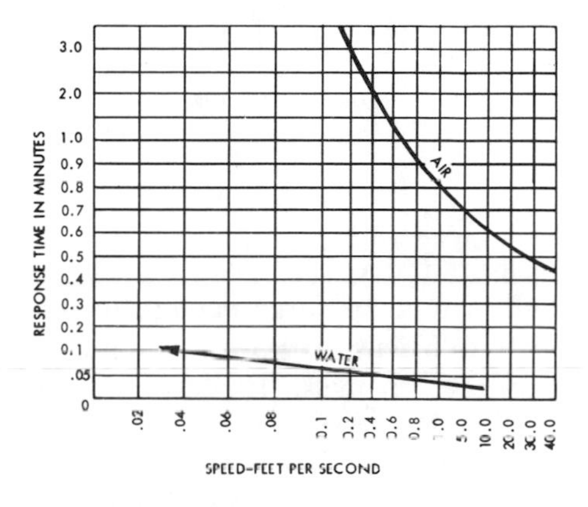

Fig. 10-20. A comparison of response rate between moving liquid and a moving gas.

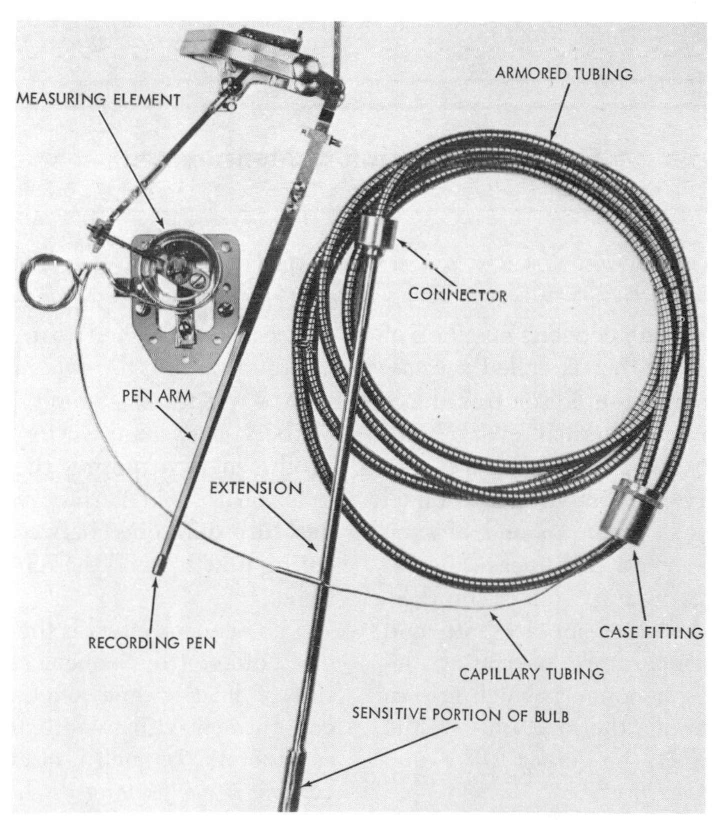

Fig. 10-21. A bulb and measuring element for process thermometer. (Bristol Division of Acco)

TABLE 3 CHARACTERISTICS OF TUBE SYSTEMS

	FILLING FLUID	LOW LIMIT	HIGH LIMIT	SHORTEST RANGE	LONGEST RANGE
CLASS 1	LIQUIDS OTHER THAN MERCURY	- 300 °F	600 °F	25 °F	300 °F
CLASS 2	VAPOR	- 300 °F	600 °F	40 °F	300 °F
CLASS 3	GAS	- 450 °F	1000 °F	100 °F	1000 °F
CLASS 4	MERCURY	- 48 °F	1000 °F	40 °F	1000 °F

measured process involves a liquid, the bulb should be large enough to provide sufficient area for sensing temperature changes. However, if the bulb is larger than required, its size can negatively affect response time. Again the manufacturer can determine the best pressure spring thermometer system and size for any process, if a complete description of the operating conditions is provided in advance. Table 3 shows the characteristics of the various tube systems.

Electrical Temperature Measurement

Thermocouples

A device which converts one form of energy into another is called a *transducer.* A thermocouple is a transducer which converts thermal energy into electrical energy. It is a device consisting of two dissimilar wires joined at their ends. When an end of each wire is connected to a measuring instrument (as in Fig. 10-25), the thermocouple becomes an accurate and sensitive temperature measuring device. Three phenomena which govern the behavior of a thermocouple are the *Seebeck Effect,* the *Peltier Effect,* and the *Thomson Effect.*

The Seebeck Effect. The joined ends of a thermocouple form a junction called the *hot junction* (measuring junction). The opposite ends of the wires are connected to a meter or a circuit and form the *cold junction* (reference junction). Simply stated, the Seebeck electromotive force (voltage) produced by exposing the measuring junction to heat varies with the temperature difference between this measuring junction and the reference junction.

The Seebeck effect is the production of a voltage (the Seebeck emf) as a result of heating one junction of a circuit formed with two dissimilar metal conductors. In such a circuit, electric current flows as long as the two junctions have different temperatures. See Fig. 10-22.

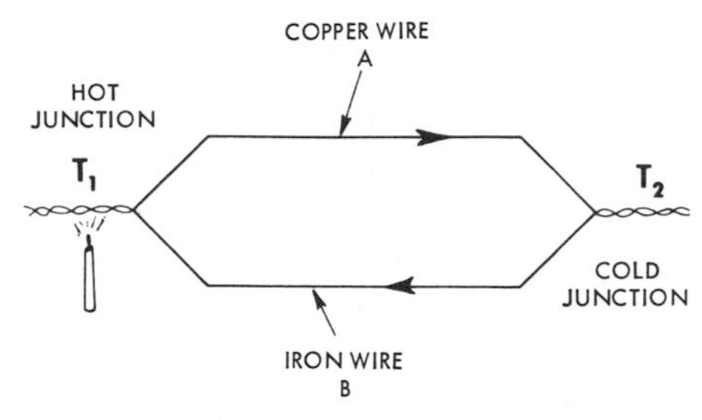

Fig. 10-22. The heat source at one junction enables electric current to flow through the circuit. (The Seebeck Effect)

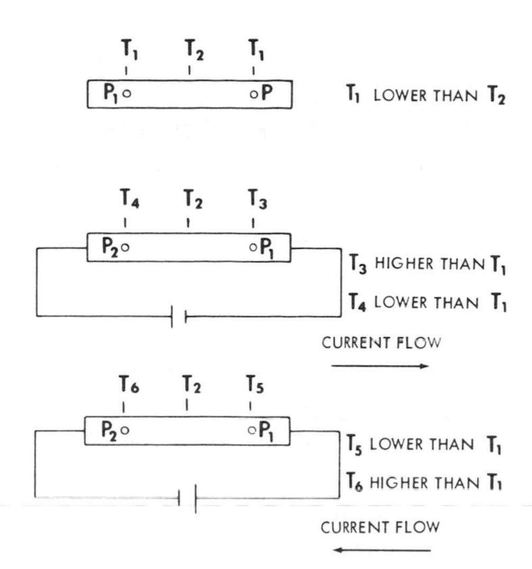

Fig. 10-24. With the Thomson Effect, heat is released when the current flows in the same direction as the heat.

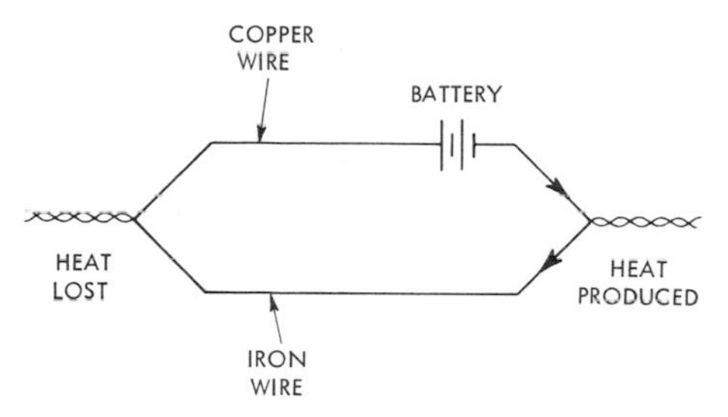

Fig. 10-23. An example of the Peltier Effect.

The Peltier Effect. If a current flows across the junction of two dissimilar metal conductors which have the same temperature, the heat is either released or absorbed, depending on the direction of current flow. If the current flow is in the same direction as that produced by the Seebeck Effect, heat is released at the hot junction and absorbed at the cold junction. See Fig. 10-23.

The Thomson Effect (the Reversible Heat Effect). The Thomson Effect is the result when an electric current passes through a conductor in which there is a temperature gradient. The temperature of a copper wire carrying current can vary along its length. Heat is released at any point where the current flows in the same direction as the heat, that is, from hot to cold. See Fig. 10-24. When an iron wire is used, however, heat is absorbed at any point where the current flows in the same direction as the heat. In a circuit comprised of iron and copper wires, with the cold junction at 32°F, the emf (voltage) increases as the temperature of the hot junction increases. However, the rate of increase diminishes until it finally reaches zero, and the emf reaches its maximum. Once it is past the maximum, the emf decreases until it reaches zero. Then it reverses polarity and begins to increase again.

The voltage which produces the Seebeck current is the sum of the Peltier *emf* at the junctions and the two Thomson *emfs* along the dissimilar wires. This is the true basis of thermoelectric thermometry.

The Law of Intermediate Temperatures. As shown in Fig. 10-25, the law of intermediate temperatures states that the sum of the voltages generated by two thermocouples, one with its reference junction at T_1 and its measuring junction at a higher temperature (T_2), and the second with its reference junction at T_2 and its measuring junction at an even higher temperature (T_3), is equal to the voltage generated by a single thermocouple with its reference junction at T_1 and its measuring junction at T_3. Because of this law, it is possible to use a reference temperature, with any fixed value (T_2). The voltage V_2 only differs from the voltage V_3 by the amount of the constant voltage V_1. V_1 remains constant as long as T_1 and T_2 do not change. Also, the voltage in the circuit remains at a fixed value, if a temperature sensitive resistor, which automatically eliminates voltage changes at the reference junction due to ambient temperatures, is used in the measuring circuit. Hence, only a change in the temperature of the measuring junction can effect a voltage change in the circuit.

The Law of Intermediate Metals. The law of intermediate metals states that the use of a third metal in a thermocouple circuit does not affect the voltage, as long as the temperature of the three metals at the point of junction is the same. See Fig. 10-26. Therefore, metals different from the thermocouple materials can be used as extension wires in the circuit. This is a common practice in Industry. For example, since platinum is very expensive, the extension lead wires used with Platinum/Platinum-Rhodium thermocouples are often made of copper. Another application might use one copper wire and a second made from an alloy. The extension lead wires for Chromel-Alumel thermocouples can be copper and constantan, or iron and an alloy.

Of all the materials used for the thermocouples, platinum is no doubt the most important. The Platinum/Platinum-10%Rhodium thermocouple is the primary standard for temperatures between 630.5°C and 1063°C. The useful range of this thermocouple actually is from 0°C to 1480°C.

The performance of other thermocouple materials is generally determined by using them as the second

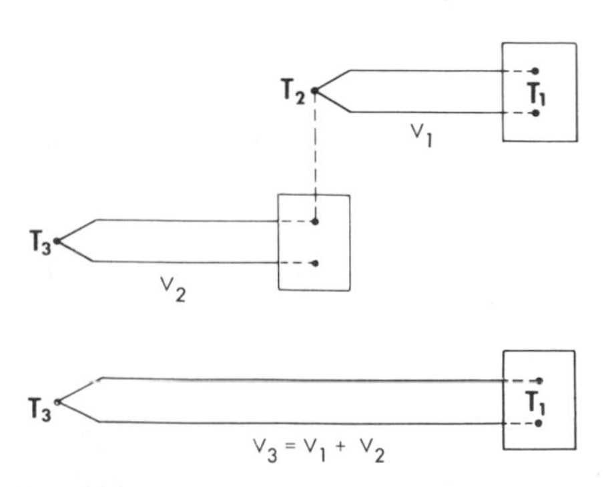

Fig. 10-25. An example of the Law of Intermediate Temperatures.

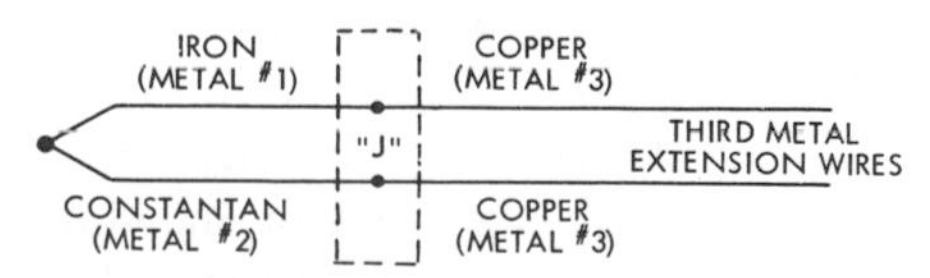

Fig. 10-26. The temperatures of the three metals is the same at junction J.

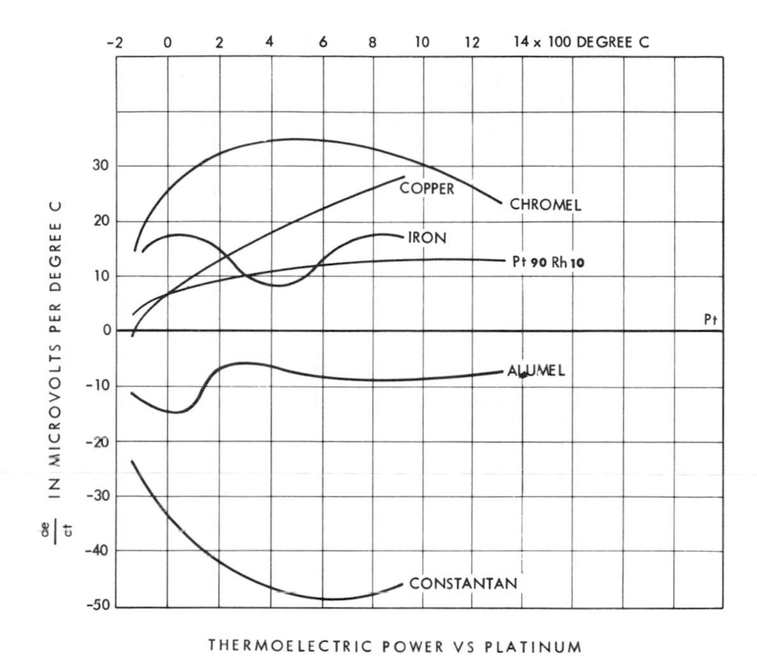

Fig. 10-27. Characteristics of thermocouple materials when used with platinum.

material with platinum. One of the most important factors in selecting a pair of materials for use as a thermocouple is the thermoelectric difference between them. A significant difference between the two materials results in a better performance. Fig. 10-27 illustrates the characteristics of common thermocouple materials when used in conjunction with platinum.

Other materials, in addition to the more common ones shown in Fig. 10-27, are used in thermocouples. Chromel-Constantan, for example is excellent for temperatures up to 2000°F. Nickel/Nickel-Molybdenum sometimes replaces Chromel-Alumel. Tungsten-Rhenium is used for temperatures up to 5000°F. Other combinations for special applications are Chromel-White Gold, Molybdenum-Tungsten, Tungsten Iridium, and Iridium/Iridium-Rhodium.

The size of the wire and the protection of the thermocouple are also important factors to consider. The wire size affects both the sensitivity and the maximum operating temperature of the thermocouple. Thermocouple wire sizes range from fine, 40 AWG, to heavy, 8 AWG. (AWG indicates American Wire Gage. It is used interchangeably with B. & S., Brown and Sharpe.) While a 20 gage wire requires only two minutes to reach 80% of its final temperature reading, an 8 gage wire requires nearly nine minutes. Table 4 is a summary of the maximum temperature for various thermocouples.

In most cases, thermocouples can-

TABLE 4 MAXIMUM TEMPERTATURE FOR THERMOCOUPLES

ANSI Type	THERMOCOUPLE	GAGE	MAXIMUM OPERATING TEMP. °F
T	COPPER-CONSTANTAN	14 20 24 28	700 600 500 400
J	IRON-CONSTANTAN	8 14 20 24 28	1400 1000 900 700 700
K	CHROMEL-ALUMEL	8 14 20 24 28	2500 2000 1800 1600 1600
S	PLATINUM-10% RHODIUM -PLATINUM	24	3000
	CHROMEL-CONSTANTAN	8 14 20	1800 1200 1000
	TUNGSTEN 5% RHENIUM-TUNGSTEN 5% RHENIUM	18 20 24 28	4200 1800 1600 1600

TABLE 5 PROTECTING TUBE MATERIALS

TYPE	RECOMMENDED MAX. TEMP.	COMMENTS
WROUGHT IRON	1200°F	FOR GENERAL USE EXCEPT CORROSIVE ATMOSPHERES
CAST IRON	1500°F	FOR ACID AND ALKALINE SOLUTIONS
304 STAINLESS STEEL	1800°F	FOR CORROSIVE ATMOSPHERES AND SOLUTIONS
NICKEL	2000°F	FOR SPECIAL CHEMICAL APPLICATIONS
INCONEL	2200°F	SUBSTITUTE FOR NICKEL WHEN SULFUR IS PRESENT
CORUNDUM	3000°F	FOR STEEL INDUSTRY, WHERE THERMAL SHOCK MAY BE HIGH
CARBOFRAX	3000°F	FOR APPLICATIONS WITH HIGH THERMAL AND MECHANICAL SHOCK
CHRONIUM & ALUMINUM OXIDE	3000°F	FOR BRASS AND BRONZE FOUNDRIES HIGH THERMAL CONDUCTIVITY

not be used without protection from the environment in which they are used. The environment might be a reducing atmosphere (where oxygen content is low and hydrogen and carbon monoxide are present) or an oxidizing atmosphere (where oxygen and water vapor are present). For example, iron corrodes in an oxidizing atmosphere, and chromel becomes contaminated in a reducing atmosphere. Platinum/Platinum-Rhodium thermocouples always require protection.

The devices used for protecting thermocouples are called thermocouple wells and protecting tubes. They may be made of metals such as iron, steel, nickel, or inconel; silica compounds such as corundum or carbofrax; or metal ceramic compounds such as chromium oxide and aluminum oxide. Sometimes a thermocouple must be enclosed in a primary metal protector as well as a secondary silica protector. Table 5 lists some of the more common protecting tube materials.

The selection of the complete thermocouple assembly should be made carefully. Manufacturers provide a considerable amount of data for this purpose.

The Millivoltmeter. The millivoltmeter used to measure temperature is made with a permanent magnet and a moving coil. See Fig. 10-28. When a voltage is applied across the coil, the coil becomes a magnet and rotates because of the interaction between its magnetic force and the magnetic force of the permanent magnet. The coil rotation is opposed by the action of spiral springs. These springs also act as conductors, bringing current into the meter coil. Because the millivoltmeter requires power to move the coil, the voltage due to temperature must be sufficient to produce such power. For this reason, the temperature range of a millivoltmeter type thermometer must be wide enough to produce nearly full-scale deflection electrically. For example, if an iron-constantan thermocouple is used, a temperature range of 0°F to + 200°F or greater is needed to provide a five millivolt deflection.

Ambient temperatures which occur at the meter must be compensated for because they affect the temperature of the reference junction located at the meter. Variations in the meter temperature can also cause the electrical resistance of the meter to change. Compensation for changes in the reference junction temperature is generally accomplished by using a bimetallic strip in the coil spring mechanism. See Fig. 10-29. The bimetallic strip corrects the tendency of the electrical resistance to move the coil when the temperature changes. A meter is said to be temperature-compensated if a change of 10°F causes less than 0.25% error in its reading.

Variations in the internal electrical resistance of the millivoltmeter are kept at a minimum by using wire

Fig. 10-28. Mechanism with a permanent magnet and moving coil. (Weston Instruments, Inc.)

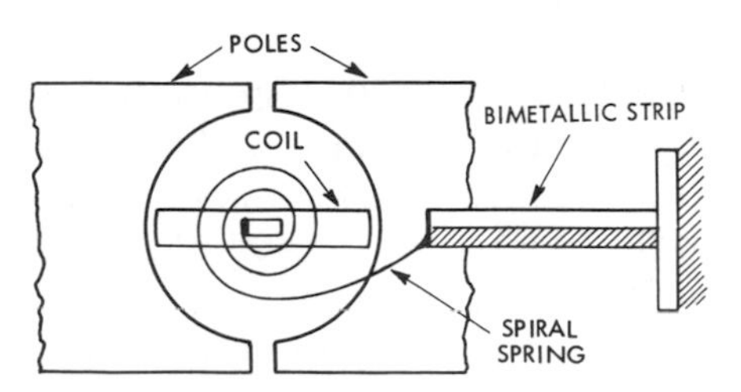

Fig. 10-29. A bimetallic strip used as compensation in the reference junction millivoltmeter.

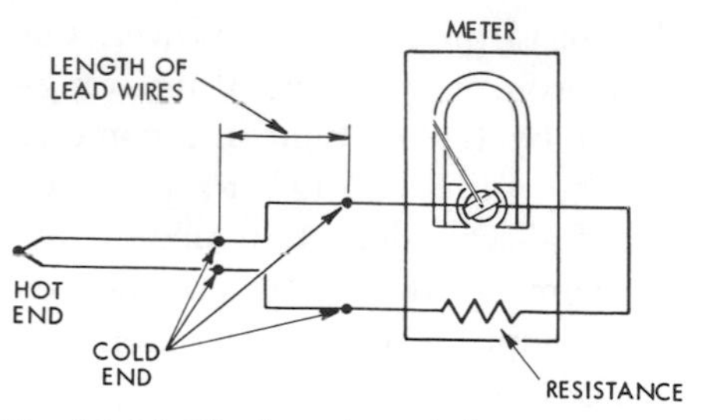

Fig. 10-30. The length and diameter of the lead wire determines the wire's resistance in a thermocouple pyrometer circuit.

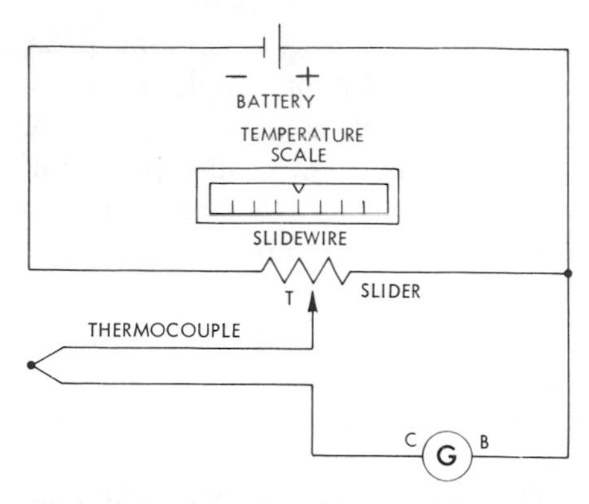

Fig. 10-31. In the potentiometer, the thermocouple voltage is measured between points T and C. The voltage supplied by the battery is measured between T and B. A scale can be used with the slider to indicate temperature.

which does not vary its resistance as the temperature changes. Manganin wire is frequently used for the internal meter resistors.

The external resistance of the millivoltmeter circuit likewise affects the accuracy of the system, including the resistance of the lead wire and the thermocouple. Inaccuracy is generally overcome by making the resistance of the meter as high as possible so that changes in external resistance will have little effect. A fixed external resistance is used when the instrument is calibrated. This fixed value is usually printed on the meter scale. Best accuracy is obtained if the external resistance is as near to the fixed value as possible during installation. The length and diameter of the lead wire are important factors in determining resistance. See Fig. 10-30. A greater length will increase resistance, and a greater thickness will decrease the resistance.

The usual static accuracy of the millivoltmeter thermocouple is $\pm$ 1%. The speed of response of a millivoltmeter is generally better than that of the thermocouple if the meter has been well-designed.

The Potentiometer. Unlike the millivoltmeter, which uses the voltage of the thermocouple to activate the meter mechanism, the potentiometer compares the thermocouple voltage with voltage supplied by a battery. The battery-supplied voltage is adjusted until it is equal and opposite the thermocouple voltage. The battery-supplied voltage is applied to a slidewire resistor calibrated in degrees of temperature. See Fig. 10-31. Thus there is a temperature reading for each position of the slider. Because a battery does not maintain voltage, it is frequently necessary to adjust the voltage to a particular value so that the relationship between the temperature and the battery-supplied voltage remains constant. This adjustment of the voltage is made using a standard cell. See Fig. 10-32. The battery voltage is temporarily disconnected from the thermocouple and connected to the standard cell. Adjustable resistors are used to

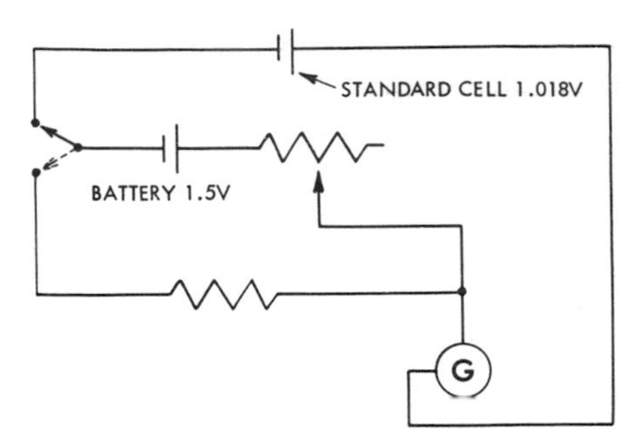

Fig. 10-32. The galvanometer compares the voltages of the standard cell and the battery.

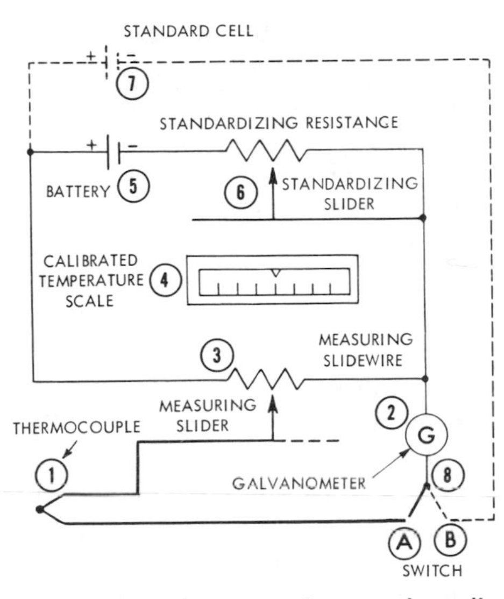

Fig. 10-33. Schematic of a simple null-balance potentiometer circuit. (1) Thermocouple, (2) Galvanometer, (3) Slidewire and Slider, (4) Scale, (5) Battery, (6) Standardizing Slider, (7) Standard Cell, (8) Standardizing Switch with (A) Thermocouple Position and (B) Standard Cell Position.

make the battery voltage equal to the voltage of the standard cell. The standard cell maintains a constant voltage (1.018V) for long periods of time. The comparison of battery and standard cell voltage is called *standardization* and can be performed manually or automatically. During standardization, the measuring circuit of the instrument is adjusted, and the standard cell replaces the thermocouple.

It is perhaps simpler to explain the operation of a potentiometer which contains a galvanometer, although this type is seldom used in industry. A galvanometer, like the millivoltmeter, has a permanent magnet and a moving coil. In fact, the millivoltmeter is actually a galvanometer used for measuring voltage in units of 1/1000 of a volt. The galvanometer used in the potentiometric circuit ordinarily has few or no scale graduations. A zero point indicates when no electrical current is passing through. There can be marks which indicate if the current is above or below the zero point.

Fig. 10-33 shows a simple potentiometer circuit. It is called a null-balance circuit since the value to be measured is correct only when the galvanometer is at its zero (null) point. At this point, the measuring circuit voltages are in balance, since the thermocouple and battery voltages are equal and opposite.

The standard cell is cut out of the circuit when the standardizing switch is at the thermocouple position. The position of the slider determines the amount of the battery voltage that is tapped. The thermocouple voltage is automatically and instantly subtracted from this amount. The remaining difference in voltage is applied to the galvanometer. The circuit is at the null point when the voltages of the battery and the thermocouple are equal and opposite. A pointer attached to the slider then indicates the temperature of the thermocouple.

When the standardizing switch is at the standard cell position, the slider and the thermocouple are no longer a part of the circuit. They are replaced by the standard cell. The entire slidewire is in the circuit, connected to the positive terminal of the battery and the standard cell at one end, and to the galvanometer at the other end. The negative terminal of the standard cell is directly connected to the side of the galvanometer not hooked up to the slidewire. The standardizing resistor slider is adjusted until the battery voltage is equal and opposite to that of the cell. The voltage across the entire measuring slidewire is at the value required for the scale to be correctly read in units of temperature. The thermocouple voltage can be read as temperature, as long as the battery maintains proper voltage.

The simple circuit shown in Fig. 10-33 does not include the resistors necessary to establish the operating range of the potentiometer or to make the instrument suitable for use with a particular type of thermocouple. In modern industrial instruments, the measuring resistors are supplied as a prewired unit for plug-in, or by another simple method of installation.

Temperature Variations. Ambient temperatures at the instrument must be compensated for to eliminate changes in resistance due to temperature, and changes in the reference temperature.

The resistors used in the measuring circuit should be carefully selected so that their resistance remains constant over the range of temperatures for the particular process application. The resistors can be made of manganin, which has exceptional stability, if the temperature range is extensive, and high accuracy is required.

Reference junction compensation in measuring circuits, such as the one shown in Fig. 10-34, is usually accomplished automatically by using a special nickel resistor in the measuring circuit. Variations in the temperature of the reference junction are compensated for by separately enclosing the resistor and the reference junction terminals inside the instrument. The nickel resistor must be selected so that its variation in resistance causes a voltage equal and opposite to that caused by the change in reference junction temperature.

Balancing Mechanisms. In the simple circuit of the potentiometer, the balancing is a manual operation, involving an instrument of the type shown in Fig. 10-35. This is not satisfactory for most industrial applications because the instrument must be in continuous balance to provide an accurate record and precise control.

In previous galvanometer-type instruments many different automatic balance mechanisms were employed. Most manufacturers have adopted a design that eliminates the galvanometer and substitutes a converter electronic amplifier and balancing motor. A typical circuit is shown in Fig. 10-36.

In this circuit the dc voltage of the measuring circuit is converted to ac. Then the ac voltage is amplified to control the balancing motor which moves the slidewire slider to maintain balance in the measuring circuit. The

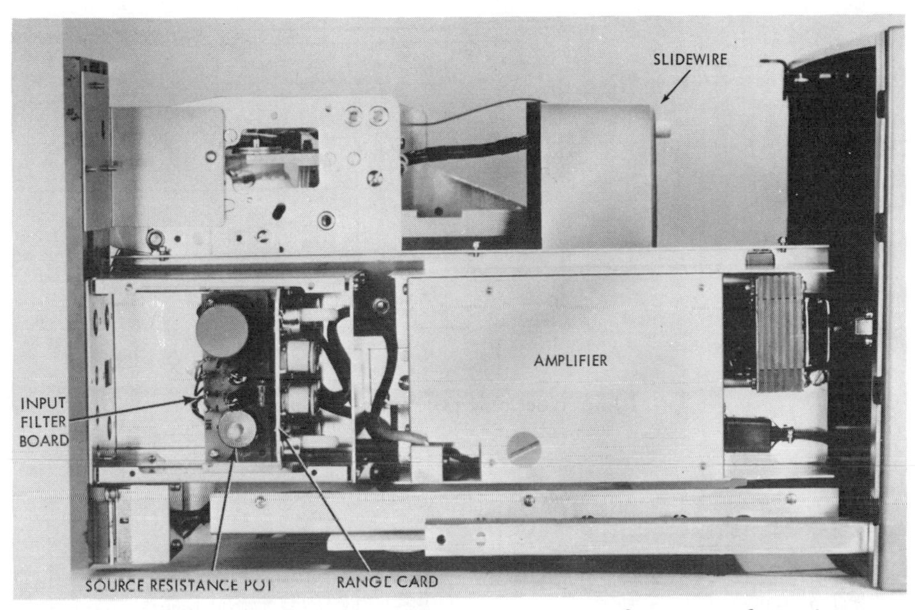

Fig. 10-34. Interior of a measuring circuit. (Bristol Division of Acco)

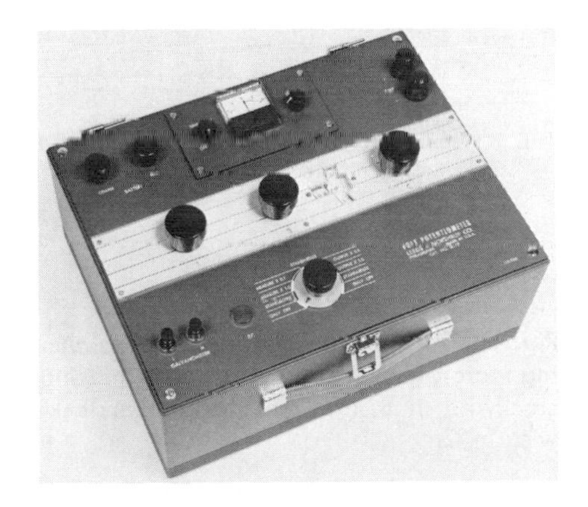

Fig. 10-35. Manual balance indicator. (Leeds & Northrup)

galvanometer is not used. Its function of detecting measuring circuit unbalance is performed by the converter. The converter changes from dc to ac the current that flows in the measuring circuit due to the voltage imbalance. The ac current passes through a transformer and an amplifier and then to the balancing motor causing the motor to drive the slidewire slider to a position that will balance the measuring circuit. When the measuring circuit is

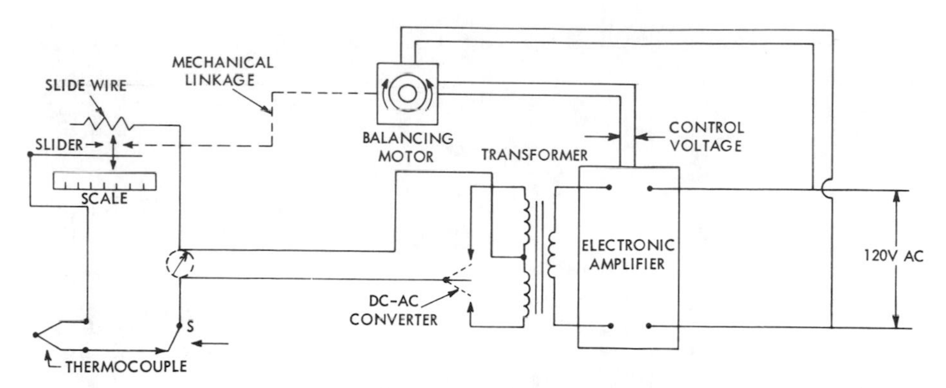

Fig. 10-36. Electronic potentiometer circuit.

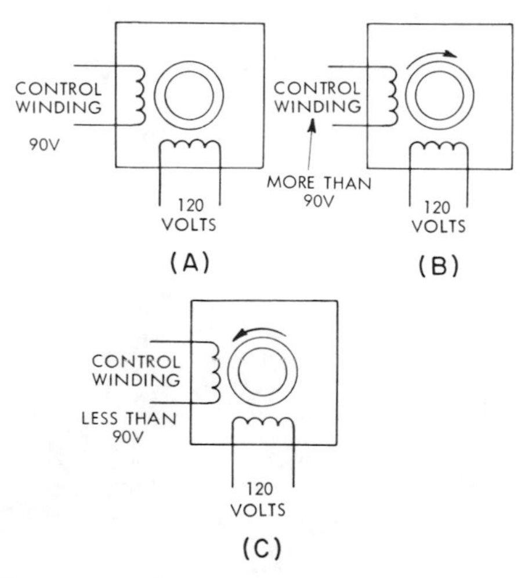

Fig. 10-37. The characteristics of a balancing motor. Motor is stationary in A, moving clockwise in B, and moving counterclockwise in C.

Fig. 10-38. A high-speed single-point strip chart recorder. (Honeywell Process Control Div./Fort Washington, Pa.)

again in balance no dc current is fed into the converter. Hence there is no ac current and the motor stops. The slider then indicates the correct temperature.

The balancing motor contains two windings, one of which is supplied with line voltage. The other winding is called the *control winding*. The motor remains stationary at a specific voltage on the control winding. When the voltage decreases the motor drives in one direction. When the control voltage increases, the motor drives in the opposite direction. See Fig. 10-37.

The static accuracy of automatically balanced potentiometers is usually better than ± ¼ of 1%, which is, of course, considerably better than the accuracy of most industrial thermocouples. Some potentiometers of this type respond to a full scale temperature change in a fraction of a second. This is faster than the temperature response of the thermocouple itself. See Fig. 10-38.

Resistance Thermometers. When precise temperature measurement by electrical means is desired, the resistance thermometer is used in a bridge circuit.

The thermometer itself is generally manufactured in the form of a bulb. The bulb consists of a fine wire wrapped around an insulator and enclosed in metal. See Fig. 10-39. The most common shape for a resistance thermometer resembles a bimetallic thermometer bulb. See Fig. 10-40.

Platinum wire is the best material for a resistance thermometer since it is useful over a wide range of temperatures (−400°F to +1200°F). Nickel is frequently used because it is economical and, over its useful range (−250°F to +600°F), its resistance per degree of temperature change is greater than that of platinum. Copper is generally restricted to temperatures below that of nickel. Its entire useful range is −328°F to +250°F.

The important considerations in the selection of resistance thermometer wire are:

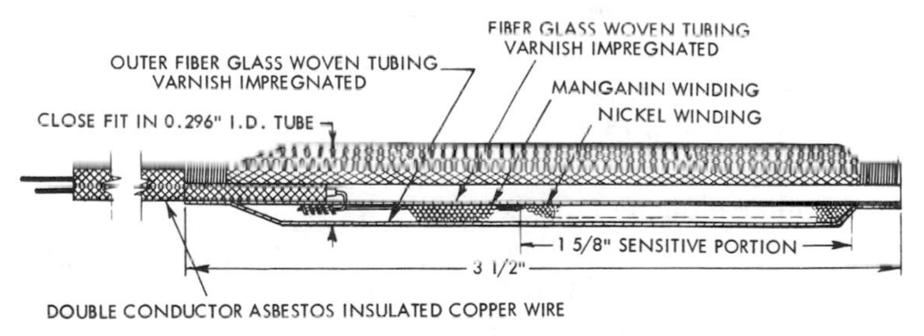

Fig. 10-39. Resistance thermometer bulb element. (Honeywell Process Control Div./Fort Washington, Pa.)

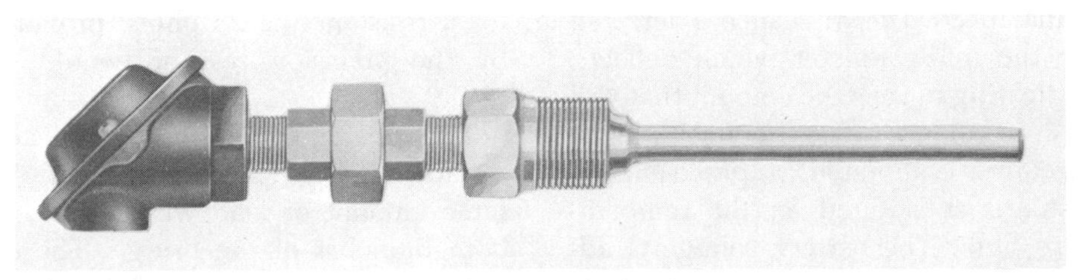

Fig. 10-40. High-speed resistance thermometer bulb. (Honeywell Process Control Div./ Fort Washington, Pa.)

1. Purity
2. Uniformity
3. Stability
4. High resistance change per degree temperature change
5. Good contamination resistance.

Resistance thermometers made from the same material should be interchangeable without requiring recalibration of the instrument being used. For this reason, resistance thermometers are manufactured to have a fixed resistance at a certain temperature. Platinum resistance thermometers generally have a resistance of 25 ohms at 32°F. Nickel thermometers generally have a resistance of 100 ohms at 77°F. Copper thermometers generally have a resistance of 10 ohms at 77°F.

Since temperature measurement with resistance thermometers is actually a measurement of resistance, the Wheatstone bridge, with variations, is used. The bridge may be dc or ac. In this discussion, we deal with the dc type in its simplest form.

An indicating millivoltmeter is found in the simplest forms of resistance thermometers. Fig. 10-41 shows the Wheatstone bridge circuit used with this type of meter.

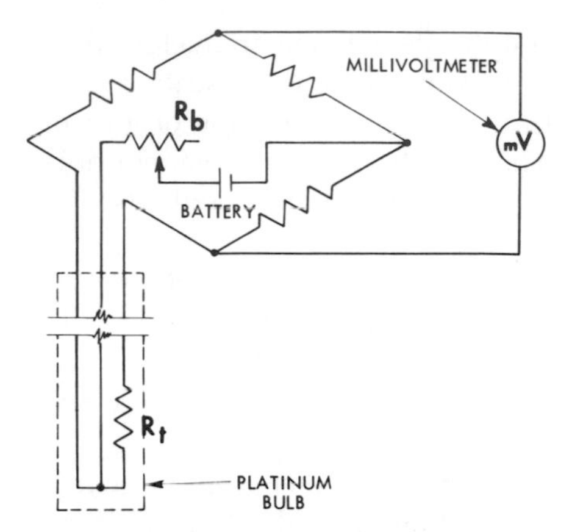

Fig. 10-41. Schematic of a resistance thermometer using a Wheatstone bridge circuit and platinum bulb.

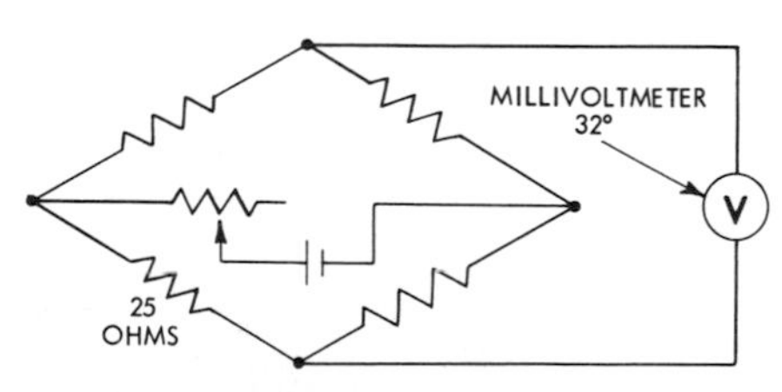

Fig. 10-42. Wheatstone bridge circuit with resistor used in place of the platinum bulb.

When the temperature (R_t) of the resistance bulb rises, its resistance increases. As a result of this increase in resistance, the Wheatstone bridge is unbalanced. The imbalance is detected by the millivoltmeter which deflects, indicating in volts the amount that will restore balance to the circuit. When the circuit is calibrated, a precision resistor is substituted for the temperature bulb. The battery voltage is adjusted by varying R_b until the meter indicates the temperature that is correct for this resistance. Fig. 10-42 shows an example. If a platinum bulb (replaced at the point of calibration) is to be used, the correct meter reading for a resistance of 25 ohms, provided by the precision resistor, would be 32°F.

The resistance bulb has three leads, as shown in Fig. 10-43. In this way, the same amount of lead wire is used in both branches of the bridge. The arrangement allows the lead length to be increased or shortened, without af-

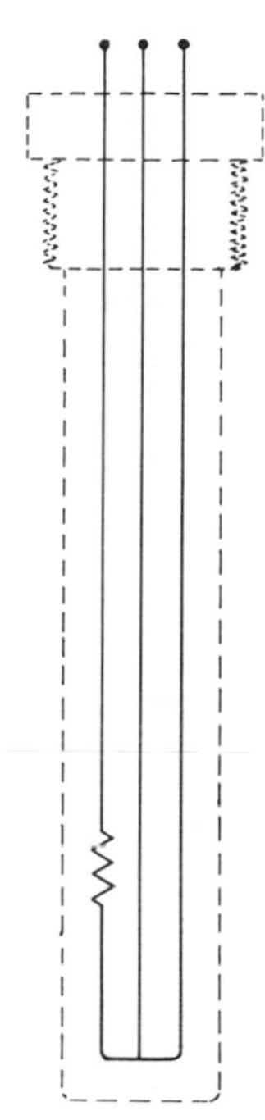

Fig. 10-43. Resistance bulb showing the three lead wires.

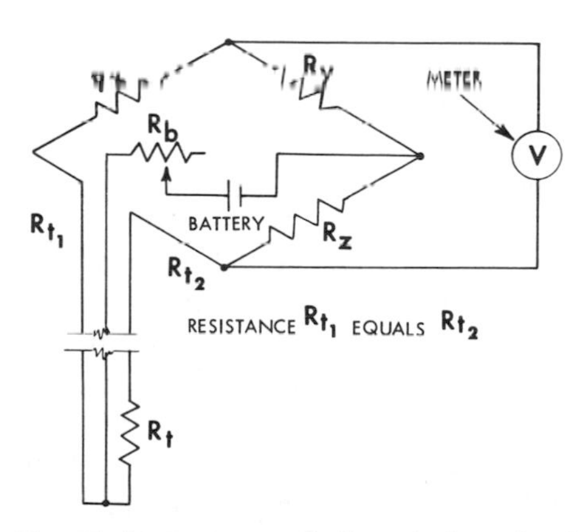

Fig. 10-44. Resistance bulb with three lead wires in a Wheatstone bridge circuit.

fecting the meter reading. See Fig. 10-44.

Resistance thermometers are also used with slidewire instruments which have automatic null balance. Except for the circuitry and the lack of standardizing equipment, these instruments are basically the same as the self-balancing potentiometers. They eliminate the galvanometer, using instead a dc-ac converter or an ac power supply in place of a battery. Fig. 10-45 shows a dc bridge circuit typically used with a resistance thermometer. It should be noted that the position of the slidewire here is different from that in a simple Wheatstone bridge. The slidewire is in the position shown (Fig. 10-45) so that the effect of a resistance change between the slidewire and the slider is eliminated. Hence, any change in resistance due to poor contact affects both arms of the bridge in the same manner. Thus any effect on the accuracy of the measurement is negligible.

The null balance principle in this type of instrumentation depends on a balance between the thermometer bulb resistance and the division of the slidewire resistance. The amount of slidewire resistance in the two arms changes with each position of the slider. If the resistance of the thermometer bulb increases, the resistance in its arm is reduced, and the resistance in the opposing arm increased, to obtain a balance. When

$$\frac{(R_1 + R_x)}{R_t} = \frac{(R_r + R_y)}{R_z}$$

the circuit is in balance. Because there is no voltage flow to the converter, amplifier, and motor, the pointer remains stationary, and the temperature reading is accurate.

The accuracy of the thermometer using a millivoltmeter is about $\pm 1\%$.

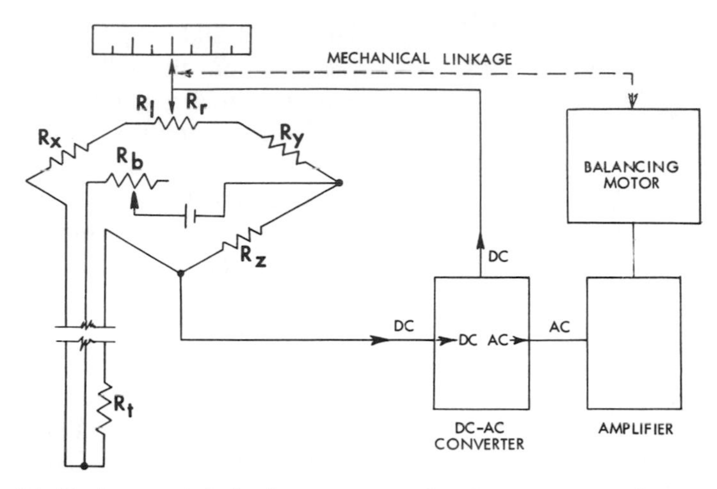

Fig. 10-45. A typical dc bridge circuit used with a resistance thermometer.

The accuracy of the null-balance type thermometer is greater, generally better than $\pm 0.25\%$.

The response of the resistance thermometer is about the same as the response of the thermocouple if it is used under the same conditions and in similar enclosures. However, the resistance thermometer and the thermocouple are usually enclosed in some type of well which slows down response.

Thermistors. Thermistors are very small, solid-state semi-conductors, made from various metal oxides. They are available in several shapes, such as rods, discs, beads, washers, and flakes. See Fig. 10-46. The electrical resistance of a thermistor decreases with an increase in temperature. Thus the thermistor has a negative temperature coefficient of resistance. For temperature measurement, they are used in bridge circuits like the resistance thermometer. Since thermistors are very small (to prevent overheating), the bridge current must be maintained at a low level. Their accuracy is slightly inferior to that of the conventional resistance bulb.

Radiation Thermometry. All bodies (or surfaces) which are not at absolute zero give off radiation, depending on the activity of their molecules. The reference standard for radiation is called a *blackbody*. A blackbody is an ideal body which completely absorbs all radiant energy of any wavelength falling upon it and reflects none of this energy. Hence, its reflectivity would be zero, and its absorbency would be 100%. It would be a complete radiator, radiating a greater amount of thermal energy at a faster rate than any other body at the same temperature and under the same conditions. The ability of a body to radiate thermal energy is called *emissivity*. Emissivity is a comparison of the relative emissive power of any radiating surface with the emissive power of a blackbody radiator at the same temperature. It is expressed by the following ratio or fraction:

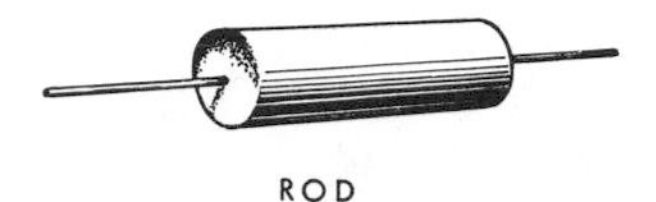

ROD THERMISTORS—are made by an extrusion process and sintered at a high temperature. Lead wires are soldered to the ends of the element.

DISC THERMISTORS—are produced by pressing thermistor powders in a mold and sintering at a high temperature.

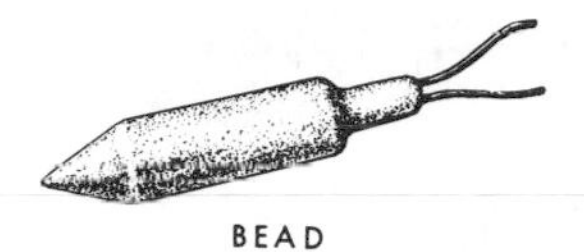

BEAD THERMISTORS—are made by utilizing the surface tension of a slurry of mixed oxides applied to closely spaced platinum wires. Shrinkage of the oxides occurs at the high sintering temperature used and results in a permanent electrical contact.

Fig. 10-46. Common types of thermistors.

TABLE 6 EMISSION FACTORS OF OXIDIZED METALS

MATERIAL	100C	200C	500C
BRASS		0.61	0.59
COPPER (CALORIZED)	0.6		0.6
CAST IRON		0.95	0.84
IRON		0.74	
MONEL		0.43	0.43
NICKEL		0.37	
STEEL		0.79	0.79
WROUGHT IRON		0.94	0.94
STAINLESS STEEL		0.62	0.73

$$\text{Emissivity} = \frac{\text{Total radiation from a non-blackbody}}{\text{Total radiation from a blackbody}}$$

Radiation thermometry is based on this concept. See Table 6.

It should be noted that thermal radiation, like light, travels in waves. The wavelengths range from the longest infrared to the shortest ultraviolet rays. The color of a body can change with temperature. For example, as a steel billet is heated, its color changes from red (at about 1112°F / 600°C) to yellow (at about 2012°F / 1100°C) to white (at about 2732°F / 1500°C). Thermal wavelengths range from 4500 angstrom units for red heat to 6500 angstrom units for white heat. (An *angstrom unit* equals one billionth of a centimeter.)

When the radiation from a hot ob-

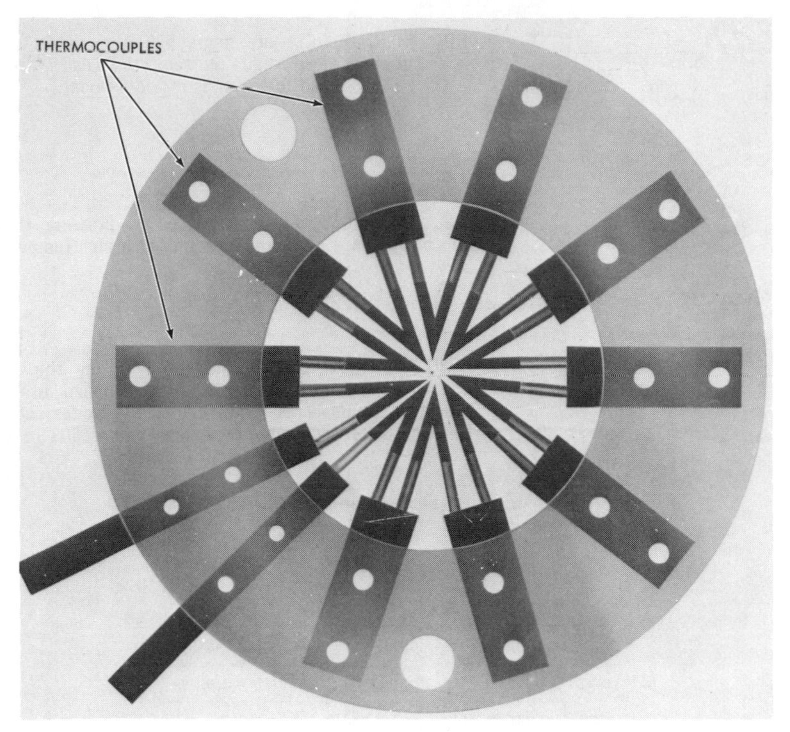

Fig. 10-47. A thermopile showing the thermocouples. (Honeywell Process Control Div./ Fort Washington, Pa.)

ject is visible, as is the case of a steel billet, the temperature can be measured using a lens system and a thermopile. A *thermopile* consists of several thermocouples connected in series to add up their outputs. See Fig. 10-47. Chromel and Constantan are the two metals commonly used in the individual thermocouples because of their physical properties and high voltage output. The thermopile, like the thermocouple, is used with potentiometric circuits. These radiation thermometers, such as the one shown in Fig. 10-48, are used for measuring temperatures from 125°F to 4000°F.

Heat rays that are invisible are infrared rays. Just as with visible rays, infrared rays can be detected using a thermopile. However, since infrared rays have greater thermal energy than visible rays, the temperature of objects emitting these rays is often measured by another type of radiation thermometer. This type uses a semiconductor crystal. A material commonly used for the crystal is Indium. All thermal rays consist of energy units called *photons*. A lens system is used to focus these photons on the crystal. The bombardment of the crystal causes an energy change in the form of electrical voltage. This voltage is proportional to the temperature of the

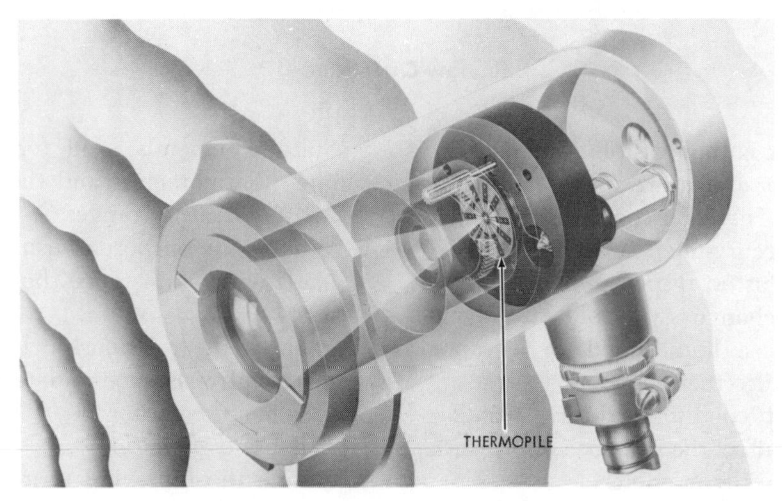

Fig. 10-48. A radiation thermometer showing thermopile position. (Honeywell Process Control Div./Fort Washington, Pa.)

object being measured. A voltmeter with its scale calibrated in temperature units can be used for this measurement. Such infrared radiation thermometers are used for temperatures ranging from 100°F to 5000°F. A lead sulfide cell is used instead of the crystal in some applications.

Words to Know

thermal capacity
specific heat
coefficient of linear expansion
coefficient of volumetric expansion
Charles' law
transducer
Seebeck effect
Peltier effect
Thomson effect (reversible heat effect)
cold junction (reference junction)
hot junction (measuring junction)
emf
law of intermediate temperature
law of intermediate metals
AWG
standardization
balancing motor
thermistor
blackbody
emissivity
angstrom unit
photon
thermopile

Review Questions

1. What is the possible source of error for a pressure spring thermometer? How is it compensated for in a mercury-filled system?
2. The fastest response to a temperature change is possible with a gas-filled, a liquid-filled, or a vapor pressure thermometer?
3. What conditions during installation affect the response of pressure spring thermometers?
4. State the two laws of thermoelectricity.
5. State the metals used for resistance thermometers and their useful temperature ranges.
6. What is the purpose of using three leads with a resistance thermometer?
7. What is the relationship between electrical resistance and temperature in a thermistor?
8. What is *emissivity?*
9. Define the following terms: *angstrom unit, thermopile,* and *photon.*

Pressure

Chapter

11

There are three scales for pressure measurement:

1. Gage pressure scale
2. Absolute pressure scale
3. Vacuum scale

The difference between the gage pressure scale and the absolute pressure scale is the location of the zero point. On the gage pressure scale, the zero point is at atmospheric pressure. On the absolute pressure scale, the zero point is at the absolute zero pressure point.

The vacuum scale has its zero at atmospheric pressure and its maximum point at the absolute zero pressure point. Thus, the vacuum scale is used to indicate negative gage pressure. See Fig. 11-1.

The measurement of atmospheric pressure is essential to the establishment of the gage pressure scale and the vacuum scale. Atmospheric pressure is the pressure exerted by the air surrounding the earth. This pressure varies with altitude, since the air nearer

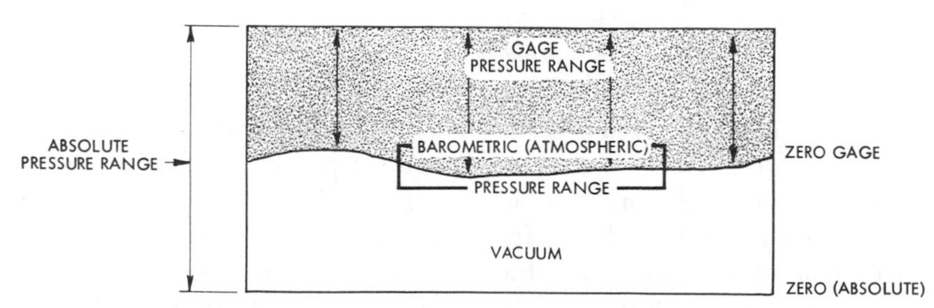

Fig. 11-1. Gage pressure is pressure above atmospheric pressure. The location of zero on a gage pressure scale depends on the applicable barometric reading. Absolute pressure is pressure above absolute zero pressure. A vacuum is any pressure less than atmospheric pressure.

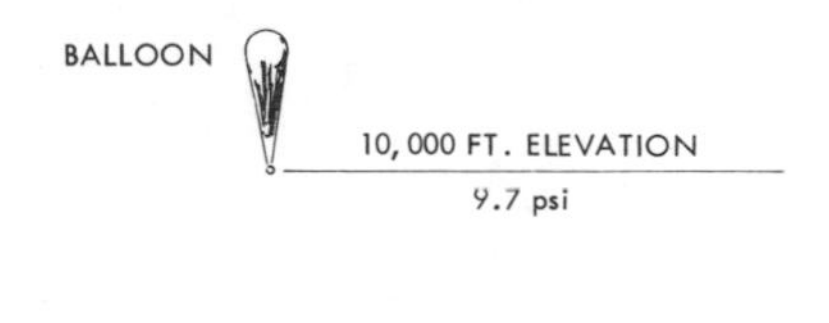

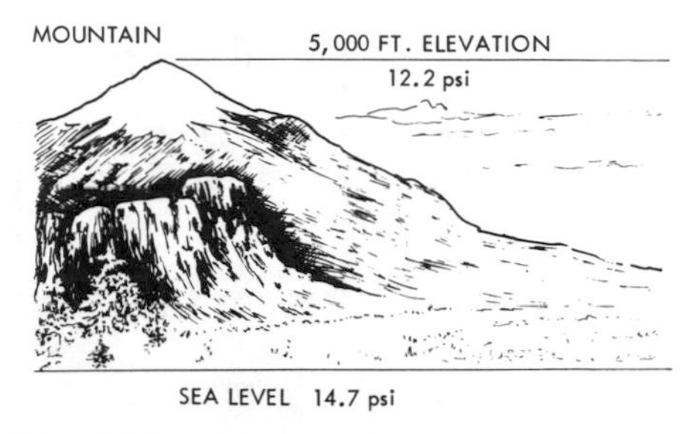

Fig. 11-2. Atmospheric pressure at sea level, at 5,000 feet, and at 10,000 feet.

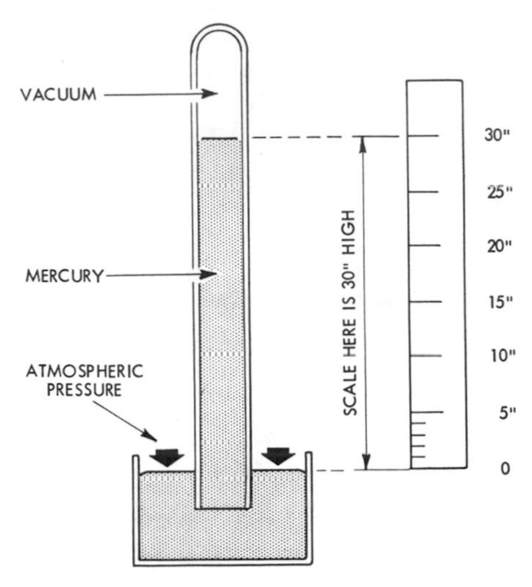

Fig. 11-3. Barometer used for measuring atmospheric pressure.

the earth is compressed by the air above. At sea level, the atmospheric pressure is 14.7 pounds per square inch (psi). At 5,000 feet elevation the atmospheric pressure is 12.2 psi. At 10,000 feet elevation, the atmospheric pressure is 9.7 psi. See Fig. 11-2. The instrument used for measuring atmospheric pressure is the barometer. The simplest barometer consists of a long glass tube which is sealed at one end, filled with mercury, then inverted and placed in a pan of mercury. See Fig. 11-3. The mercury in the tube settles, leaving a vacuum above it. The height of the mercury in the tube above the level of the mercury in the pan indicates the atmospheric pressure in inches of mercury. Another device for measuring atmospheric pressure is the *aneroid barometer.* See Fig. 11-4. In this type, a pressure capsule, from which all air has been removed, is linked to a pointer. The pointer then indicates atmospheric pressure on the scale in pounds per square inch.

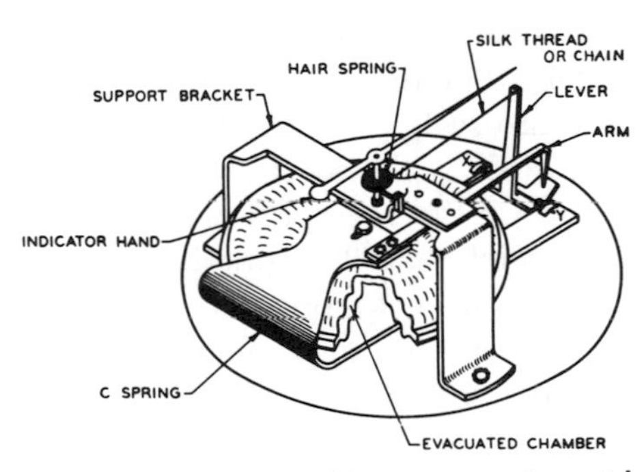

Fig. 11-4. An aneroid barometer. Aneroid means using no fluid. (Taylor Instrument)

For measuring absolute pressures, the principle of the aneroid barometer is used. However, instead of having atmospheric pressure act upon an evacuated pressure element, the pressure to be measured is used. In Fig. 11-5, the pressure inside the evacuated

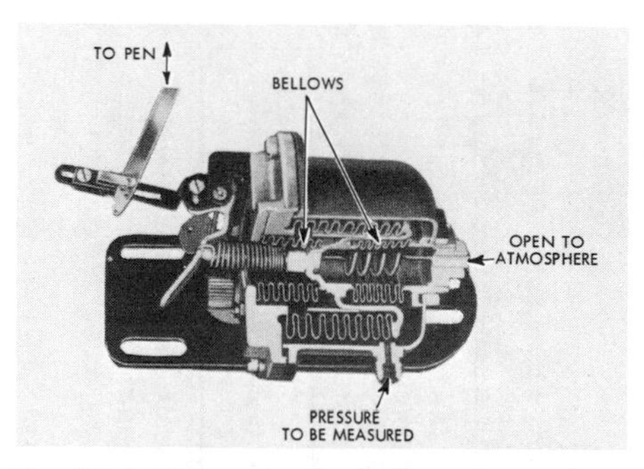

Fig. 11-5. Cutaway of a bellows-type absolute pressure gage. (Taylor Instrument)

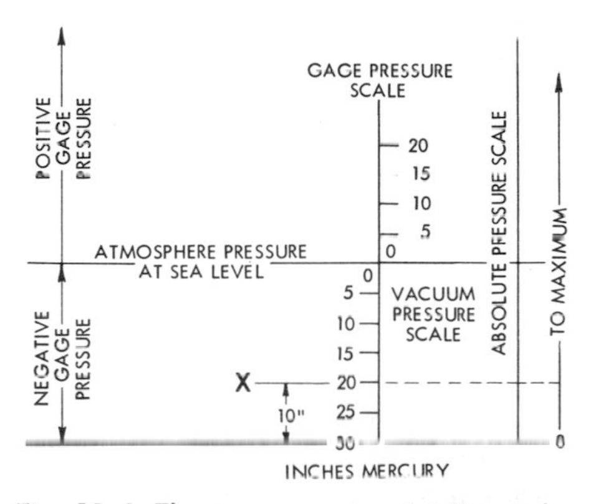

Fig. 11-6. The pressure at point X can be expressed as 10 inches of mercury absolute, or as 20 inches of mercury vacuum, or as —20 inches of mercury gage.

bellows is at the absolute zero point. Therefore, the pressure indicated by the instrument is in absolute units.

Vacuum measurement is the measurement of pressures below atmospheric pressure. The vacuum scale is graduated in inches of mercury or inches of water. Elastic deformation elements, such as the Bourdon tube, bellows, or diaphragm, are used to measure vacuum just as they are used to measure pressure.

The three pressure scales are shown in Fig. 11-6. Notice that the absolute pressure scale includes the gage pressure scale and the vacuum scale. Thus, a pressure of 10 inches of mercury absolute can also be expressed as vacuum of 20 inches of mercury or a gage pressure of —20 inches of mercury.

Pressures which do not exceed 1 psi above atmospheric pressure are generally regarded as low pressures. Instruments for measuring such low pressures are usually calibrated in inches of water (1 pound per square inch above atmospheric pressure equals 27.7 inches of water).

Manometers

The manometer can be used for measuring low pressures. In addition to the simple U-tube manometer, several special types of manometers have been devised to measure low pressures. The direct reading well-type manometer is one of these. In this instrument, the pressure to be measured is admitted to the well.

Assume that a pressure of 2 psi is applied to the well side of the manometer depicted in Fig. 11-7. This causes the water in the tube side to rise 55.15 inches above the zero line. At the same time, the water in the well side is lowered. Since the area of the well is 300 times the area of the tube, this movement is considerably less (0.185 inches). In order for the pressure to be read directly on a scale attached to the tube side, each scale inch must be corrected for the movement of water in the well. Thus, each inch of water read

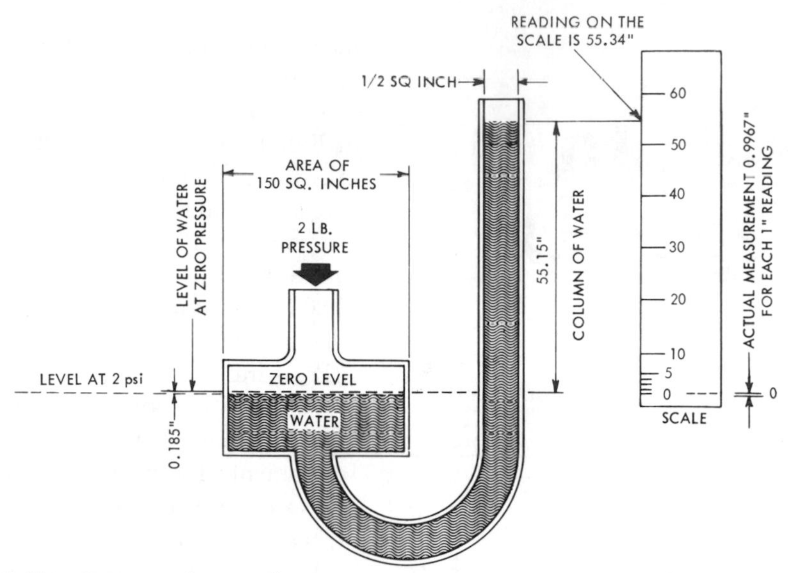

Fig. 11-7. This direct reading well-type manometer uses water as the liquid for measuring low pressure.

on the manometer scale is actually less than an inch in length.

In the example in Fig. 11-7, each inch mark on the scale is $1 - \frac{0.185}{55.34}$ inches in length, or .9967 inches, after adjustment. Note that the sum of the change in heights of the water in the two sides of the manometer (55.15 + 0.185) equals 55.34 inches. This is proven, since 2 psi is equivalent to the pressure exerted by a column of water 55.34 inches in height.

Another form of manometer is called a *draft gage*. In this well-type manometer, the tube is almost horizontal. This has the effect of lengthening the graduations of the scale. A vertical rise of two inches in the tube will actually show up as a six-inch diagonal movement of the water in the tube. See Fig. 11-8. Note that this type of manometer must be installed in a level position. Levelling devices are usually built into the instrument. Again, the scale must be corrected for the lowering of the water in the well.

Water and mercury are the most common liquids used in manometers. Since the measurements produced are in common units of pressure, other fluids can be used, as long as their specific gravities remain constant under the operating conditions. The specific gravity of liquid actually determines the pressure range of the manometer. When one liquid is substituted for another, the reading must be corrected accordingly. For instance, if a liquid with a specific gravity of 3.0 is substituted for water (specific gravity = 1.0), each inch shown on the manometer scale is actually equal to 3 inches of water in terms of pressure measurement. There are manometer fluids available whose specific gravities are

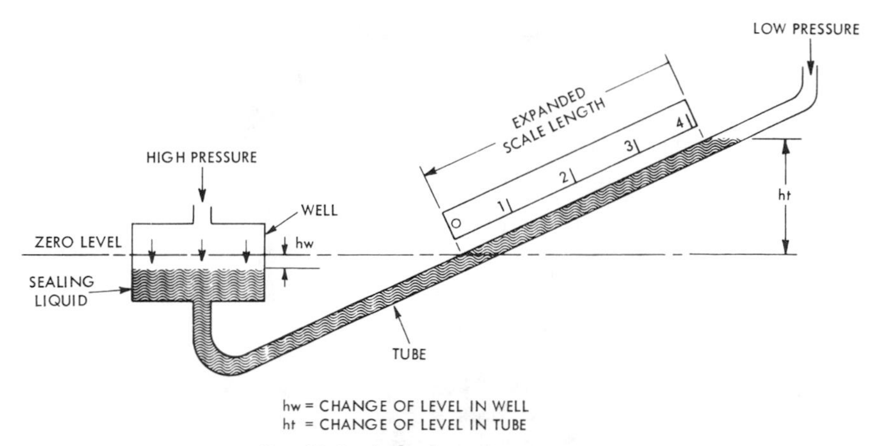

Fig. 11-8. A draft gage manometer.

less than 1.0 as well as others with specific gravities from 1.0 to 3.0. In addition to their usefulness in extending or limiting the manometer scale, some liquids have physical or chemical properties which make them superior to water or mercury.

Elastic Deformation Elements

The diaphragm, among the elastic deformation elements available, is best suited for low pressure measurement.

One type of diaphragm is non-metallic. It is made of such materials as leather, teflon, and neoprene. Generally, diaphragms of this type are large and non-circular. See Fig. 11-9. They are satisfactory for pressure ranges from 0 to .5 inches of water to 0 to 10 inches of water. The diaphragm is opposed by a light spring. The principal applications of such diaphragms are for indicating, recording, or controlling draft gages.

Another type of diaphragm is a single, circular, metallic disc, either

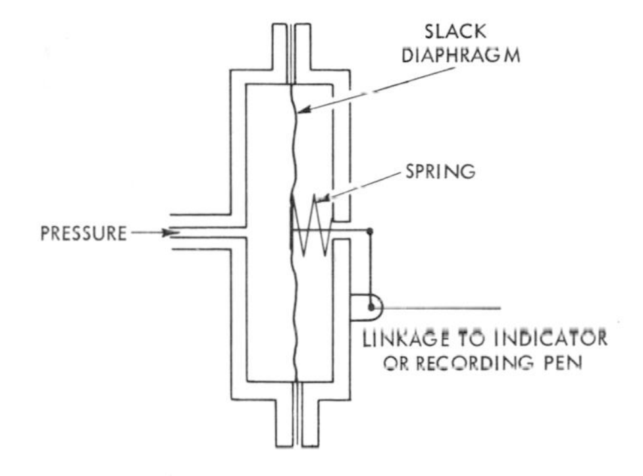

Fig. 11-9. A diaphragm deformation element suitable for low pressures.

flat or corrugated. It depends on its own elasticity for operation.

Pressure Capsules. Frequently, two circular diaphragms are welded or soldered together to form a pressure capsule. These capsules can be used singly or stacked, as shown in Fig. 11-10, depending on the pressure range to be measured. The deflection of a pressure capsule depends on the diameter of the

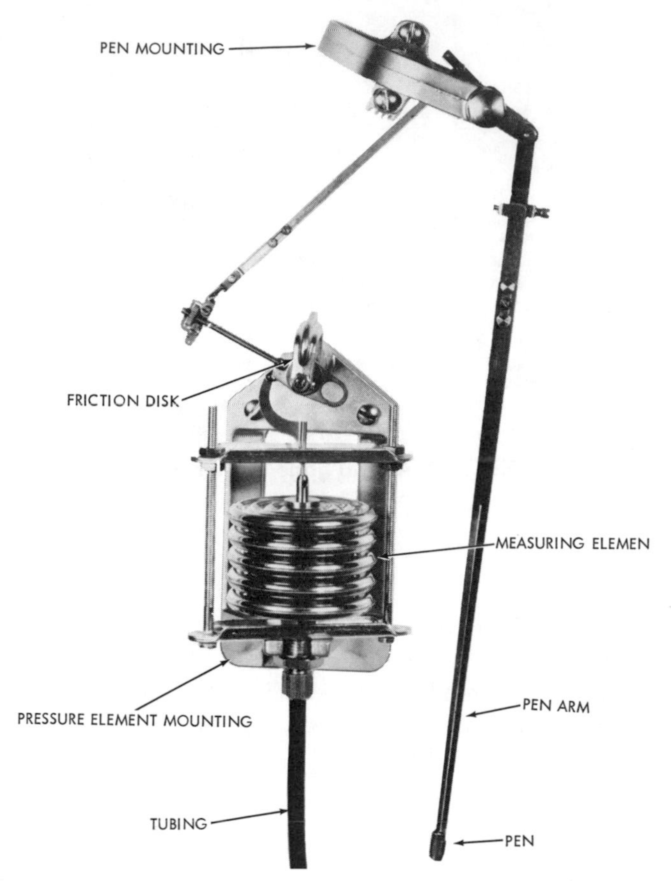

Fig. 11-10. Stacked pressure capsule measuring element. (Bristol Division of Acco)

capsule, the thickness of the material, the elasticity of the material, and the design of the capsule. The design of the capsule includes the shape and the number of corrugations.

It is important that the deflection-pressure relationship be linear. The capsule must be made of material which maintains its accuracy after many deflections. Phosphor bronze, stainless steel, and Ni-Span-C (an iron-nickel alloy) are the most common materials used for pressure capsules. Fig. 11-11 shows the deflection-pressure curves of these three materials.

Phosphor bronze capsules are suitable for most applications with ranges of from 0 to .5 inches of water to 0 to 30 psi. They should not be used if chemical corrosion might result, or if wide temperature variations are expected.

Stainless steel capsules are used for corrosive applications. They are available for ranges from 0 to 8 inches of water to 0 to 50 psi.

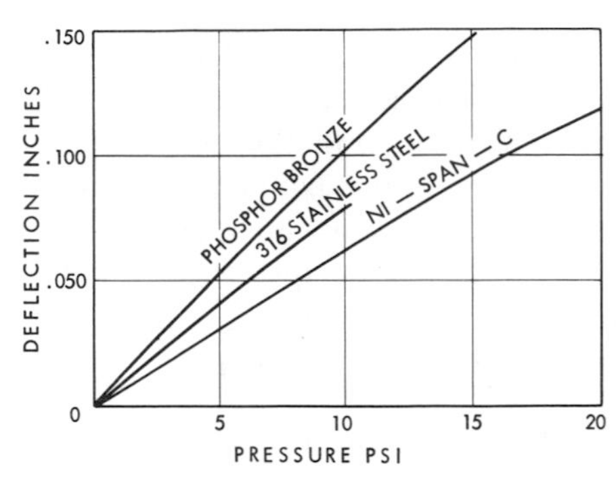

Fig. 11-11. The deflection characteristics of three materials most widely used for low pressure capsules. (Bristol Division of Acco)

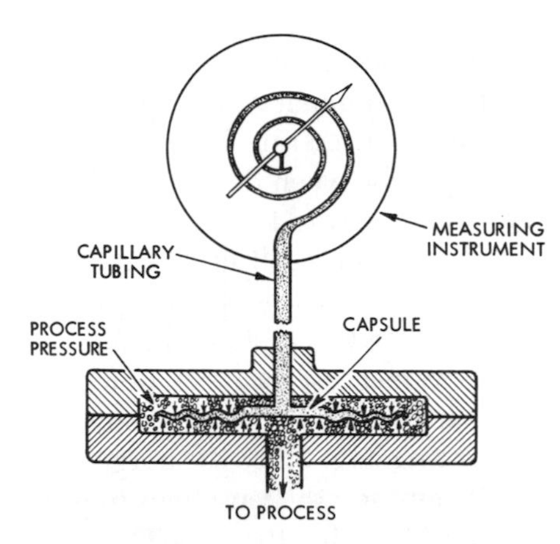

Fig. 11-12. Measuring process fluid pressure with capsule-type seal. Process fluid acts on a secondary fluid which moves up the capillary tubing, forcing the coil to expand or unwind. (Bristol Division of Acco)

Ni-Span-C capsules are virtually unaffected by temperature changes from −50° to 150°F. They are available for ranges from 0 to 4 inches of water to 0 to 30 psi.

Response of all pressure elements to a full scale pressure change is extremely rapid. This response is, of course, affected by the length and diameter of pipe or tubing used. A long sensing line will provide a slower response because of line resistance. When the diameter of the line is large, the response is also slower. These dimensions must be carefully selected to avoid excessively slow response. For low pressure measurement, the length of the sensing line must be short and the diameter small. For high pressure measurement, the sensing line can be long and the diameter larger. If the pressure to be measured pulsates, a small needle valve should be inserted in the line as a pulsation damper.

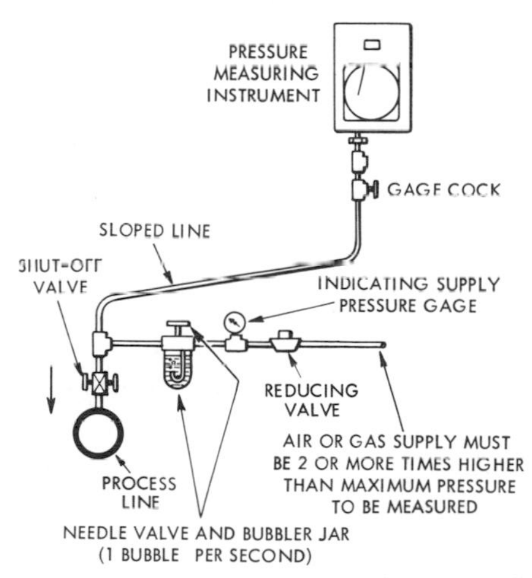

Fig. 11-13. With a constant air or gas purge pressure, a change in the reading of the combined pressures indicates a change in the process pressure. (Bristol Division of Acco)

Seals are available for protecting the measuring elements from corrosive fluids. Fig. 11-12 shows a typical capsule-type seal. The process fluid acts on the pressure capsule. The capsule, along with the measuring element and connecting capillary tubing, is filled with a secondary fluid. Sometimes it is possible to eliminate the capsule and provide a seal merely by inserting an inert liquid between the process liquid and the measuring instrument.

Still another sealing method employs an air or gas purge, as shown in Fig. 11-13. In this system, the measuring instrument senses the difference between the purge pressure and the process pressure. Since the purge pressure is constant, the variations in process pressure indirectly actuate the measuring instrument.

Bellows. For slightly higher pressures, the bellows is a serviceable elastic deformation element. A bellows is a one-piece, collapsible, seamless metallic unit with deep folds formed from very thin-walled tubing. Metals such as brass, phosphor bronze, and stainless steel are among the type used for such bellows. The diameter may be as small as 0.5 inches or as large as 12 inches. A bellows can have as many as 24 folds.

Generally, only a portion of the total available motion of bellows is used. This prevents the bellows from taking a set due to frequent expansion to its limit. A spring is often used to limit the motion of the bellows. The larger the diameter of the bellows, the lower the pressure that it can measure. Instruments with bellows elements are used for pressure ranges from 0 to 5 inches of water to 0 to 800 psi.

Pressure Springs. Pressure springs of helical or spiral shape, or of the simple C-shaped Bourdon type, are used for measuring higher pressures. They are formed from seamless metal tubing with wall thickness, varying from .01 inches to .05 inches. Beryllium copper, steel, and stainless steel are among the metals used for such elements. The wall thickness and the material itself determine the maximum pressure to which such an element can be subjected. A thick-walled steel Bourdon tube is available for measuring pressures as high as 100,000 psi.

Pressure Transducers

Any of the elastic deformation elements can be joined to an electrical device to form a pressure transducer. These transducers can produce a change of resistance, inductance, or capacitance. It is, of course, essential that a corresponding change of electrical characteristic take place for each unit change of pressure.

Resistance-Type. Resistance-type devices used in pressure transducers include strain gages and moving contacts.

A strain gage is simply a fine wire in the form of a grid. See Fig. 11-14. When the grid is distorted, the resistance of the wire changes according to the formula:

$$R = K\frac{L}{A}$$

when

K = a constant for the particular kind of wire
L = length of wire
A = cross-sectional area.

As the strain gage is distorted by the elastic deformation element, its length is increased and its cross-sectional area is reduced. Both of these changes increase the resistance. Little distortion is required to change the resistance of a strain gage through its total range. This type of transducer can be used

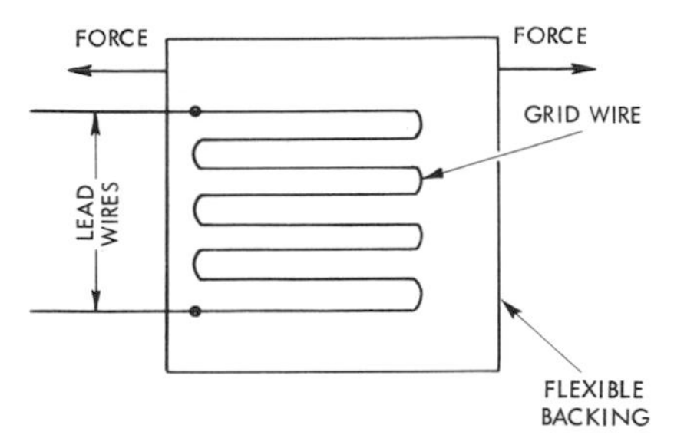

Fig. 11-14. A schematic of a strain gage. A strain gage consists of a wire grid bonded to impregnated paper or ceramic backing.

to detect very small movements and, therefore, very small pressure changes.

Fig. 11-15 shows a simple strain gage pressure transducer. Increased entering pressure causes the bellows to expand. This expansion moves a flexible beam to which a strain gage has been attached. Deflection of the beam alters the shape of the strain gage. This causes its resistance to change. The temperature compensating gage counteracts the heat caused by the current flowing through the very fine wire of the strain gage. The compensating gage must be quite similar to the strain gage and located close to it. Strain gages, which are resistors, are used with bridge circuits. Fig. 11-16 shows a typical circuit.

Alternating current is provided to the bridge by an exciter. An ac bridge is similar to a dc bridge. When a change or resistance in the strain gage causes an unbalance, an error signal enters the amplifier. This actuates the balancing motor which moves the slider along the slidewire, thus restoring balance to the bridge. As the slider moves, its position is noted on a scale marked in units of pressure, provided the strain gage resistance is changed due to pressure.

Another strain type pressure transducer uses an elastic disc. A strain sensing grid has been deposited on the disc by a process called *crystal diffusion*. The silicon strain element is sealed in a silicone fluid and protected from the process fluid by a diaphragm. See Figs. 11-17 and 11-18. Fig. 11-19 shows the bridge circuit used with this transducer.

Other resistance type transducers combine a bellows or Bourdon tube with a variable resistor. See Figs. 11-20 and 11-21.

Inductance-Type. Inductance-type pressure transducers consist of three parts: a coil, a movable magnetic core, and the elastic deformation element.

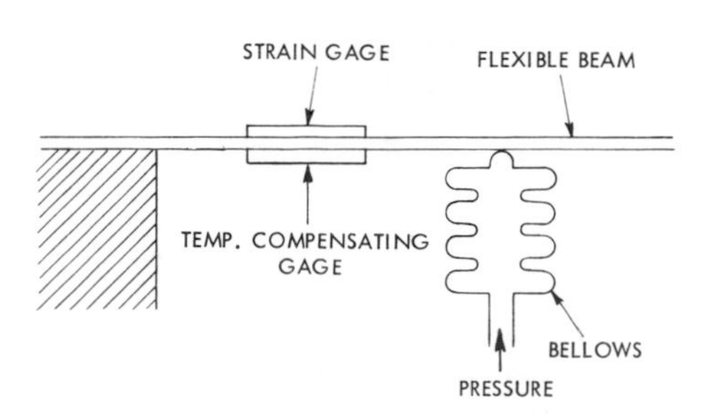

Fig. 11-15. A strain gage pressure transducer.

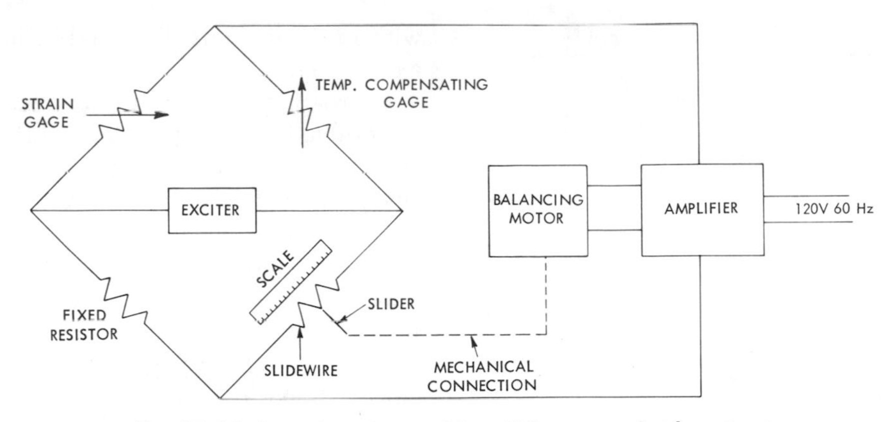

Fig. 11-16. A strain gage used in a Wheatstone bridge circuit.

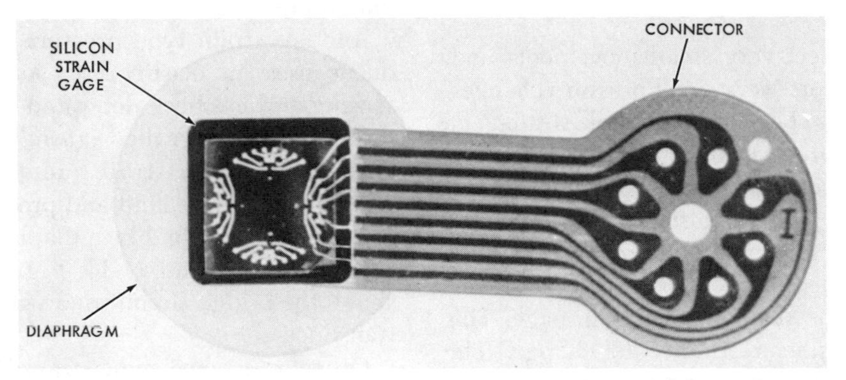

Fig. 11-17. Silicon strain gage. (Honeywell Process Control Div./Fort Washington, Pa.)

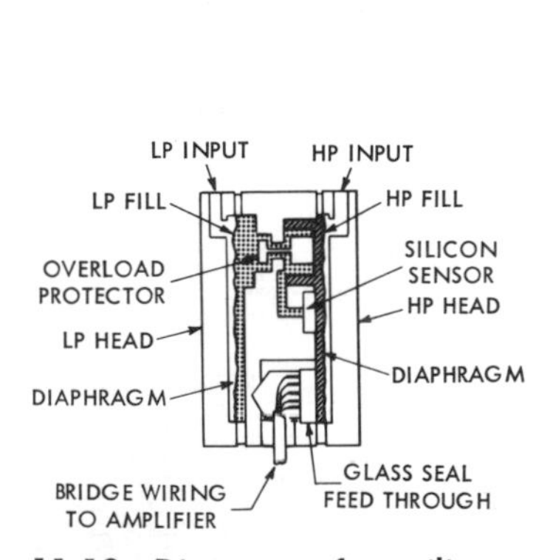

Fig. 11-18. Diagram of a silicon strain gage. (Honeywell Process Control Div./ Fort Washington, Pa.)

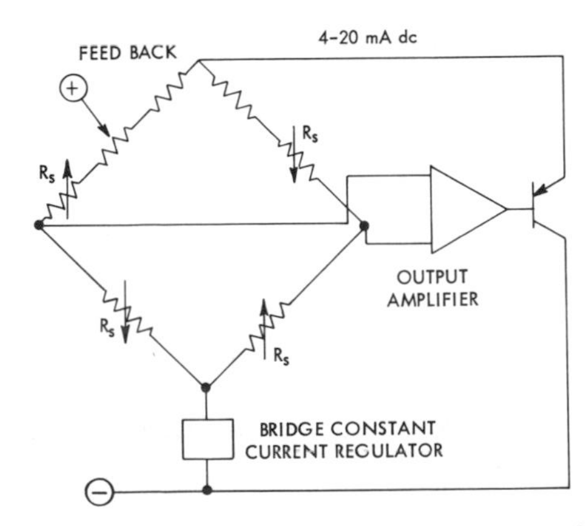

Fig. 11-19. Wheatstone bridge circuit used with the silicon strain gage. (Honeywell Process Control Div./Fort Washington, Pa.)

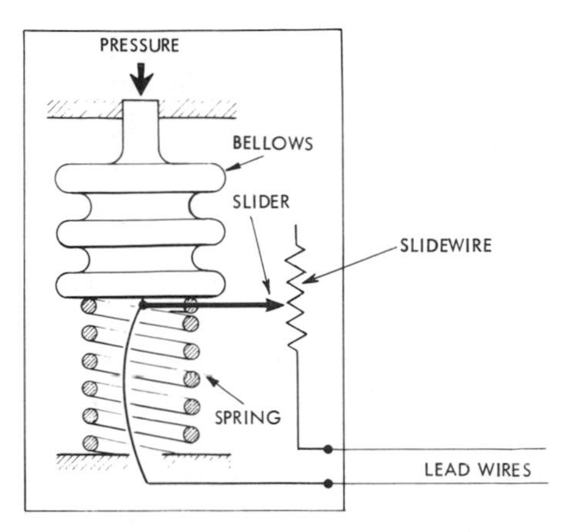

Fig. 11-20. Moving contact-type of resistance transducer.

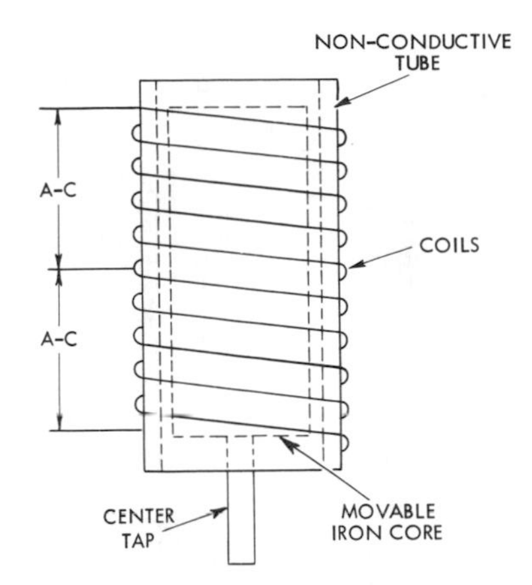

Fig. 11-22. Coil for an inductance-type pressure transducer. Center tap provides greater sensitivity.

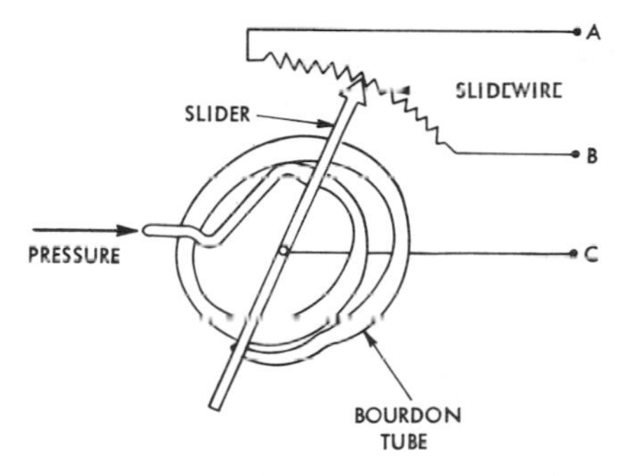

Fig. 11-21. Resistance transducer using a Bourdon tube.

The element is attached to the core. As the pressure varies, the element causes the core to move within the coil. An alternating current is passed through the coil, and, as the core moves, the inductance of the coil changes. The current passing through the coil increases as the inductance decreases. This type of transducer is used in a current-sensitive circuit.

For increased sensitivity, the coil can be divided in two, using a center tap. See Fig. 11-22. This actually provides two coils. As the core moves inside the coils, the inductance of one coil decreases as the inductance of the other increases.

In the type of inductance unit shown in Fig. 11-23, two coils are wound on a single tube. The primary coil is wrapped around the middle of the tube. The secondary coil is divided with one half wrapped around each end of the tube. The two halves of the secondary coil are wound in opposite directions which causes the induced voltages to oppose one another. A core, positioned by a pressure element (for example, a diaphragm), is movable within the tube. When the core is in an upper position, the upper half of the secondary coil provides the output. When the core is in the lower position, the lower half of the secondary coil provides the output. The direction and magnitude of the output signal depend

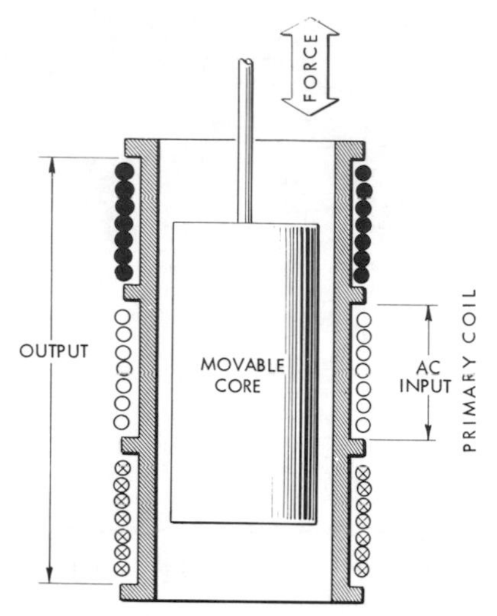

Fig. 11-23. Movable core for an inductance-type pressure transducer.

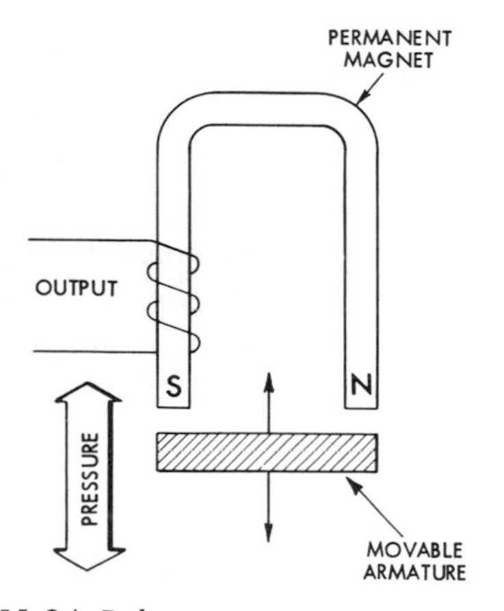

Fig. 11-24. Reluctance-type pressure transducer.

on the amount the core is displaced from its center position. When the core is at midposition, there is no output from the secondary coil. An amplifier is used to detect the output. The amplifier produces a signal of a strength sufficient to actuate an indicator, a recorder, or a controller. See Fig. 11-23.

The reluctance-type transducer is similar to the inductance-type transducer. See Fig. 11-24. The principal difference is in the source of magnetic energy. The reluctance unit has a permanent magnet. An ac input is used in the inductance type. To obtain a pressure measurement, the elastic deformation element is moved *to* or *from* the permanent magnet. The inductance of a coil wound on the permanent magnet varies due to reluctance changes in the magnetic circuit with armature movement.

Capacitive-Type. Capacitive pressure transducers consist of two conductive plates and a dielectric. See Fig. 11-25. As the pressure increases, the plates move farther apart, changing the capacitance. The fluid being measured serves as the dielectric.

Several other types of transducers with elastic deformation elements are used for pressure measurement. Fig. 11-26 shows a carbon pile transducer. As pressure is applied to a confined

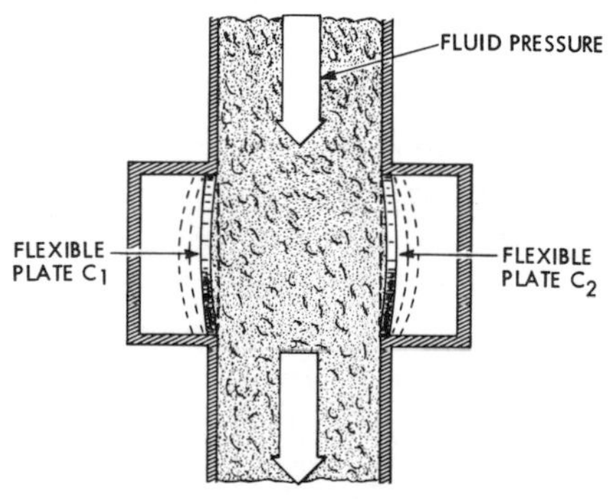

Fig. 11-25. Capacitive pressure transducer.

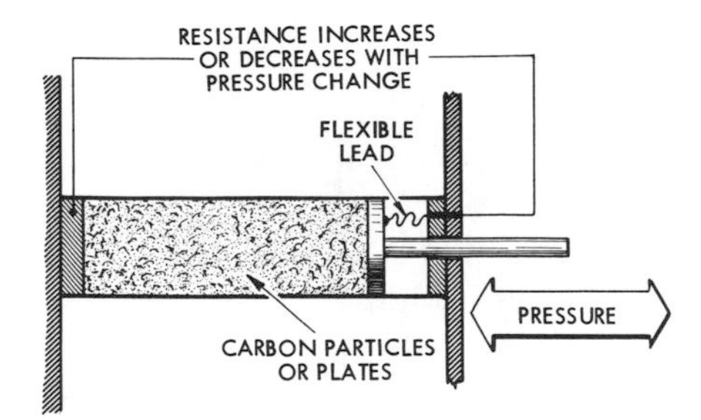

Fig. 11-26. Carbon pile pressure transducer.

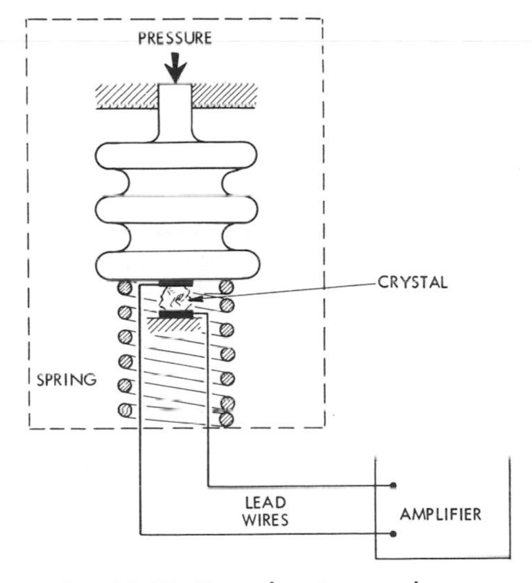

Fig. 11-27. Piezoelectric transducer.

mass of carbon particles, the electrical resistance of the carbon pile changes. Fig. 11-27 depicts a piezoelectric transducer. When pressure is applied to crystals of certain materials, a difference in voltage across particular points of their structure occurs.

Differential Pressure Measurement

Because of its importance in flow rate and level measurement, precise differential pressure measurement is of prime importance in industrial processes.

All the pressure sensitive elements already described are adaptable to differential pressure measurement. Manometers of all kinds can be used with the differential pressure (ΔP) being proportional to the difference in vertical height ($h_1 - h_2$) of the two columns multiplied by the density of the manometer liquid (d_m). Expressed as a formula, this is:

$$\Delta P \sim (h_1 - h_2)\, d_m$$

Density is defined as the amount of matter (mass) per unit volume and is often stated in pounds per cubic foot. Water has a density of 62.4 pounds per cubic foot, mercury a density of 848.4 pounds per cubic foot, and glycerine 78.6 pounds per cubic foot.

As an example, consider a U-tube manometer with water as the fluid. A difference in height of the columns of 1 inch produces this relationship:

$$\Delta P = 1 \times \frac{62.4}{1728} = .03611 \text{ psi}$$

(Note: 62.4 lb./cu. ft. must be converted to lb./cu. in.
1 cu. ft. = 1728 cu. in.

$$62.4 \text{ lb./cu. ft} = \frac{62.4}{1728} \text{ lb./cu. in.)}$$

Using mercury in the same manometer, a 1 inch difference in column height gives:

$$\Delta P = 1 \times \frac{848.4}{1728} = .49097 \text{ psi}$$

In many differential pressure applications, the density of the measured fluid must be taken into account. This is particularly true when the density

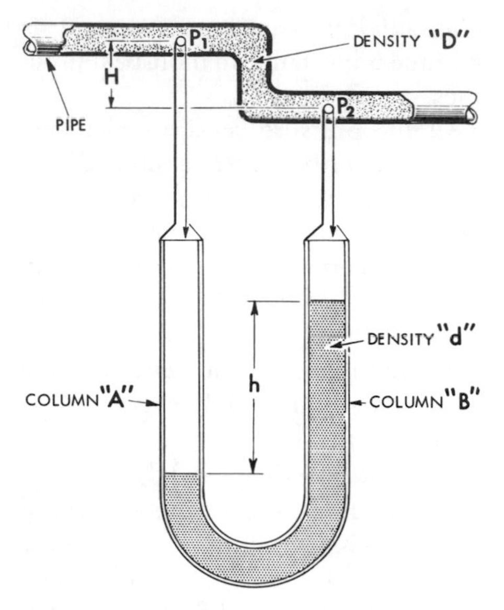

Fig. 11-28. Differential pressure installation for a dense liquid with pressure connections at different heights.

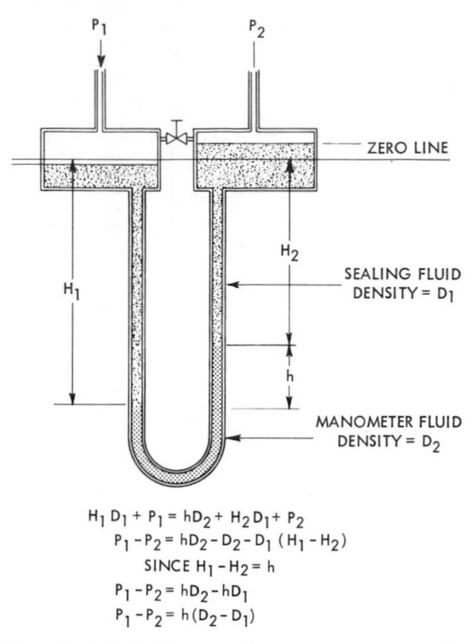

Fig. 11-29. Differential pressure meter. The formulas indicate a method of compensating for the difference of fluid height in the various columns.

is quite high, and there is a difference in height between the two pressure points. Such an installation is illustrated in Fig. 11-28.

The difference in height of the liquid columns results from the fact that P_1 is greater than P_2, and there is an unbalance caused by the difference in height (the column *H*) of the measured fluid on the two manometer columns. The manometer reading must, therefore, be corrected for the pressure caused by the difference. The pressure equals $H \times D$. Therefore, the difference between P_1 and P_2 is actually $P_1 - P_2 = \Delta P =$ hd-HD.

Sometimes this correction is more difficult because of a sealing fluid between the manometer fluid and the measured fluid. A sealing fluid is used when the measured fluid might damage the manometer. In addition to the correction required to overcome the unbalance caused by the difference in height of the columns of measured fluid, any difference in height of the sealing fluid columns must also be corrected. See Fig. 11-29.

Such corrections are made by adjusting the zero of the manometer with measuring fluid and the sealing fluid, if used, at their respective zero levels.

The well-type manometer, when used for differential pressure measurement, provides readings on one scale. The higher pressure is admitted to the well, and the lower pressure to the tube. See Fig. 11-30. The reading is made directly on the scale which is corrected for the movement of the

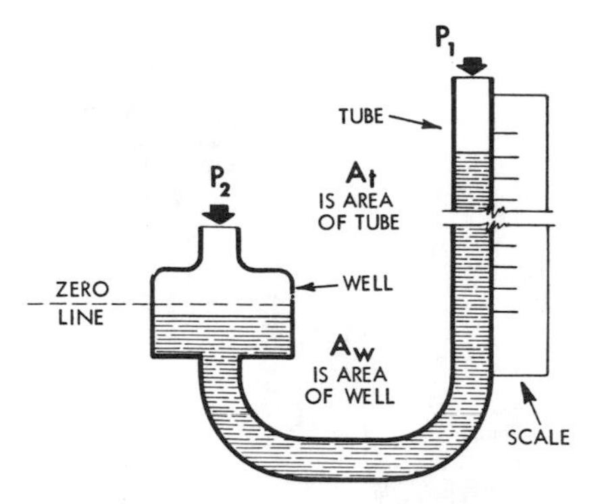

Fig. 11-30. Well-type manometer used for differential pressure measurement.

fluid in the well, as mentioned previously. Using the same units as before, the relationship between differential pressure and the height of the fluid in the tube is as follows:

$$P_2 - P_1 = (1 + \frac{A_t}{A_w})\, d \times h$$

when

d = density of the liquid

h = height of the fluid in the tube as read on a standard, non-corrected scale from scale zero.

$\frac{A_t}{A_w}$ = ratio of the areas of the tube and well.

$\frac{A_t}{A_w}$ is usually between $\frac{1}{300}$ and $\frac{1}{600}$.

With this ratio fixed at $\frac{1}{500}$, each scale inch read actually corresponds to $1 - \frac{1}{500}$ inch in length or .998″.

When water or mercury is used as the manometer fluid, the readings are in standard units. When some other fluid is used, the readings must be corrected to allow for the density of the substitute fluid.

The inclined-tube manometer permits the scale to be expanded so that lower differential pressures can be measured precisely.

The relationship between differential pressure and the position of the fluid in this type manometer is expressed:

$$P_2 - P_1 = d\,(1 + \frac{A_t}{A_w})\, h \sin \alpha$$

when

d = density of the fluid

$h \sin \alpha$ = height of the manometric fluid

α = angle of the inclined tube with horizontal

h = length of fluid along the inclined tube measured from level in well

The mercury float manometer is a modification of the well-type manometer. Because the buoyant force on the float is sufficient to move a recording mechanism, the mercury float manometer can be used as a differential pressure recording instrument. See Fig. 11-31. The manometer fluid does not have to be visible. Therefore, the manometer may be made of metal, permitting its use at high pressures, up to 5,000 psi. Another feature of this kind of manometer is the interchangeability of the tube columns (range tubes), which permits a change in the differential range of the instrument. The relationship between the differential pressure and height of the

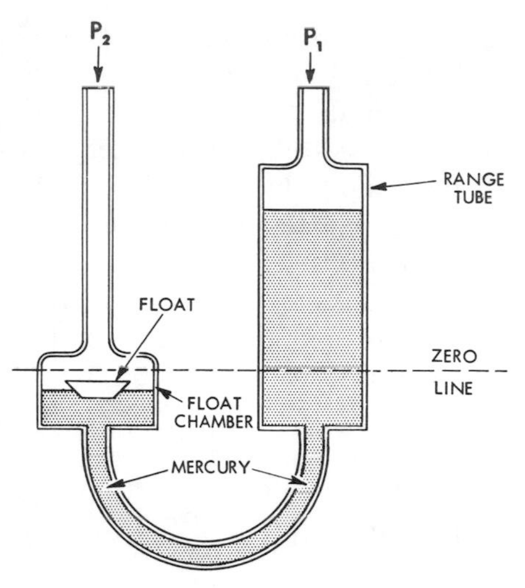

Fig. 11-31. Mercury-float manometer used for differential pressure measurement.

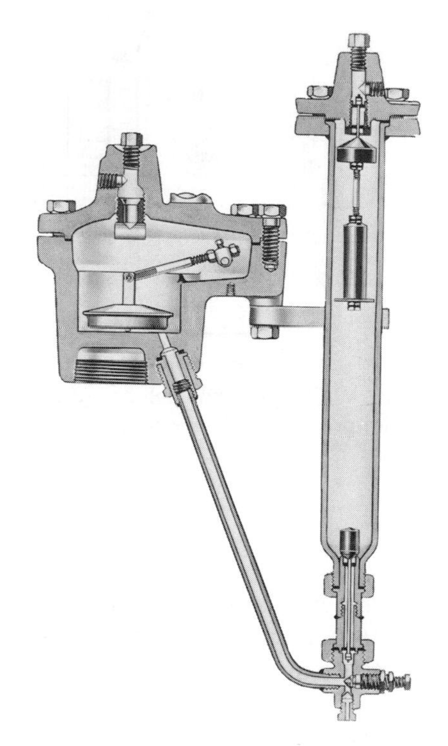

Fig. 11-32. A cross-section of the mercury chamber of a mercury-float manometer. (American Meter Co.)

fluid in the well, or height of the float, as follows:

$$P_2 - P_1 = (1 + \frac{A_t}{A_w})\, d \times h$$

It can be seen that, by changing the range tubes, the ratio $\frac{A_t}{A_w}$ is changed, thus changing the range of the meter. The differential pressure ranges available with such instruments vary from 0 to 10 inches of water to 0 to 200 inches of water. Very low differential pressure ranges from 0 to .2 inches of water to 0 to 1.5 inches of water can be accommodated by substituting oil for mercury. When such a substitution is made, special oil seals are required inside the meter body, and the instrument can be used only at low pressures.

The kind of instrument shown in Fig. 11-32 usually includes check valves to prevent excessive movement of the mercury when the instrument is subjected to differential pressure above its normal range or when occasional reversal of pressures occurs. Accuracy of such a meter is excellent.

The bellows or dry-type differential meter uses two matched bellows enclosed in separate chambers of the meter body. See Fig. 11-33. The two pressures are admitted separately to each chamber, thus causing each bellows to be compressed. Because of the difference in the pressures, their compression is not equal. The two bellows are attached to a common shaft which moves an amount equal to the differ-

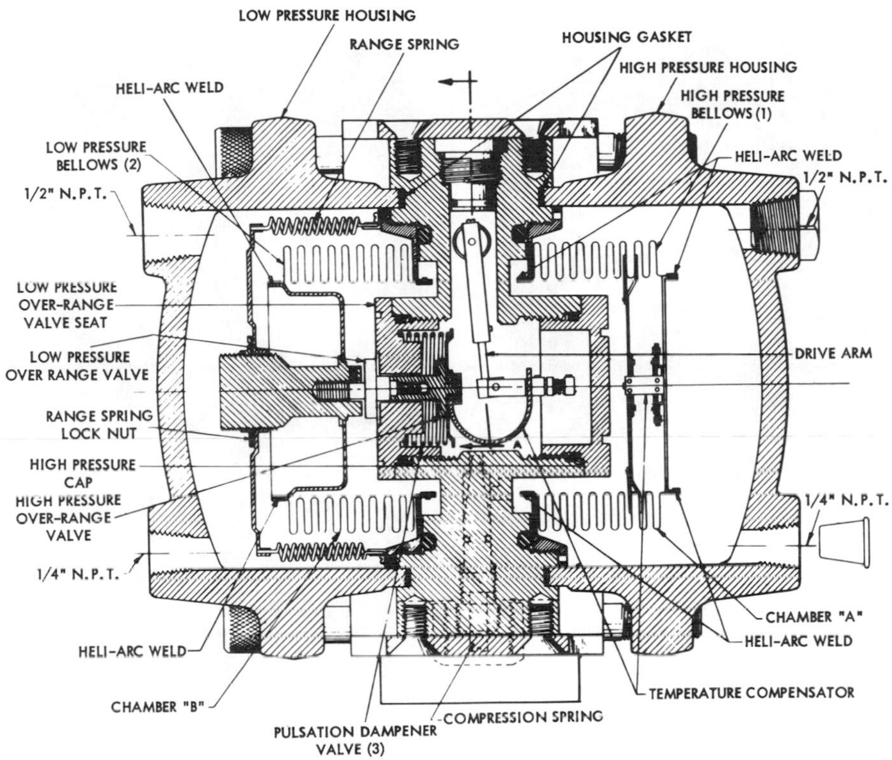

Fig. 11-33. Cutaway of chambers for a dry-type differential meter, showing the matched bellows. (Meriam Instrument Co.)

ence in the movement of the two bellows. The shaft movement is opposed by a spring which determines the amount of bellows unbalance necessary to produce the shaft motion. The spring determines the range of the meter.

In a typical meter of this type, the bellows are filled with a hydraulic fluid which is permitted to flow from one bellows into the other. This provides a smooth damped motion of the bellows. By including a valve in the hydraulic system which shuts off the passage of the fluid between the bellows, the meter is protected against differential pressures above its normal range. In addition, an adjustable restriction in the hydraulic line permits slowing or speeding up the action of the bellows. Such a meter is available for pressures up to 10,000 psi and differential ranges from 0 to 20 inches of water to 0 to 400 psi. Accuracy of such a meter is $\pm$ .5% of range.

The bell-type differential pressure meter is a simple instrument that can be used for measuring low range differential pressures from about 1 to 20 inches of water. Meter bodies suitable for pressures up to 250 psi are available. In the kind of meter shown in

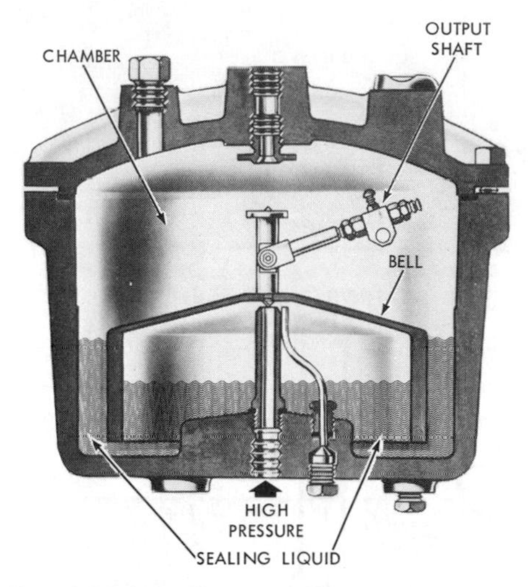

Fig. 11-34. Bell-type differential pressure meter. (American Meter Co.)

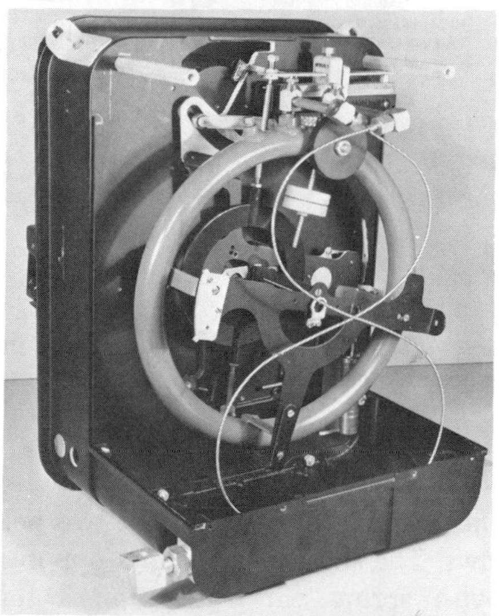

Fig. 11-35. Ring Balance Meter. The top view shows the front of the recording instrument. The bottom view shows the interior, including the ring with flexible connections.

Fig. 11-34, a bell is floated in a shallow pool of liquid in the meter body. The higher pressure is admitted to the underside of the bell and the lower pressure to the outside. The movement of the bell is opposed by a spring which determines the range of the meter for any particular bell. The weight and the thickness of the bell also determine the range since the weight affects its buoyancy, and its thickness establishes the ratio of the area of the under surface to the area of the outer surface. The heavier the bell, the more differential pressure required to cause it to move upward.

There are several kinds of weight-balance type differential pressure meters. One of these is the ring balance meter. See Fig. 11-35. This is not really a manometer but rather a weight balance. In this type of meter, the mercury contained in the ring is displaced by the differential pressure. The ring is balanced on a knife edge and counterbalanced by a weight. The pressure connections to the ring are flexible, allowing the ring to rotate. When rotation of the ring due to pressure displacement of mercury has been balanced out, using the counterweight, the equation for balance is:

$$P_2 - P_1 = \frac{R}{r}\left(\frac{W}{A} \sin\theta\right)$$

when

R = radius of the arc described by the weight
r = the radius of the ring
W = weight of the counterweight
A = area of the tube
θ = angle through which the ring moves due to differential pressure.

Note that the density of the fluid does not appear in this equation. The liquid is used only to provide a seal. Mercury is most commonly used.

The bellows, ring balance, and tilting U-tube instruments discussed above are motion-balance devices. In addition to motion-balance instruments, force-balance differential pres-

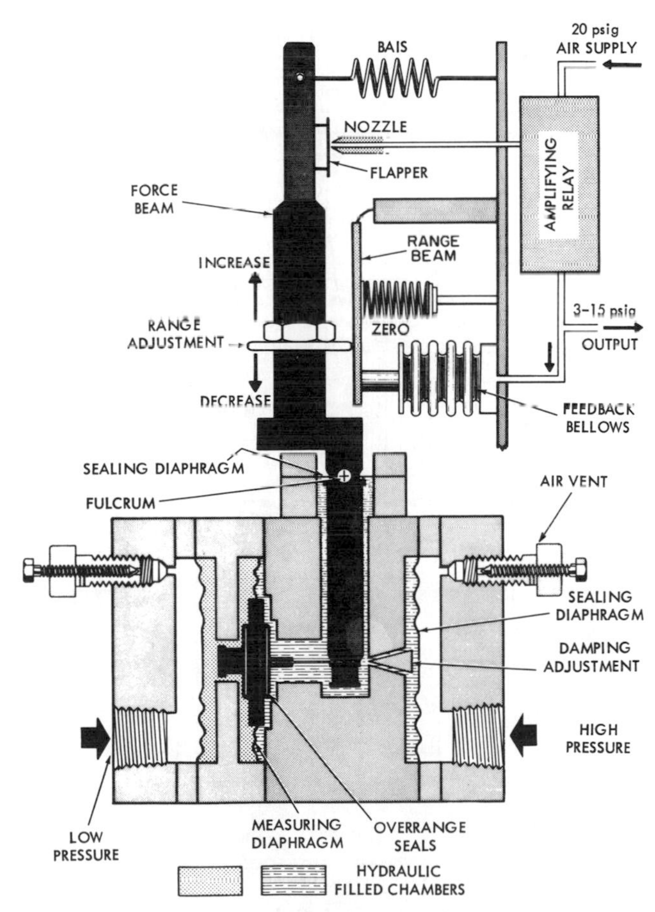

Fig. 11-36. Diagram of a pneumatic differential pressure meter. (Fischer & Porter)

sure-sensing instruments are also common, used mostly as transmitters. They can be pneumatically or electrically actuated. A diaphragm capsule or Bourdon tube serves as the sensing element.

Both the pneumatic and electric instruments contain a similar mechanical assembly. The force beam, pivoted at the top of the measuring section, is the principal part. As the process pressure changes, the pressure element deflects one end of the beam. The total movement is only a few thousandths of an inch.

In the pneumatic version of the instrument, shown in Fig. 11-36, a flapper is attached to the opposite end of the beam. The flapper acts against a pneumatic nozzle, causing a change in the signal air pressure. The signal is amplified in a pneumatic relay and be-

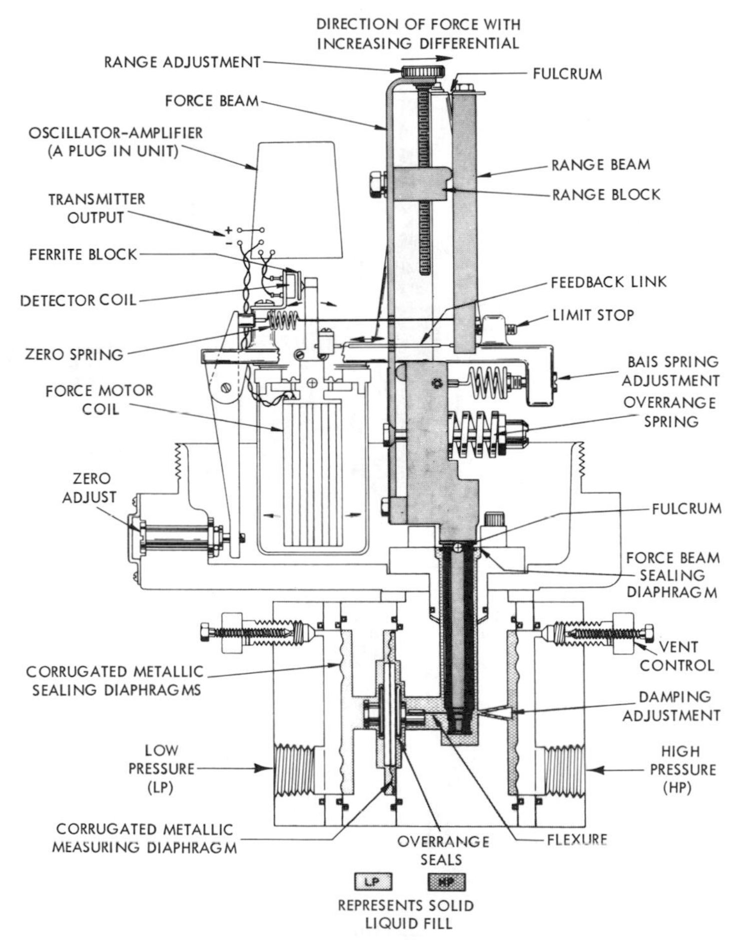

Fig. 11-37. Diagram of an electrical differential pressure meter. (Fischer & Porter)

comes the 3 to 15 psig (gage pressure) output signal of the instrument. The same output signal acts to restore the beam to its balance position.

In the electronic version, shown in Fig. 11-37, the opposite end of the beam is attached by linkage to an inductance type transducer. The change in inductance is amplified and becomes the 4 to 20 mA dc output signal of the instrument. This same output signal serves as the feedback to a force motor which restores the beam to its balance position.

Words to Know

manometer	draft gage	strain gage
deformation elements	specific gravity	inductance
atmospheric pressure	pressure capsule	differential pressure

Review Questions

1. What is the absolute pressure, if the gage pressure at sea level is 26 psi? What is the reading on the vacuum scale in inches of mercury if the gage pressure at sea level is −5 psi?
2. If the well area of a well-type manometer is 50 square inches (400 times greater than the tube area), what is the actual length of each scale inch?
3. Does the substitution of a liquid with a specific gravity of .75 for water reduce or extend the range of a U-tube manometer? Does it make the instrument more sensitive or less sensitive?
4. List the useful ranges of the following elastic deformation pressure sensitive elements: slack diaphragm, metal bellows, Bourdon tube.
5. What is a strain gage? How can a strain gage be used to measure pressure?
6. What happens to a mercury manometer if one pressure line to the instrument is suddenly suptured? What devices are usually included in such an instrument to prevent this?
7. What is the function of the hydraulic fluid contained inside the bellows of a typical dry-type differential pressure measuring element?
8. How do the dimensions of the bell-type manometer affect the range of the meter?
9. To measure a differential pressure in a range of 0 to 200 inches of water, when the operating pressure is 2000 psi, and there is a possibility of the loss of one of the pressures, what type of instrument is used?

Chapter

12

Flow

The physical properties of fluids important in flowmetering are pressure, density, viscosity, and velocity.

Pressure has already been defined as force divided by area. Density is weight divided by volume, and is usually expressed in pounds per cubic foot.

Viscosity. The viscosity of a fluid refers to its physical resistance to flow. Fuel oil is more viscous than water, and water more viscous than gasoline. There are several viscosity units, the most widely used being the centipoise. The viscosity of water at 68°F is 1.0 centipoise, and the viscosity of kerosene at 68°F is 2.0 centipoises. The viscosity of liquids decreases as the temperature rises. The opposite is true of gases.

Velocity. The velocity of a flowing fluid is its speed in the direction of flow. It is an important factor in flowmetering because it determines the behavior of the fluid. When the average velocity is slow, the flow is laminar. See Fig. 12-1. This means that the fluid flows in layers with the fastest moving layers toward the center and the slowest moving layers on the outer edges of the stream. As the velocity increases, the flow becomes turbulent. The layers disappear, and the velocity across the stream is more uniform. See Fig. 12-2. In this discussion the flow is as-

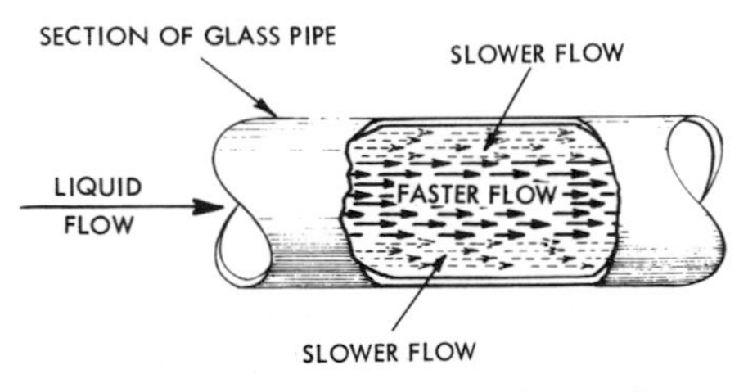

Fig. 12-1. Laminar flow. The faster flow is in the center, with slower flow on the outer edges.

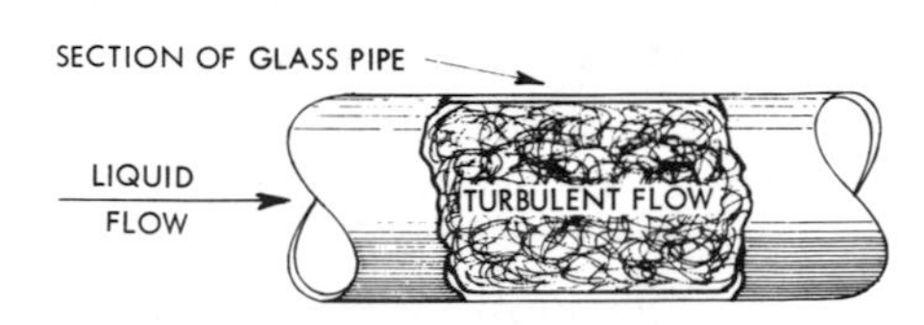

Fig. 12-2. Turbulent flow.

sumed to be turbulent. The term *velocity* refers to the average velocity of a particular cross-section of the stream. Velocity is expressed as follows:

$$\text{Velocity} = \frac{\text{rate of flow (cu. ft./sec.)}}{\text{area of pipe (sq. feet)}}$$

Reynolds Number. In flowmetering, the nature of flow can be described by a number, called a Reynolds number. A Reynolds number is the average velocity × density × internal diameter of pipe ÷ viscosity. In equation form, this is expressed:

$$R = \frac{v D \rho}{\mu}$$

when

v = velocity (feet/sec.)
D = inside diameter of pipe (feet)
ρ = fluid density (lbs./cu. ft.)
μ = viscosity (centipoise).

Although the Reynolds number has no dimensions of its own, it is important that the dimensions, used in the equation determining it be consistent with each other (e.g., all metric, etc.).

From the Reynolds number, it can be determined whether the flow is laminar or turbulent. If the Reynolds number is less than 2000, the flow is laminar. If the number is greater than 4000, the flow is turbulent. Between these two values, the nature of the flow is unpredictable. In most industrial applications, the flow is turbulent.

Although measurement can be made without consideration of the Reynolds number, greater accuracy is possible when the correction is based on it.

Flow Calculations

The greatest number of rate-of-flow meters for fluids are those which measure the differential pressure across a restriction in the pipe line. The most common restriction is the concentric orifice plate.

Fig. 12-3 is a cross section of a typical orifice plate installation, showing the variation in pressure that occurs upstream and downstream of the plate. Notice that the main flow stream takes the shape of a Venturi tube with the

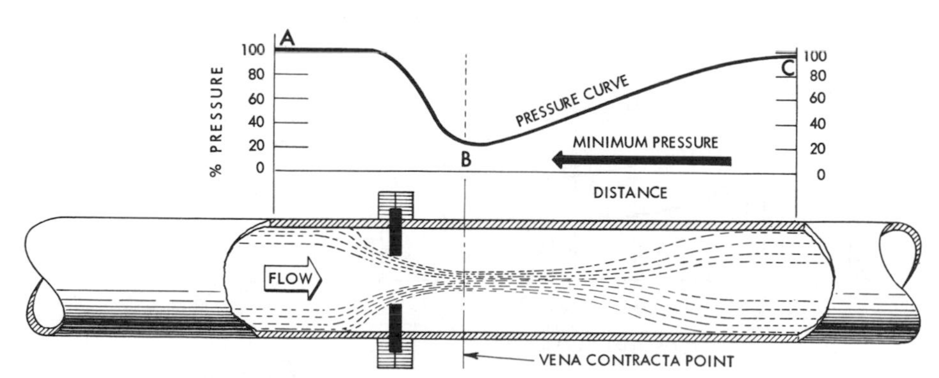

Fig. 12-3. Orifice plate installation. Maximum pressure is at point A, minimum pressure is at point B. Because of the loss of pressure across the plate, the downstream pressure at point C is not equal to the pressure at A.

smallest stream cross-section occurring slightly downstream of the plate. This point is called the *vena contracta.* At this point, the pressure is at its minimum. From this point on, the fluid again begins to fill the pipe, and the pressure rises. The pressure, however, does not rise fully to its upstream value. There is a loss of pressure across the plate. The principal consideration in selecting an orifice plate is the ratio of its opening (d) to the internal diameter of the pipe (D). This is called the *beta ratio.* If the d/D ratio is too small, the loss of pressure becomes too great. If the ratio is too great, the loss of pressure becomes too small to detect and too unstable. Ratios from .2 to .6 generally provide best accuracy.

Several procedures have been developed for calculating what size orifice is suitable for measuring a particular range of flow rate. The fundamental equation on which all these procedures are based is:

$$Q = E A_0 \sqrt{2gh}$$

when

Q = flow rate (volume per unit of time)

E = efficiency factor (dimensionless)

A_0 = area of orifice (square feet)

g = acceleration due to gravity (32 feet/sec/sec)

h = differential pressure across orifice (feet).

The efficiency factor (E) is required since the actual flow through an orifice is not the same as the theoretically calculated flow. Values of E have been determined by tests and are found in tables and graphs. See Fig. 12-4. The value is different for each combination of d/D ratio and Reynolds number. (In some equations, the letter *K* is used to express this factor, in others the letter *C*. It can also be called the *flow coefficient.*)

Concentric orifice plates are ⅛

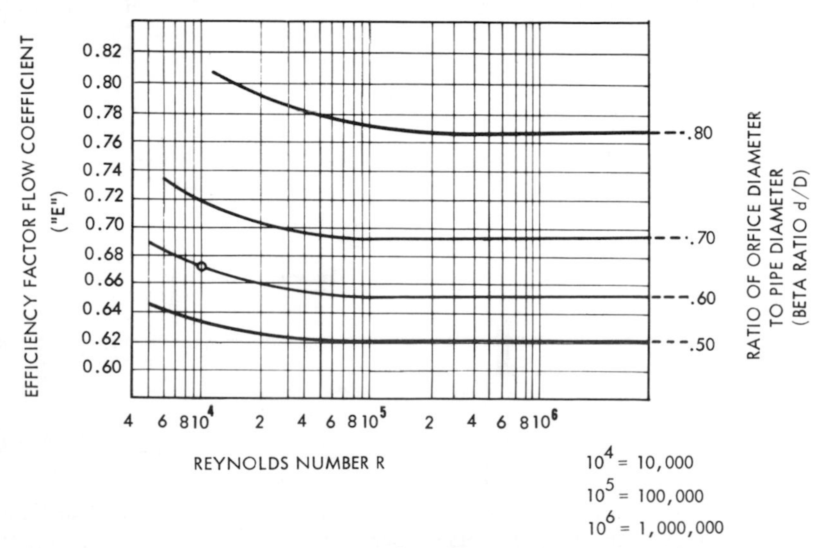

Fig. 12-4. Graph for determining the efficiency factor (E) of flowmeter.

inch thick in the small sizes, up to 10 inches, and ¼ inch thick for large sizes. The upstream edge should be square and as sharp as possible, and the upstream face as smooth as possible. The plate should be flat, and the orifice exactly centered in the pipe.

The fundamental equation cannot be used in calculating either flow rates or orifice sizes. It must be changed to a working equation. A typical working equation for liquids is:

$$Q = 19.65 \times d^2 \times E \times \sqrt{h}$$

when

Q = gallons/minute
19.65 = units constant
d = orifice diameter in inches
E = efficiency factor
h = differential pressure in inches.

As stated above, values of E are found in tables or on graphs, such as Fig. 12-4.

The following is an example. For a Reynolds number of 10,000 and a beta ratio of 0.6, the value of E is .673.

The orifice plate, flow nozzle, and venturi tube operate on the same principle, and the same equation is used for all three. In addition to the difference in E, other factors determine which element should be used for each.

Venturi Tube. The venturi tube is the most expensive and the most accurate. High beta ratios (above .75) can be used with good results. The pressure recovery of the venturi tube is excellent, which means that there is little pressure drop. Functionally, the venturi tube is good, since it does not obstruct abrasive sediment. In fact, it resists wear effectively because of its shape.

Flow Nozzle. The flow nozzle is simpler and cheaper than the venturi tube. It is slightly less accurate and does not provide as fine a pressure recovery. The flow nozzle can be used with higher beta ratios (above .75), but it is not quite as wear resistant as the venturi.

Orifice Plate. The orifice plate is quite simple and inexpensive. It is the easiest to install and replace. There is considerable data available for calculation of correct sizes. The orifice plate is not as accurate as either the venturi or flow nozzle and does not have as fine a recovery. It cannot be used with high beta ratios. Functionally, the orifice plate is subject to erosion and damage, but it is easy to replace. It is the most widely used of the primary flow elements.

There must be a long continuous run of straight pipe leading up to any of the primary elements. See Fig. 12-5. Considerable information is available regarding the length of straight pipe needed between such devices as elbows and valves and the primary ele-

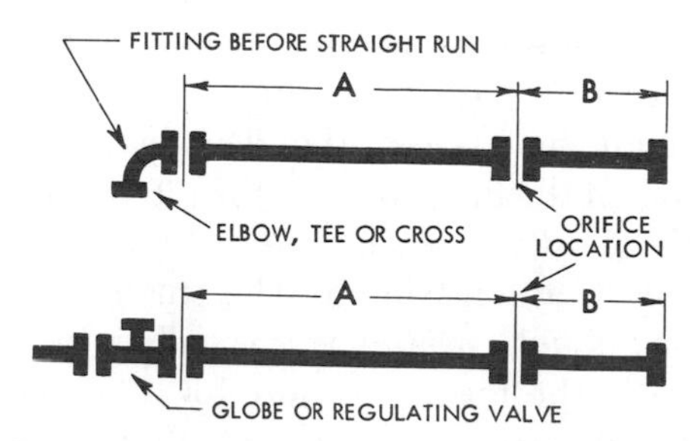

Fig. 12-5. Diagram showing elbow and valve and section of straight pipe leading up to the orifice plate location.

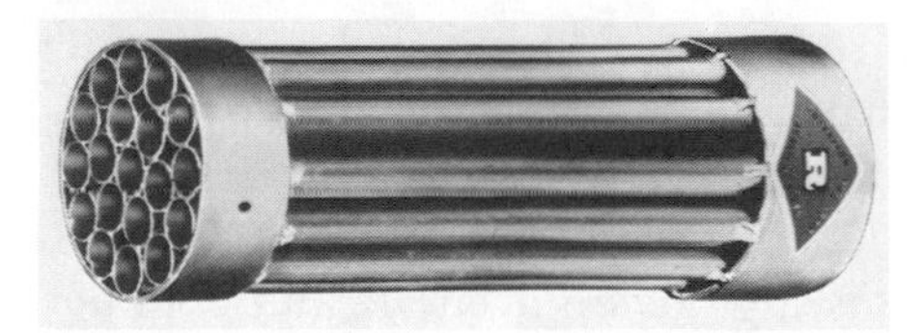

Fig. 12-6. Straightening vane used to reduce turbulence. (Roberts Mfg. Co.)

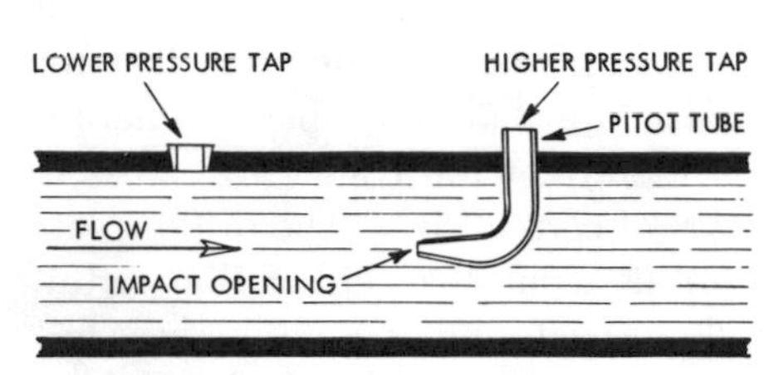

Fig. 12-7. Pitot tube for providing differential pressure reading.

ments. When insufficient straight pipe is impossible, the disturbances can be reduced or eliminated by the installation of straightening vanes. One kind of straightening vane has a bundle of tubes joined together. See Fig. 12-6.

Pitot Tube. Another primary flow element used to produce a differential pressure is the *pitot tube*. In its simplest form, the pitot tube consists of a tube with a small opening at the measuring end. See Fig. 12-7. This small hole faces the flowing fluid. When the fluid contacts the pitot tube, the fluid velocity is zero and the pressure is at a maximum. This small hole, or *impact opening* as it is called, provides the higher pressure for differential pressure measurement. While the pitot tube provides the higher pressure for differential pressure measurement, an ordinary pressure tap provides the lower pressure reading.

The pitot tube actually measures the velocity of fluid flow and not flow rate. However, the flow rate can be determined from the velocity, using the formula:

$$Q = K A V$$

when

$Q =$ flow rate (cubic ft./sec.)
$A =$ area of flow cross section (feet)
$V =$ velocity of flowing fluid (ft./sec.)
$K =$ flow coefficient of pitot tube (normally about .8).

There is no standardization of pitot tubes as there is for orifice plates, venturi tubes, and flow nozzles. Each pitot tube must be calibrated for the specific installation.

Pitot tubes can be used where the flowing fluid is not enclosed in a pipe or duct. For instance, a pitot tube can be used to measure the flow of river water, or it can be suspended from an airplane to measure air flow.

Any of the differential pressure type instruments previously described can be used with the pitot tube.

Measurement Conversion

Several methods are available for converting differential pressure measurement directly to flow measurement. The simplest method is to use a square root chart. See Fig. 12-8. The observer can read the square root of the differential pressure (flow) being recorded at any given moment. In most cases, however, a square root extracting mechanism is built into the meter.

In the Ledoux bell fluid meter,

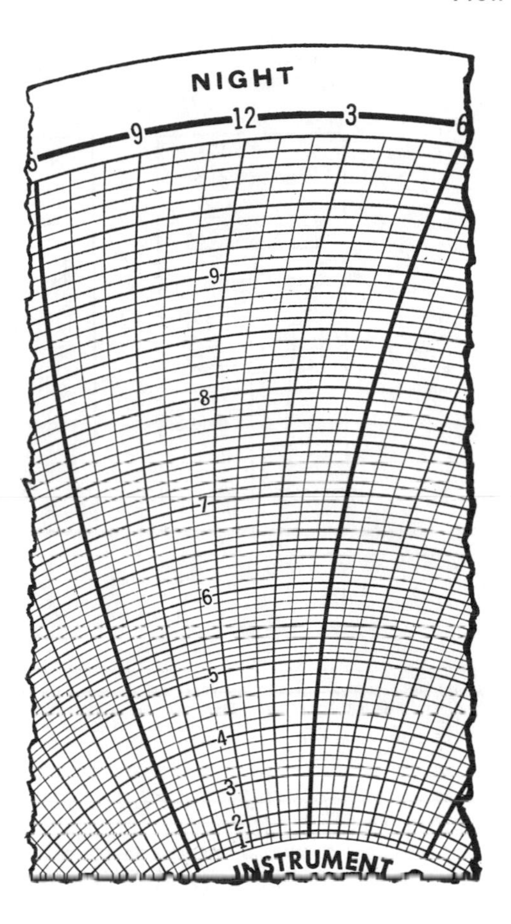

Fig. 12-8. Section of a square root chart.

shown in Fig. 12-9, conversion is achieved by shaping the inside of the bell so that it moves the same distance vertically when almost completely submerged in mercury (at low differential pressure) as it does when hardly submerged (at high differential pressure). In a mercury float-type manometer, the range tube can be similarly shaped.

In pneumatic flow transmitters, shown in Fig. 12-10, the square root extraction is performed by the pneumatic system. It is extracted in the circuitry in electric flow transmitters. See Fig. 12-11.

In order to correct the measurement of the gas flow, the operator must know the temperature and the pressure because the volume of gas varies directly with temperature and inversely with pressure. The measurement provided is in standard cubic feet per minute. A standard cubic foot of gas is the unit of volume measured at standard conditions (14.7 psig and 70°F).

Gas flow meters are most frequently used for the transmission of natural gas through pipe lines. The flow meters compensate primarily for pressure because temperature varies very little in

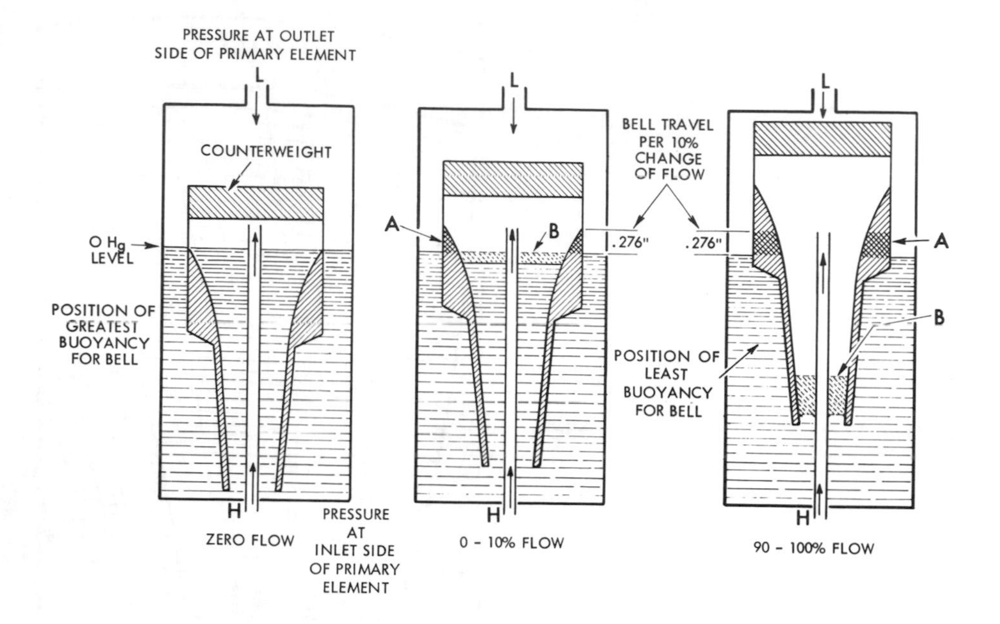

IN THE ILLUSTRATION ON THE LEFT, NO FLOW IS PASSING THROUGH THE PRIMARY ELEMENT. THE BUOYANCY GIVEN TO THE BELL BY THE MERCURY IS OVERCOME BY THE COUNTERWEIGHT, AND THE BELL SETTLES DOWN ON TOP OF THE STANDPIPE.

IN THE CENTER ILLUSTRATION, FLOW HAS STARTED THROUGH THE PRIMARY ELEMENT AT A RATE EQUAL TO 10% OF THE MAXIMUM, CONSEQUENTLY THE BELL HAS BEEN RAISED 10% OF ITS TRAVEL, OR .276". THIS IS ACCOMPLISHED BY SO SHAPING THE INSIDE SURFACE OF THE BELL THAT THE VOLUME OF BELL WALL (A) WHICH HAS EMERGED FROM THE MERCURY EXACTLY EQUALS THE DECREASE IN THE VOLUME OF THE MERCURY (B) BELOW THE ZERO LEVEL INSIDE THE BELL.

WHEN THE FLOW RATE CHANGES FROM 90 TO 100% MAXIMUM, AS IN THE ILLUSTRATION ON THE RIGHT, THE BELL MUST MOVE THE SAME AMOUNT, OR .276", IN ORDER THAT THE PEN TRAVELS THE SAME 10% ON THE CHART. HOWEVER, THE CHANGE IN THE DIFFERENTIAL PRESSURE WHEN THE FLOW INCREASES FROM 90 TO 100% IS ABOUT 19 TIMES THE INCREASE FROM 0 TO 10% RESULTING IN A MUCH LARGER VOLUME OF MERCURY DEPRESSED INSIDE THE BELL, AND THUS REQUIRING AN EQUIVALENTLY LARGER VOLUME OF BELL WALL EMERGING FROM THE MERCURY. THIS LARGER WALL VOLUME (A) CAN ONLY BE OBTAINED BY MAKING THE WALL THICKER, SINCE THE VERTICAL EMERGENCE (.276") REMAINS AS BEFORE

Fig. 12-9. Ledoux bell fluid meter. (Bailey Meter Co.)

the long pipe lines. Differential pressure and flow are measured with the same instrument, using the elements described earlier in Chapters 3 and 11. A typical instrument combines the filled bellows-type differential meter body and the helical pressure spring. The basic equation of gas flow is

$$Q = K\sqrt{hp}$$

when

$Q =$ cu. ft./hr.

$K =$ units constant

$h =$ differential pressure

$P =$ pressure

The instrument multiplies the differential pressure by the pressure and then takes the square root of the product. The mechanical linkage of the meter provides the multiplication. Fig. 12-12 is a schematic of a typical linkage. The square root is extracted by using a square root chart or a slide rule or an electronic calculator.

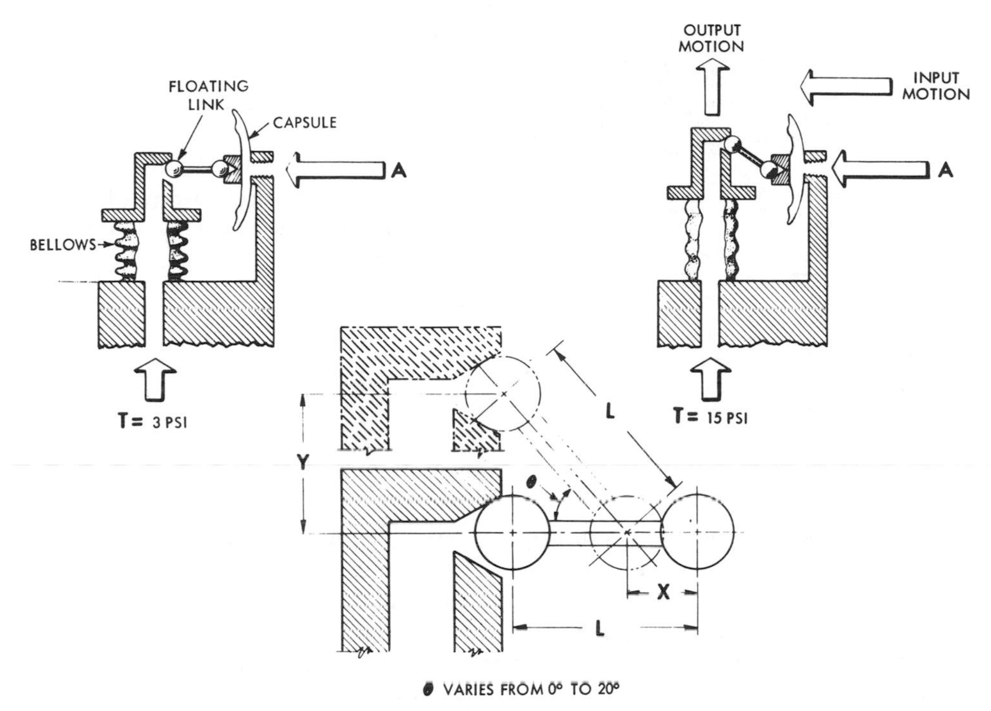

Fig. 12-10. Square root extractor. In this square root extractor, the capsule expands as the input pressure (A) increases. The floating link (L) moves against the exhaust outlet of the bellows. The bellows is now sealed, and the pressure increases inside, producing the output motion. The output motion (Y) varies as the square root of the input motion (X). (Moore Products Co.)

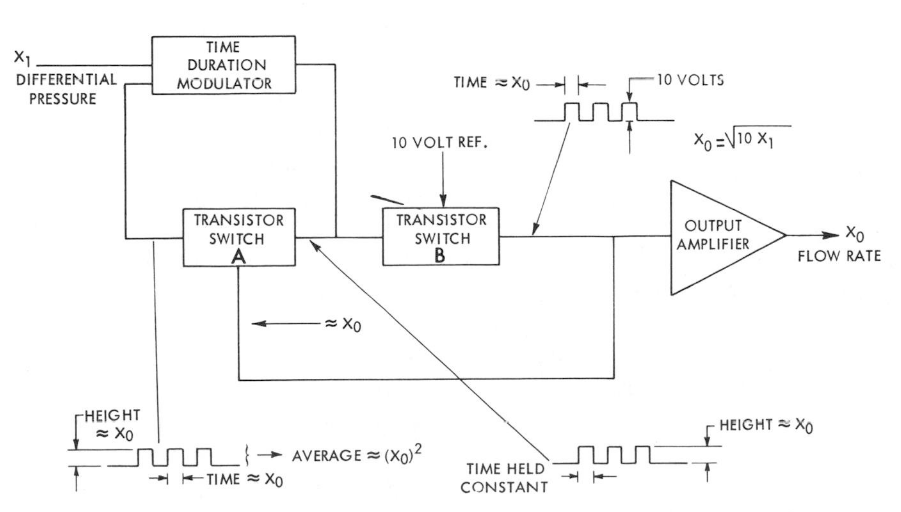

Fig. 12-11. Electrical square root extractor which provides the square root of one or two inputs.

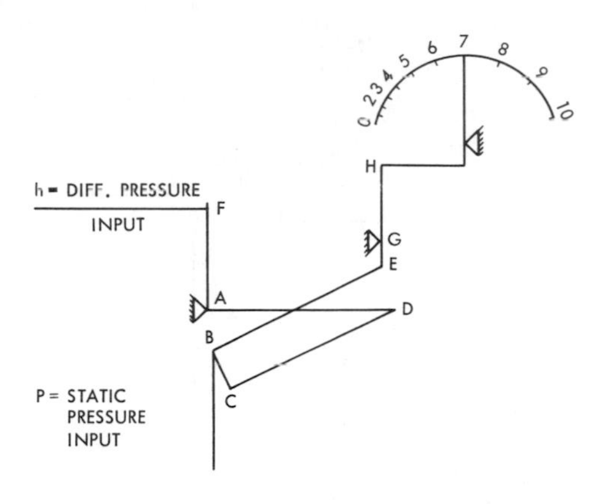

Fig. 12-12. Mechanical linkage which provides the multiplication in a square root extractor.

The Sorteberg bridge, shown in Fig. 12-14, is an unusual pneumatically operated mechanism. It can compute the flow rate which has been compensated for temperature and pressure. This system is included to show the computing capability of pneumatic systems.

This device provides simultaneous multiplication and division and can continuously solve the equation

$$Q_1 = \frac{Q \times P}{T}$$

when

Q = measured flow rate
Q_1 = corrected flow rate
P = pressure
T = temperature

The flow rate is compensated for pressure and temperature changes.

In one electrical method, a linear-differential transformer is used. See Fig. 12-14. The core of this transformer is positioned by the formula:

$$\sqrt{\frac{P \times h}{T}}$$

The square root extraction is accomplished by using a cam to position the core. The multiplication of differential pressure (h), pressure (P), and the inverse of the temperature (T) is done electrically. In a typical recorder, a resistance thermometer measures temperature, an elastic deformation inductance transducer measures pressure, and a bellows or diaphragm-actuated inductance transducer measures differential pressure. See Fig. 12-15.

Another electrical method, shown in Fig. 12-16, has similar measuring transducers. The outputs of these transducers are converted into 0 to 10 volt signals by transistorized signal-converter modules. These converted signals are then fed into transistorized multiplier-divider and square root extracting modules. The output of the whole circuit is a 0 to 10 volt dc signal, which can be the input of voltmeter-calibrated inflow units or of a compatible controller.

Variable Area Meter. In the variable area flowmeter, the differential pressure across the meter is constant. Hence, the metering area varies as the flow rate varies. The differential is maintained constant by free-moving

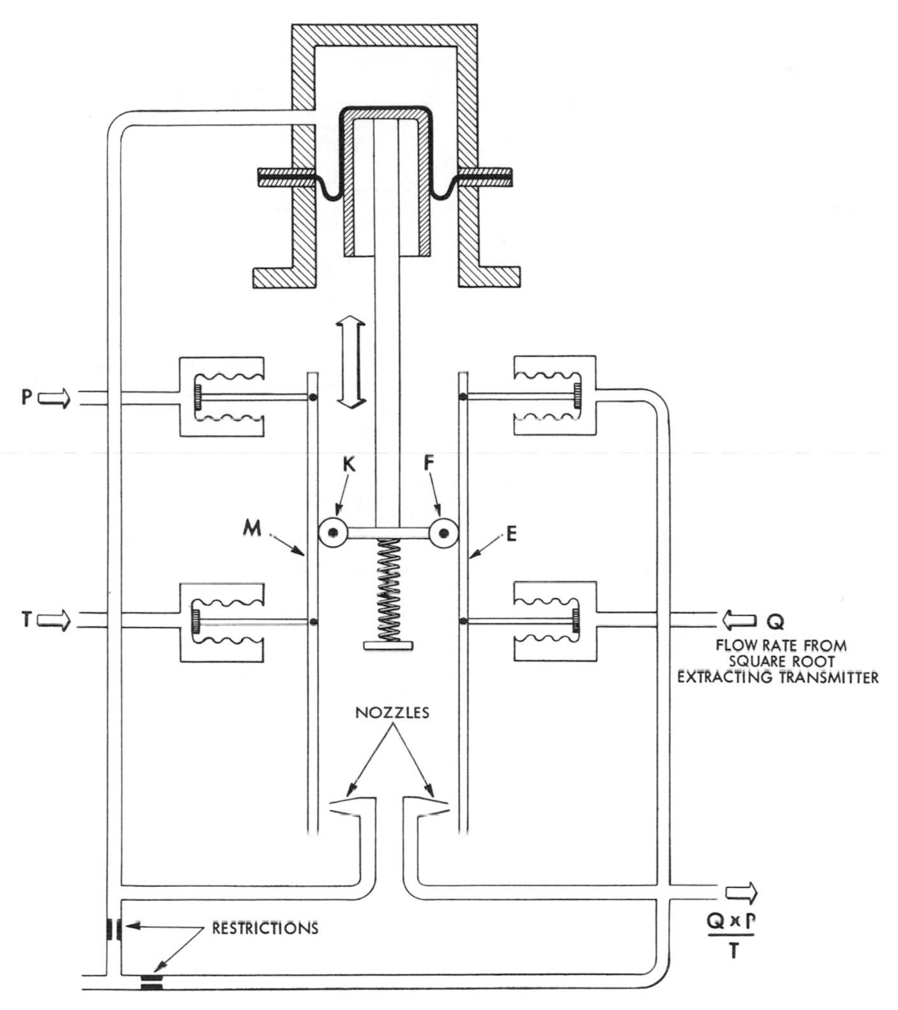

Fig. 12-13. The Sorteberg bridge. Pneumatic signals proportional to pressure (P) and temperature (T) act on the bar M. An unbalance causes wheel K to restore balance. Unbalance on bar E caused by movement of wheel F requires a change in output pressure to restore balance. Total balance is achieved when the product of signal P and signal Q, divided by signal T, equals output pressure.

element which rises and falls as the flow rate increases or decreases. As the float rises, the metering area is increased. The weight of the float and the density of the flowing fluid determine the amount of rise of the float.

Rotameter. In the rotameter, which is the most common form of variable area meter, the float moves within a tapered tube. The metering area is the annular area between the float and the tube.

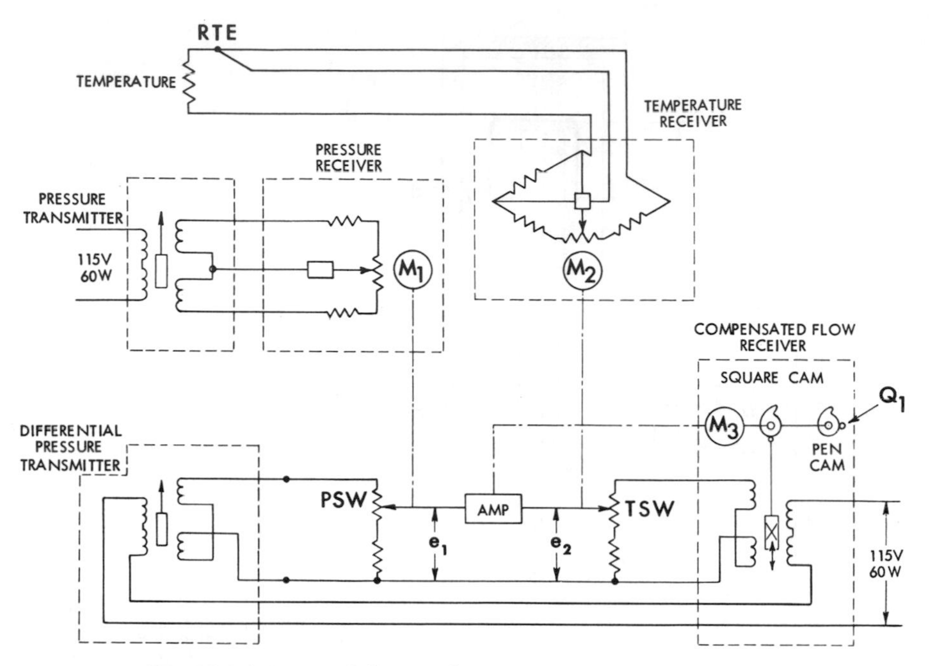

Fig. 12-14. Linear differential transformer. (Bailey Meter Co.)

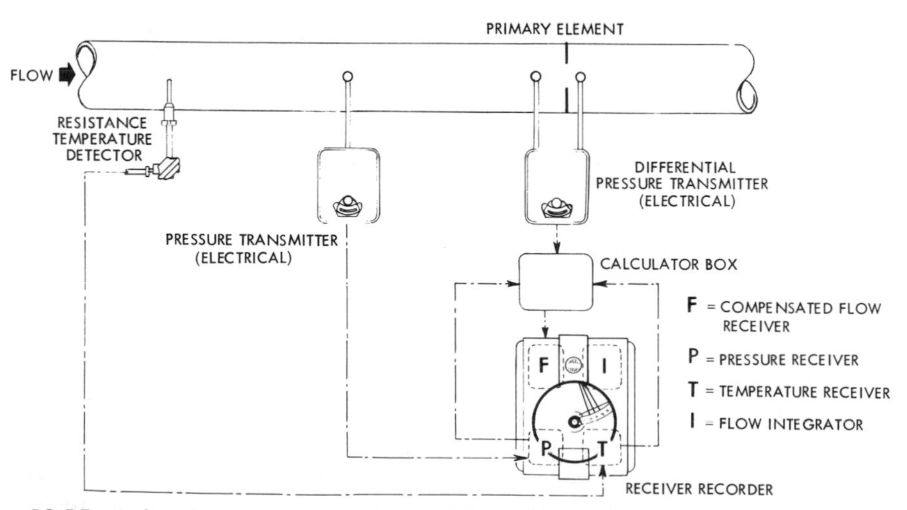

Fig. 12-15. A linedrawing of the linear differential transformer shown schematically in Fig. 12-14. The calculator box houses the pressure slidewire, the temperature slidewire, and the amplifier. (Bailey Meter Co.)

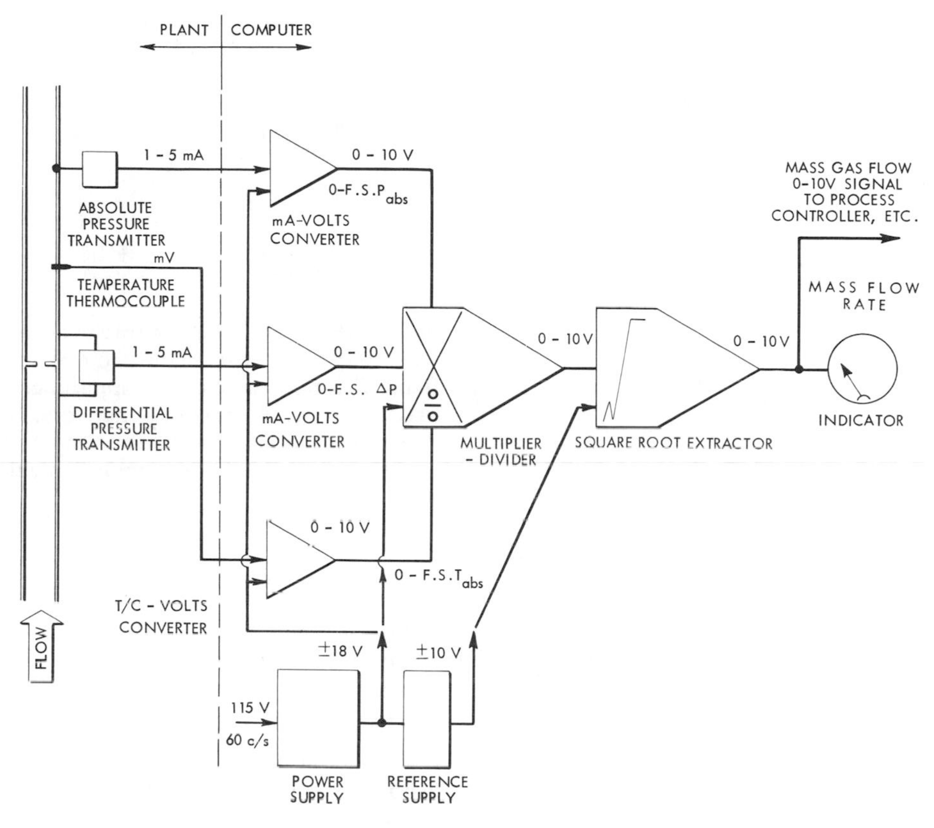

Fig. 12-16. An electrical device for square root extraction.

The same fundamental equation used for the orifice meter also applies to the variable-area meter:

$$Q = E\,A\,\sqrt{2gh}$$

In the equation for the rotameter, the differential pressure (term h) is replaced by the factor which causes it to remain constant. This factor involves the volume of the float and the area of the float, as well as the density of the float and the density of the fluid. The cross-sectional area of the float must be appropriate for the taper of the tube. The actual metering area occurs at the bottom of the float. See Fig. 12-17. The working equation becomes:

$$Q = KA_m\sqrt{2g\,\frac{V_f}{A_f}\left(\frac{\rho_f}{\rho} - 1\right)}$$

when

Q = rate of flow

K = taper constant for taper of tube

A_m = area between float and tube of float measured at indicating edge

g = 32.2 ft/sec

V_f = volume of float

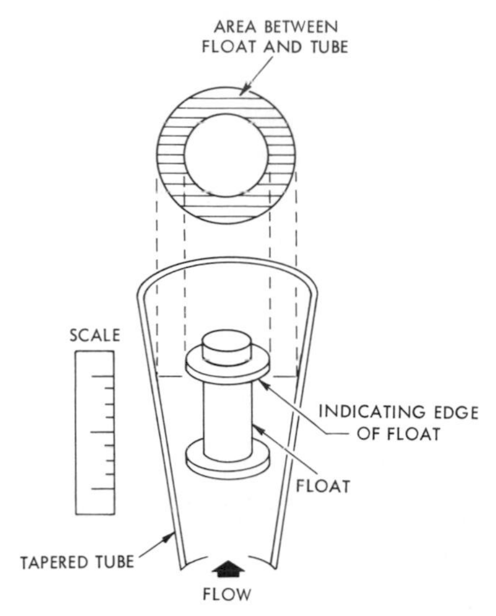

Fig. 12-17. A cutaway of a rotameter showing the float and the area between the float and the tube.

A_f = area of float
ρ_f = density of float
ρ = density of flowing fluid

The flow rate through the rotameter can be calculated for any position of the float, using this equation.

The rotameter is subject to error because of changes in the density of the flowing fluid.

Small variations in fluid viscosity do not affect measurement if the float is shaped to be insensitive to them. This makes $\frac{\rho_f}{\rho}$ equal 2. Therefore, $(\frac{\rho_f}{\rho} - 1)$ equals 1. Since

$$\sqrt{1} = 1,$$

the effect of density is eliminated.

The effect of small changes in the viscosity of the flow-in fluid is overcome, if the float is shaped so that it is insensitive to these changes. See Fig. 12-18.

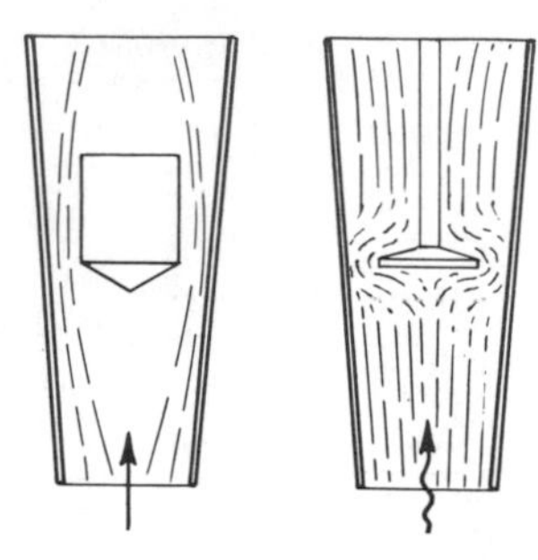

Fig. 12-18. Floats inside the rotameter vary according to the viscosity of the measured fluid. The float in the left application is used when the viscosity does not vary. The float on the right is adapted to meet the viscosity changes of the measured fluid.

The capacity of the rotameter is affected by the viscosity of the process fluid. Viscous drag on the float tends to decrease the weight of the float which decreases the differential pressure across the float. The rate of flow is related to the differential pressure. The meter can handle less flow if the viscosity of the process fluid is high. Viscosity effect is therefore a factor in the selection of the float and the calibration of the rotameter. Fig. 12-19 shows how viscosity affects float selection for a 2 inch rotameter.

Rotameters are available with glass, metal, or plastic tubes.

The glass tube can be plain or bead-guided. A plain tube has a smooth interior surface. The movement of the float is guided by a rod to avoid the float's jamming in the tube. See Fig. 12-20. A bead-guided tube has vertical glass beading along its interior sur-

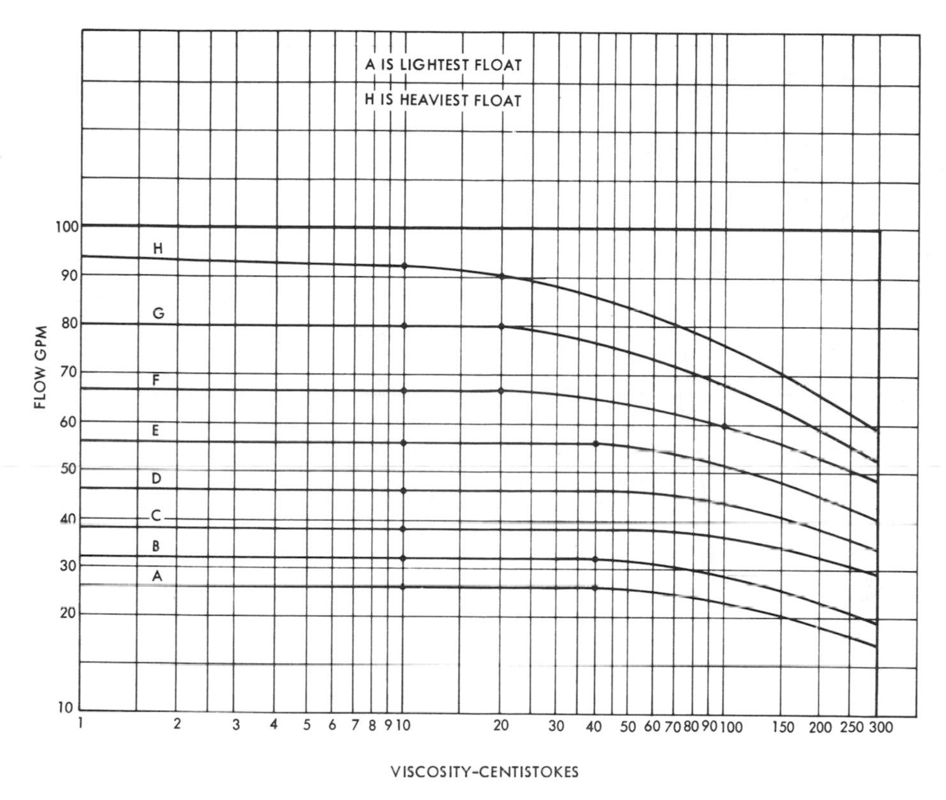

Fig. 12-19. A comparison of various floats for rotameter usage.

face, as shown in Fig. 12-21. The beading slightly reduces the area around the float. The flow capacity of a rotameter with a bead-guided float is somewhat lower than a meter with a rod-guided float.

Metal tube rotameters use rod-guided floats. The float, however, does not travel along the rod. The rod is attached to the float, and the whole assembly rises with increasing flow. The rod passes through guide bushings which are located at the top and bottom of the tube. A magnet is enclosed inside the float. A magnet attached to the indicator follows the movement of the float. See Fig. 12-22. Metal tube meters can be lined with teflon when used with corrosive fluids.

Plastic tube meters are also available. They generally have plastic floats. with encapsulated magnets. Basically, they are constructed like the metal tube-type.

By-pass meters are used when the flow rate is too great for the rotameter. The by-pass meter requires an orifice plate to be inserted in the process pipe line. The rotameter is connected across the orifice plate. A metering orifice is located in the inlet line to the rotameter. See Fig. 12-23. The function of the

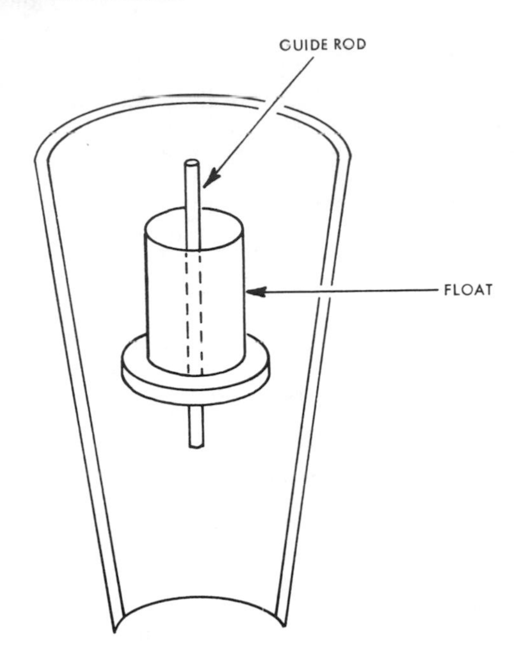

Fig. 12-20. Float in a plain tube with a guide rod.

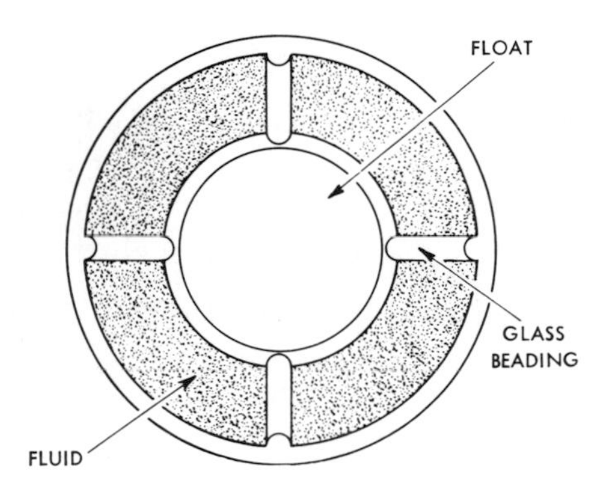

Fig. 12-21. Top view of a float in a bead-guided tube.

orifice is to establish a pressure drop in the process line. The metering orifice determines what fraction of the process flow is to enter the rotameter. Bypass rotameters can be used to accurately measure flow rate in any size pipe in which an orifice plate can be inserted. For example, flow rates as high as 20,000 gallons per minute can be measured in a 24 inch pipe. Rotameters can be equipped with alarms, and pneumatic or electric transmitters.

Turbine Meter. The pulse generating turbine flowmeter, shown in Fig.

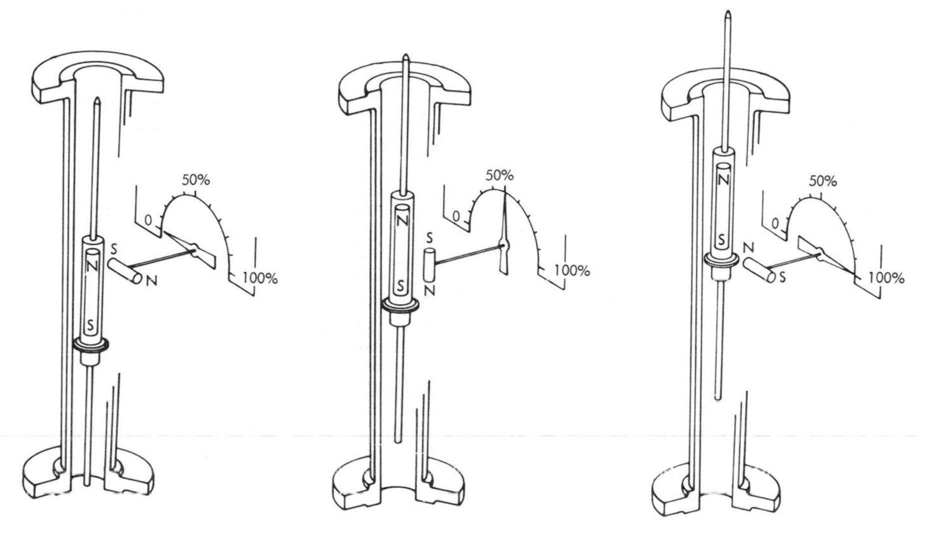

Fig. 12-22. Metal tube rotameter. The magnet connected to the indicator follows the movement of the magnet inside the float. As the float rises, the indicator shows a higher rate of flow. The float does not move up and down the guide rod. The entire assembly, including guide rod, moves with flow rate. (Wallace & Tiernan Div., Pennwalt Corp.)

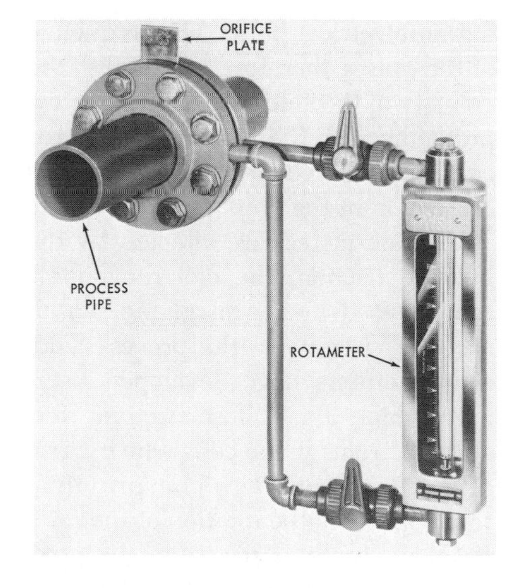

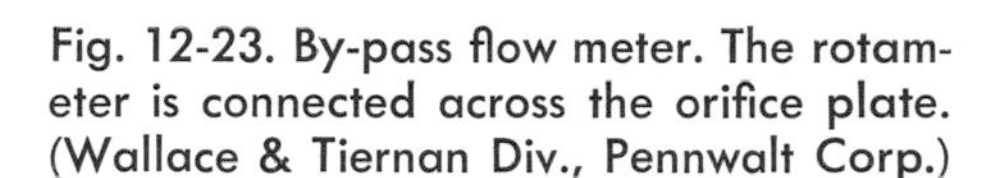
Fig. 12-23. By-pass flow meter. The rotameter is connected across the orifice plate. (Wallace & Tiernan Div., Pennwalt Corp.)

12-24, consists of a precision turbine wheel mounted on bearings inside a length of pipe, and an electromagnetic coil mounted in the pipe wall at right angles to the turbine wheel. Fluid passing through the pipe causes the

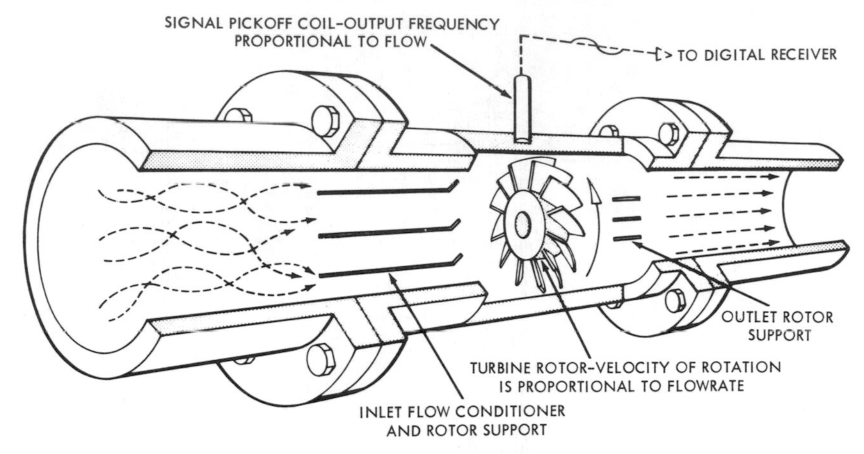

Fig. 12-24. A pulse-generating turbine flow meter. (Fischer & Porter Co.)

turbine wheel to turn at a rate which varies directly with the flow rate of the process fluid. As each blade of the turbine wheel passes the coil, it interrupts the magnetic field of the coil, producing an electrical pulse. The frequency of the pulses, therefore, varies with the velocity of the fluid flow. The electrical pulses become the input to an electrical circuit which provides continuous indication of the fluid velocity.

Turbine meters are affected by the bearing friction, the electromagnetic drag caused by the magnetic sensor, and the viscosity of the process fluid. Manufacturers have developed bearings which are almost friction free, and electromagnetic coils which virtually eliminate drag. The principal force opposing the rotation of the rotor is caused by the viscosity of the process fluid. When the rotor begins to turn, the driving force of the flowing fluid must overcome any friction, drag, and anti-rotational force caused by viscosity. As the speed of the rotor increases, the driving force is greater than those forces which oppose the rotation. When turbine meters are calibrated, several fluids of different vis-

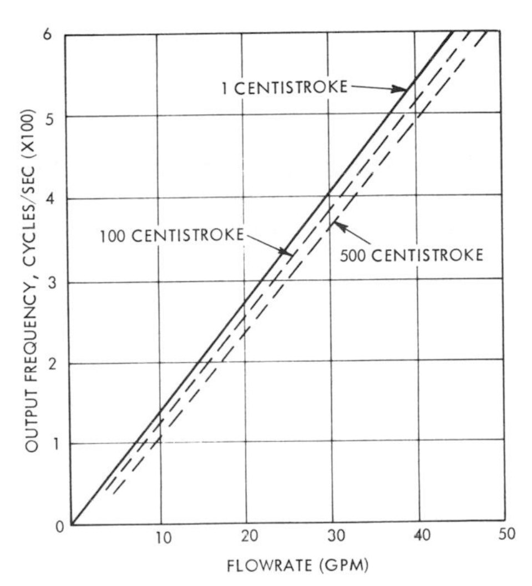

Fig. 12-25. A comparison of output frequency and the rate of flow. (Fischer & Porter Co.)

cosities are used. In a typical turbine meter, there is a reduction of the output frequency as the viscosity increases. See Fig. 12-25. Turbine meters measure flow rates of fluids with viscosities from .6 to 700 centistokes.

Electromagnetic Meter. The electromagnetic flowmeter, shown in Fig. 12-26, consists of a tube of non-conducting material with two electrodes mounted opposite one another on the tube wall. The inner ends of the electrodes are in contact with the fluid flowing in the tube. A magnet surrounds the tube, with its field at right angles to the electrodes. The flowing fluid must be an electrical conductor.

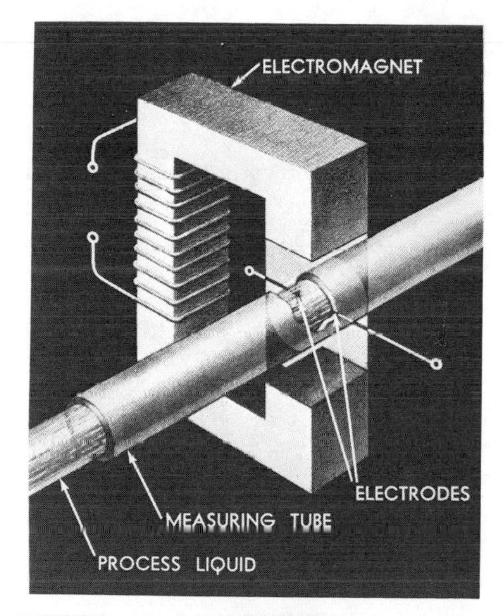

Fig. 12-26. An electromagnetic flowmeter. The top view shows the principles of operation. The bottom view shows an actual transmitter. (The Foxboro Co.)

As the moving electrical conductor (the fluid) flows through the non-conductive tube through the magnetic field, a voltage is produced which can be picked up by the electrodes. The voltage produced depends upon the strength of the magnetic field, the distance between the electrodes, and the velocity of the conductive fluid flowing through the tube. Therefore, if the strength of the magnetic field and the distance between the electrodes is held constant, the voltage produced will be entirely due to the speed at which the fluid moves through the tube. This voltage is then amplified and connected to suitable indicating or recording instruments.

The Vortex Precession Meter for measuring gas outwardly resembles a turbine meter. In the turbine meter, the flow stream causes a rotor to turn. The vortex flowmeter has a stationary swirler at one end, which causes the gas stream to rotate around the center line of the meter. A de-swirler, located at the opposite end of the meter, restores the gas stream to its original flow rate. Internally, the meter is shaped like a venturi tube, with an entrance cone, a throat, and an exit cone. See Fig. 12-27. When the gas passes through the expanded portion of the meter, the axis of rotation moves perpendicular to the center of rotation (precesses). See Fig. 12-28. The fre-

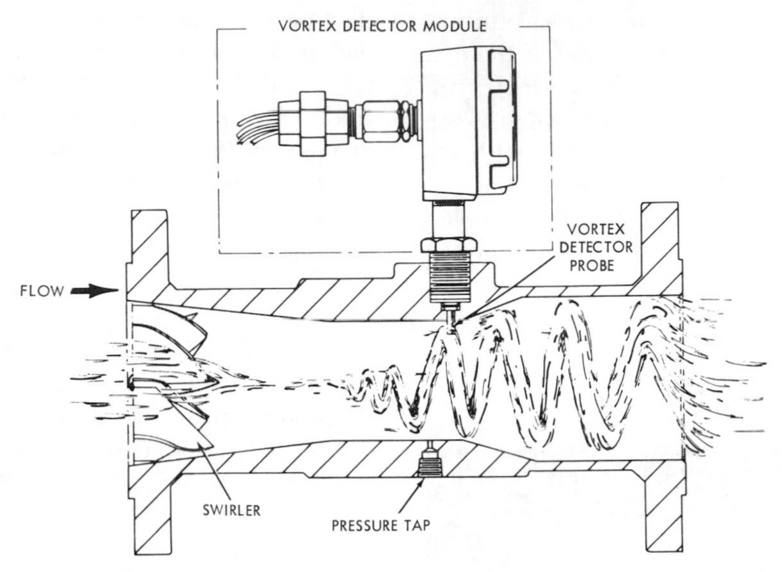

Fig. 12-27. Vortex precession meter for measuring the flow rate of gas. (Fischer & Porter Co.)

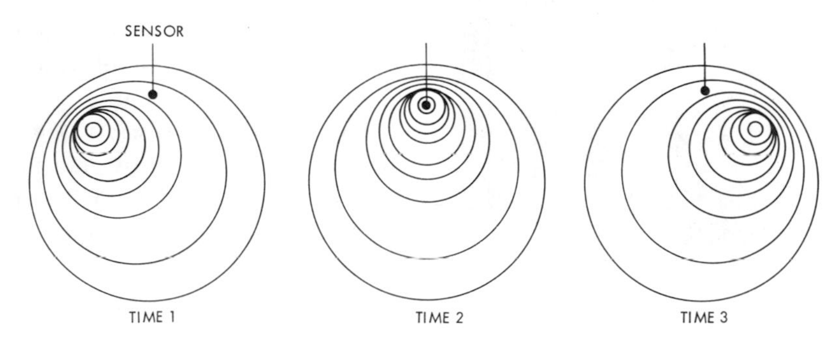

Fig. 12-28. The axis of rotation in a vortex precession meter. (Fischer & Porter Co.)

quency of the precession varies with the flow rate of the gas.

The vortex precession meter uses a thermistor as one of its sensors. A constant current is applied to the thermistor, which maintains it at a specific temperature. The gas receives thermal energy from the thermistor as it comes in contact with the thermistor. This causes a change of voltage in the thermistor circuit. The voltage change is detected, amplified, and becomes the output signal of the meter. The electronic circuit is shown in Fig. 12-29.

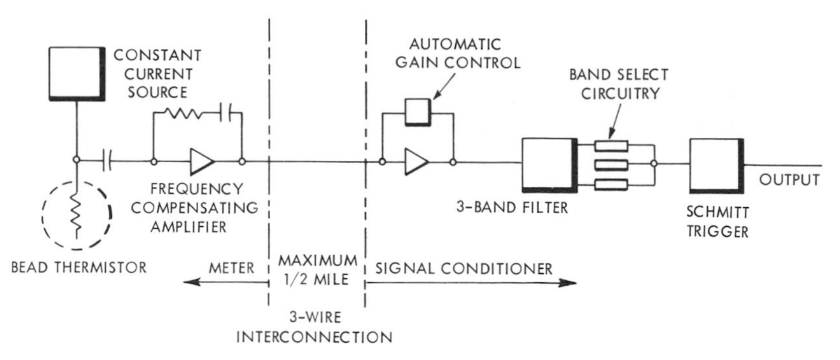

Fig. 12-29. A schematic of the electronic circuit for a vortex precession meter. (Fischer & Porter Co.)

The Vortex Shedding Flowmeter uses the formation of vortices for its principle of operation. This meter consists of a cylinder in which a V-shaped block, called a *bluff body,* is mounted. The broad surface of the cylinder faces upstream. As the gas moves around the bluff body, vortices are formed. The vortices increase in size, as the flow rate increases, until they become too large to remain attached to the bluff body. They break away and are shed downstream at a frequency which varies with the flow rate. The sensors used in the meter are two self-heated resistance elements. These elements are mounted in the front face of the bluff body. The temperature of these sensors varies as a result of velocity variations on the front face. The difference in the temperature of the sensors creates a voltage change which is amplified and becomes the output signal of the meter.

Ultrasonic is another sensing system. In this type of sensing system, the bluff body is a cylinder positioned at right angles to the fluid stream. An ultrasonic transmitter, enclosed in a stainless steel disc one-fourth inch in diameter, is located downstream from the cylinder. A piezoelectric crystal, which produces the ultrasonic signal is inside the disc. The ultrasonic signal is directed across the fluid stream and at right angles to the cylinder. The signal is in the form of a narrow ultrasonic tunnel. An ultrasonic receiver is at the other end of the tunnel. As the vortices break away from the cylindrical bluff body, they move downstream toward the ultrasonic beam. When a vortex enters the ultrasonic tunnel, the energy reaching the receiver is at its minimum. When no vortex is in the tunnel, the energy reaching the receiver is at its maximum. Since the number of vortices reaching the ultrasonic tunnel varies with the rate of flow of the process fluid, the energy developed at the receiver is proportional to the flow rate.

Mass Flow Measurement

Demands for measuring flow in weight rather than volumetric units resulted in the development of mass flow measurement. This requires continuous measurement of density as well as flow rate. One mass flowmeter consists of an electromagnetic flowmeter with a radioactive density-measuring unit. See Fig. 12-30. The outputs of these two units are fed into an electrical computing circuit to produce a single output proportional to flow in weight units (lbs./min.).

Several other specially designed in-

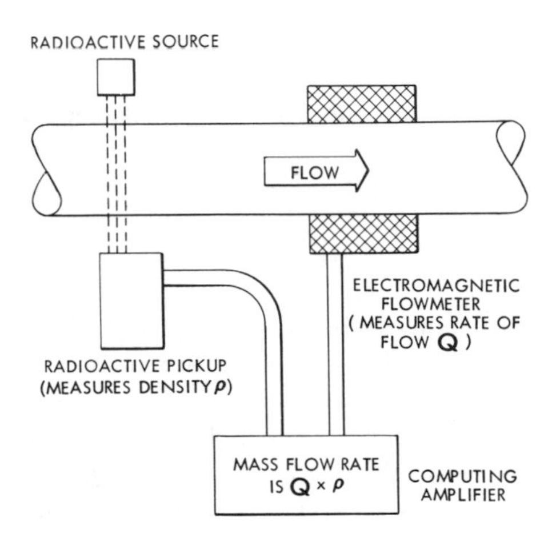

Fig. 12-30. A radioactive device added to the electromagnetic flowmeter.

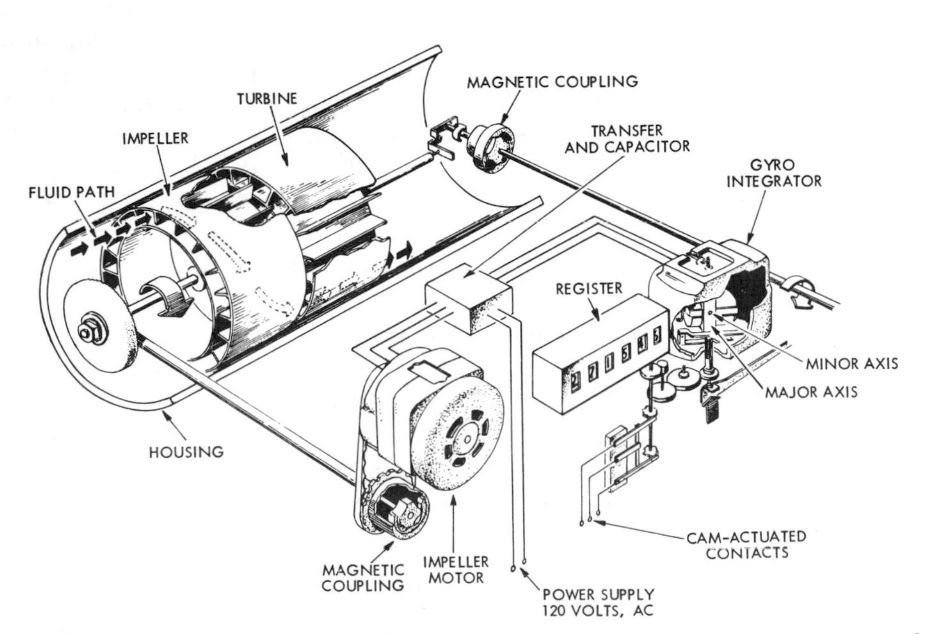

Fig. 12-31. The mass flowmeter uses an impeller to impart angular momentum to the measured fluid. (Black, Sivalls, and Bryson, Inc.)

struments are available for measuring mass flow. One of these, a constant-speed impeller, imparts angular momentum to the flowing fluid. See Fig. 12-31. A turbine wheel downstream from the impeller is displaced by this angular momentum. The turbine wheel's movement is opposed by a spiral spring. The torque exerted by the spring to restrain the movement of the turbine is proportional to mass flow. A pointer attached to the turbine shaft can be made to indicate the mass flow on a suitable scale, or the shaft movement itself can be used to actuate totalizers or controllers.

Another mass flowmeter, Fig. 12-32, senses the change in temperature of a heated thermopile due to the change in the flow rate. Two Plat-Rho-Plat thermocouples (A and B) are immersed in the fluid stream. They are heated by alternating current. A change in flow results in a change of temperature in each of the thermocouples, producing a change in total dc output of the thermocouples. A third thermocouple (C) is used, but is not heated. It merely compensates for changes in ambient temperature. All three thermocouples are combined in a dc measuring circuit. Since the change in flow rate is proportional to the change in the output of the two sensing thermocouples, and because the change in temperature is proportional to the change in density, the measurement is one of mass flow. The instrument can

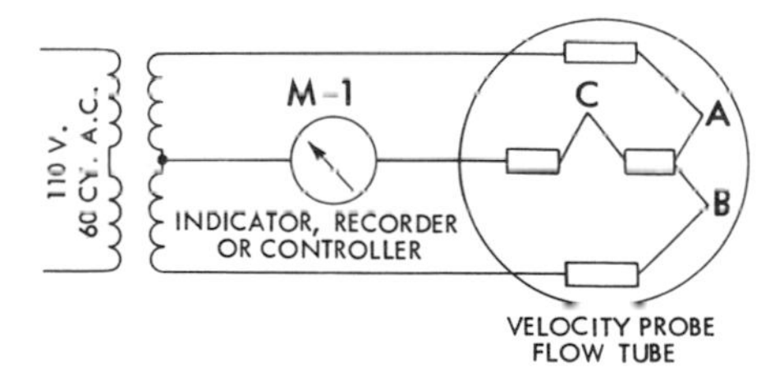

Fig. 12-32. Mass flowmeter using a heated thermopile. (Hastings-Raydist, Inc.)

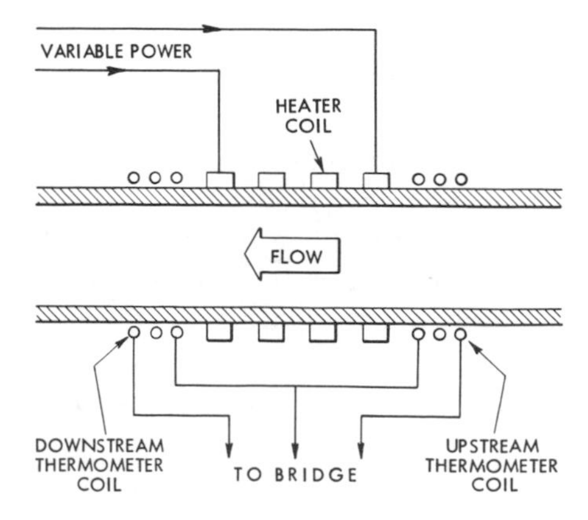

Fig. 12-33. Mass flowmeter using a resistance thermometer.

be calibrated in pounds of gas per hour, if the thermal conductivity, specific heat, and density of the gas remain constant. This requirement is satisfied over a wide range of temperature and pressure in the case of most gases.

Another type of mass flowmeter uses resistance thermometers to measure temperature difference between two points of a pipe line. As the mass flow rate increases, the process liquid carries away more heat. See Fig. 12-33. This system must include compensation for heat loss through the pipe wall, as well as the thermal characteristics of the process liquid.

Words to Know

density
viscosity
velocity
laminar flow
turbulent flow
Reynolds number
vena contracta
beta ratio
efficiency factor
venturi tube
orifice plate
pitot tube
Sorteberg bridge
bluff body

Review Questions

1. What physical properties of fluids affect their flow?
2. Describe the difference between laminar and turbulent flow. Which is more common in industrial applications?
3. What is the significance of the efficiency factor in the orifice plate equation?
4. Why is the pressure recovery of the venturi tube is better than that of the orifice plate?
5. Describe three installations in which a pitot tube might be the best primary flow element.
6. In the formula $\frac{P \times h}{T}$ what do *P, h,* and *T* stand for?
7. How are small changes in the viscosity of the flowing fluid overcome when a rotameter is used?
8. How are small changes in the density of the flowing fluid overcome when a rotameter is used?
9. What is the nature of the output in a turbine flowmeter?
10. What characteristic of water and liquid sodium make them suitable for use with an electromagnetic flowmeter?
11. What is *precession* as applied to a vortex precession meter?
12. What equipment is best suited for measuring water flow at a pressure of 1000 psi and transmitting the measurement to a remote indicator, 1000 feet away?

Analysis

Chapter

13

Continuous production processes involve the conversion of raw materials, or the combining of several ingredients, into a single final product. To be able to measure and control the physical and chemical properties of the constituent materials as they are being processed is essential to obtaining uniformly satisfactory quality in the finished product. The means of accomplishing this are provided by instruments called analyzers.

There has been a considerable increase in the number of analyzers, and applications for which they can be used. It is possible here to describe only a representative few.

Analyzers sometimes require complicated sampling systems and special sensing devices. It should be noted, however, that generally the actual measurement is temperature, electrical characteristics, or another simple variable. These can be measured using conventional instruments. In this chapter, the measurement of the following physical and chemical properties are discussed briefly:

Density and specific gravity
Viscosity and consistency
Acidity and alkalinity
Electrical and thermal conductivity
Combustibility
Chromatography

Density and Specific Gravity

The density of a material is its weight per unit of volume. The specific gravity of a liquid is the ratio of its density compared with the density of water at 4°C. At 4°C, a cubic foot of water weighs 62.4 lbs. The specific gravity of a gas is the ratio of its density compared with that of air at standard conditions (60°F and 14.7 pounds per square inch absolute pressure). Under standard conditions, a cubic foot of air weighs .076 lbs.

Liquid Density and Specific Gravity. The simplest device for measuring liquid density is the hydrometer. A *hydrometer,* shown in Fig. 13-1, is a floating instrument which displaces a volume of liquid equal to its own weight. It is usually made of hollow glass or

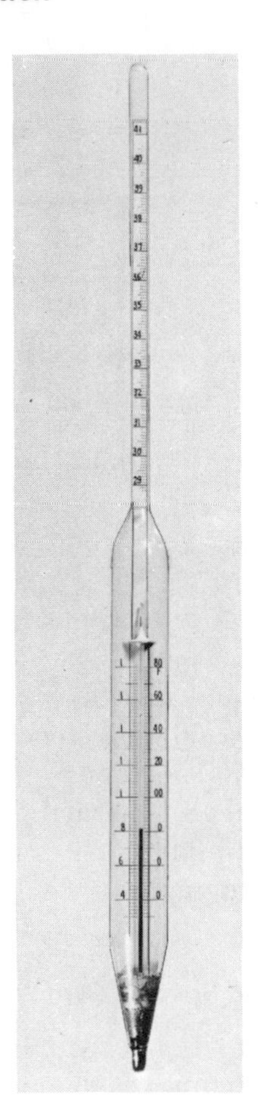

Fig. 13-1. This thermohydrometer contains a thermometer so that the effect of process liquid temperature on the density reading can be corrected. (Taylor Instrument)

metal, and is weighted at one end to make it float upright. The position of the hydrometer in the liquid depends on the density of the liquid. A less dense liquid causes the hydrometer to be positioned lower in the liquid because a greater volume of the liquid has been displaced. The density or specific gravity scale appears on the upper portion of the hydrometer. The reading is taken by noting the point on the scale to which the liquid rises. The hydrometer contains a thermometer which must be checked when a reading is taken so that any density changes due to ambient temperature can be corrected.

A metal rod can be used as a weight to permit remote reading of a hydrometer. Suspended in an electrical coil, the rod acts as the variable arm of an inductance bridge. The movement of the hydrometer and rod due to changes in liquid density is duplicated by the movement of a matching motor-positioned metal rod in the indicating or recording instrument. See Fig. 13-2. Such a hydrometer can be used for continuous measurement, but it must be placed in an enclosure which per-

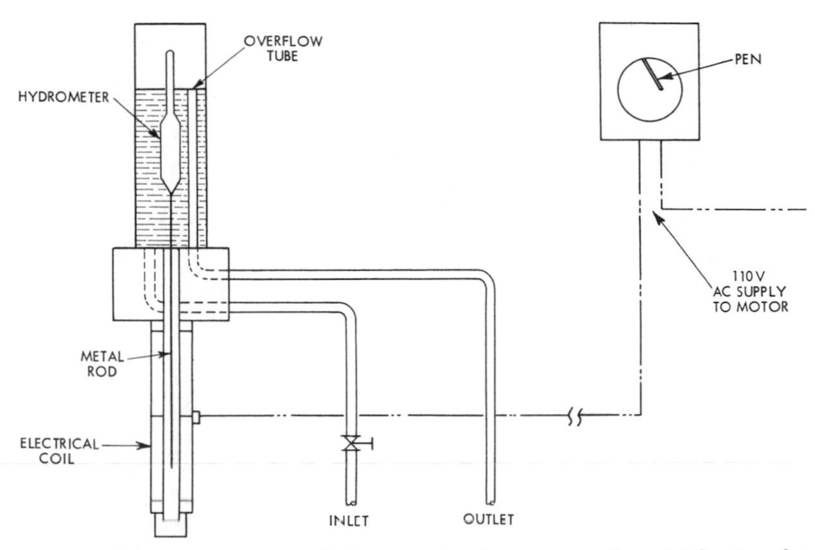

Fig. 13-2. Specific gravity meter which transmits the measured variable signal to a remote recording instrument. (Fischer & Porter Co.)

mits the liquid to flow through it. The level of the liquid in the enclosure is maintained by the use of an overflow tube. Usually the temperature of the liquid is measured continuously with a thermocouple or similar device so that density corrections due to temperature can be made.

Weight of Fixed Volume. Another method of measuring liquid density is to continuously weigh a fixed volume of the liquid. This can be done using a mechanical balance, as shown in Fig. 13-3, or an electrical load cell. See Fig. 13-4. A *load cell* is an electromechanical device containing a strain gage. When the mechanical portion is deflected due to a weight, the strain gage is distorted, causing a change in electrical resistance. The strain gage is

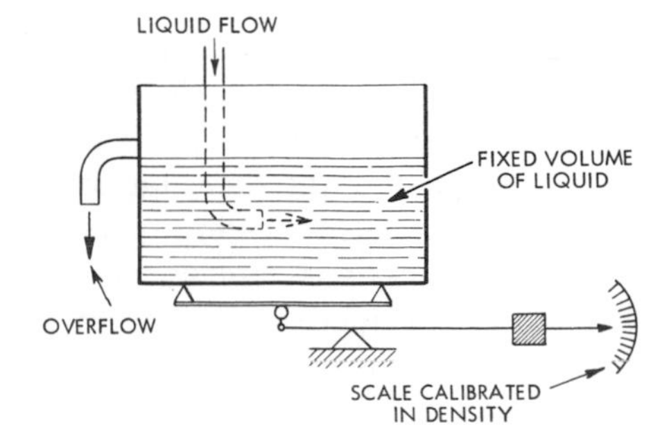

Fig. 13-3. A mechanical balance used to measure liquid density.

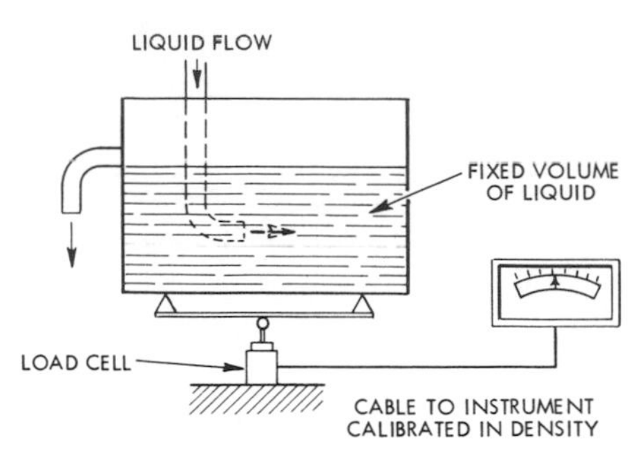

Fig. 13-4. An electric load cell used for measuring liquid density.

used as the variable resistance in a bridge network, making possible remote indication or recording of the density. For continuous measurement, the fixed volume can be enclosed in a chamber through which the liquid flows in and out at a constant rate.

Displacement. In a displacement-type liquid density meter, shown in Fig. 13-5, the displacer element is enclosed in a chamber of fixed volume. As the density of the liquid changes, the buoyant force acting on the displacer changes. The displacer is attached to a balance beam which moves slightly as the buoyant force changes. The opposite end of the balance beam acts either as the actuating mechanism of a pneumatic or an electrical measuring system.

In the pneumatic system, shown in Fig. 13-6, the beam moves the flapper in a flapper-nozzle device. When the force acting on the displacer increases due to an increase in the density of the liquid, the flapper is pressed against the nozzle. This increases the air pressure to the receiver, which can be an indicator or a recorder. The receiver

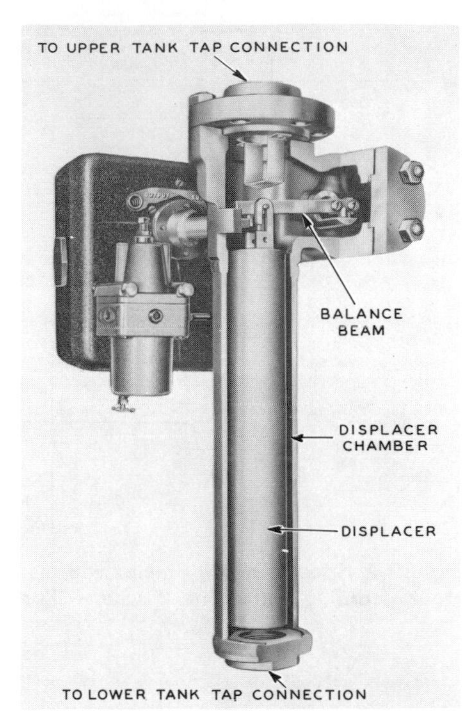

Fig. 13-5. Liquid density meter with displacer. (Masoneilan International, Inc.)

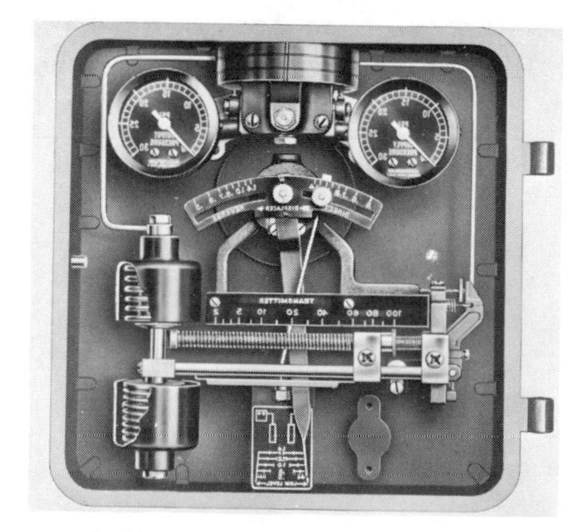

Fig. 13-6. Pneumatic transmitter used with displacer-type density element. (Masoneilan International, Inc.)

is calibrated in units of density or specific gravity.

In the electrical system, the balance beam moves a small iron rod in an inductance coil, similar to that described for use with the hydrometer. See Fig. 13-7. In this case, however, the inductance coil must provide a greater change in electrical output for a small change in position of the iron rod. This can be obtained using a differential transformer. The output of the unit is used as the variable arm of the inductance bridge.

Differential Pressure. Liquid level can be measured using pressure or differential pressure instruments. In a constant-level open tank, the pressure transmitter is installed at a measured distance below the surface of the liquid. The pressure at that point is equal to H (height of the liquid) × ρ (the density). See Fig. 13-8. The height is constant, and the pressure varies directly with the density. The transmitter can be pneumatic or electric.

In an open or closed tank which has varying level and/or pressure, the dif-

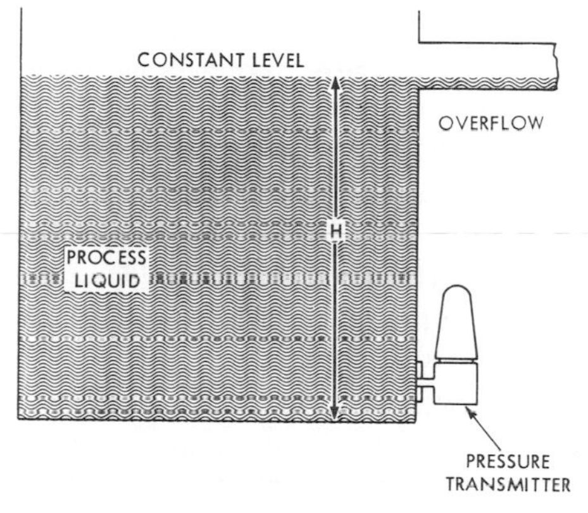

Fig. 13-8. Liquid density measurement in an open tank using pressure transmitter.

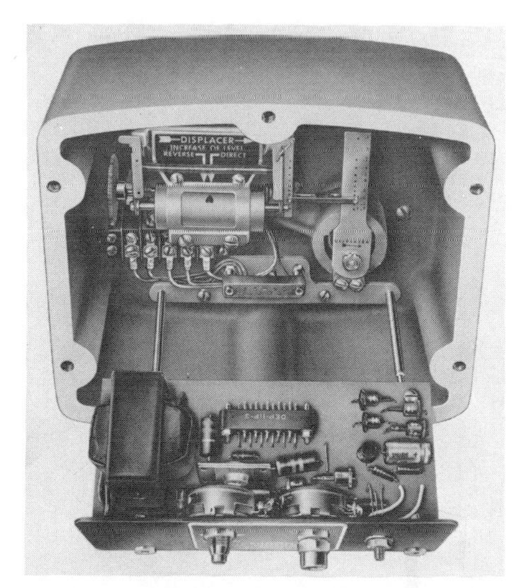

Fig. 13-7. Electronic transmitter used with displacement-type density element. (Masoneilan International, Inc.)

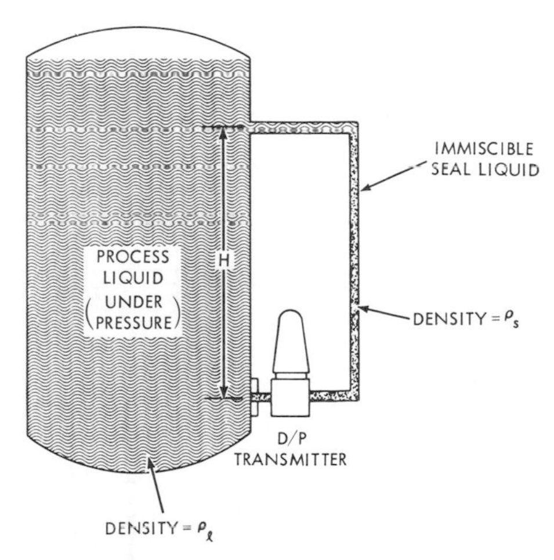

Fig. 13-9. Liquid density measurement in pressurized tank the level of which varies. The system uses a differential pressure transmitter.

ferential pressure transmitter is installed as shown in Fig. 13-9. A sealing liquid, which has a greater density than the process liquid, is enclosed in the line from the upper connection. When this method is used with hydrocarbons, water is the sealing liquid.

The differential pressure transmitter must have a suppressed range. This means the transmitter does not produce an output until a portion of the input signal is overcome. The high pressure input signal is overcome. The high pressure input to the transmitter equals H (height) × ρ_s (density of the sealing liquid). The low pressure input equals H (height) × ρ_l (density of the process liquid). The height is the same for each input, and the density of the sealing liquid remains constant. Therefore, the differential pressure signal is due solely to a change in the density of the process liquid. The amount of input suppression is equal to H (height) × ρ_s (density of the sealing liquid) − $\rho_{l\,min}$ (the minimum density of the process liquid).

A variation of the bubbler system used for level measurement can be applied to density measurement. See Fig. 13-10. Air is piped to two bubbler tubes immersed in the liquid, with their lower ends at different levels. The air pressure to each bubbler tube is regulated so that it is slightly higher than the pressure of the column of liquid in which the tubes are immersed. This permits air bubbles to be released slowly. As the density of the liquid changes, the pressures at the ends of the tubes change, since they are equal to height × density. Because the ends of the tubes are at different levels, the difference in pressures at these points is the result of the change in density of the column of liquid. The change in height is equal to the difference in the levels of the

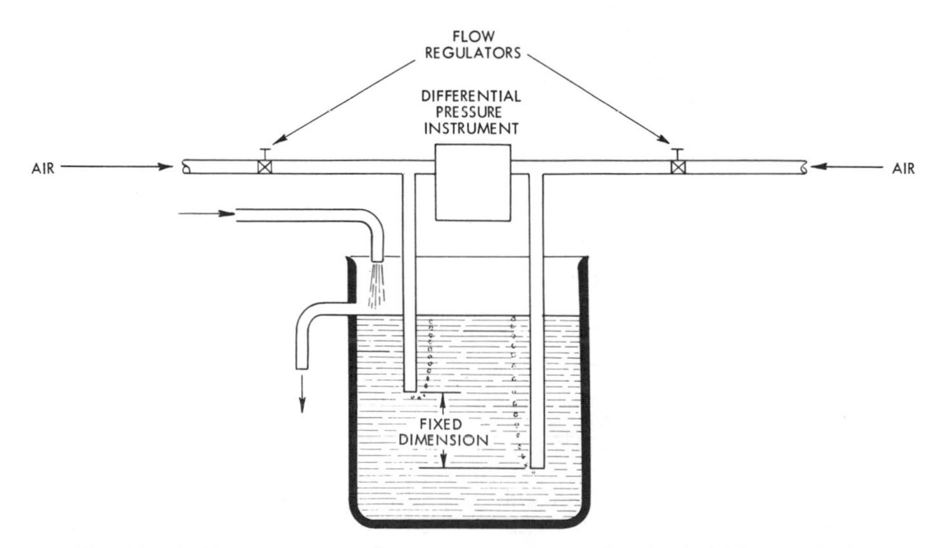

Fig. 13-10. Measuring specific gravity or density by the bubbler method.

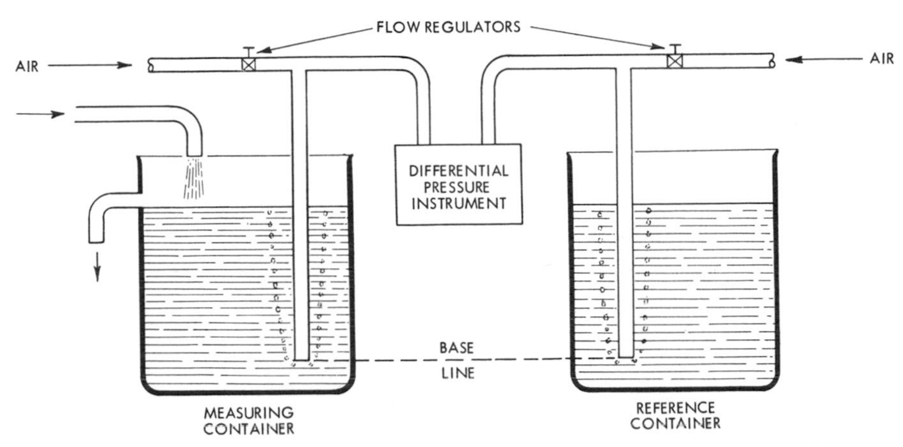

Fig. 13-11. This bubbler method for measuring specific gravity uses a second container called a reference container.

tube ends. The air pressures, in addition to passing to the bubbler tubes, enter a differential pressure measuring unit, calibrated in units of density or specific gravity. As the density of the liquid decreases, the air pressures in the bubbler tubes and the differential unit decrease. Thus a change in liquid density is measurable as a change in differential pressure.

A variation of the bubbler system differential pressure method of density measurement requires the use of two similar vessels. One contains a reference liquid (such as water), and the other the process liquid which flows through it. See Fig. 13-11. The levels of the two liquids are held constant at the same height. Bubbler tubes are immersed in each vessel. The ends are at precisely the same level. Since the liquid levels above the tube ends are equal, any difference in the pressures to the bubbler tubes and the differential unit are due to the change in density of the process liquid. The differential units are calibrated in units of density or specific gravity.

Liquid density can be measured using a hairpin-shaped tube. The tube is pivoted on flexures located at the open ends of the tube. See Fig. 13-12. The weight of the tube and the liquid contents of the tube are transferred to a weigh beam. A counterweight can be adjusted by moving it along the weigh beam. A change in the density of the liquid produces a directly proportioned change in force on the weigh beam. The force is measured by a force balance transmission system. This system can be either pneumatic or electrical.

In the pneumatic system, shown in Fig. 13-13, *top*, the force of the weigh beam acts against a deflection bar. The deflection bar serves as the flapper moving against the nozzle. A bellows provides the feedback signal for proportional measurement. The measurable density span in this system is .025 to .5 grams per millimeter.

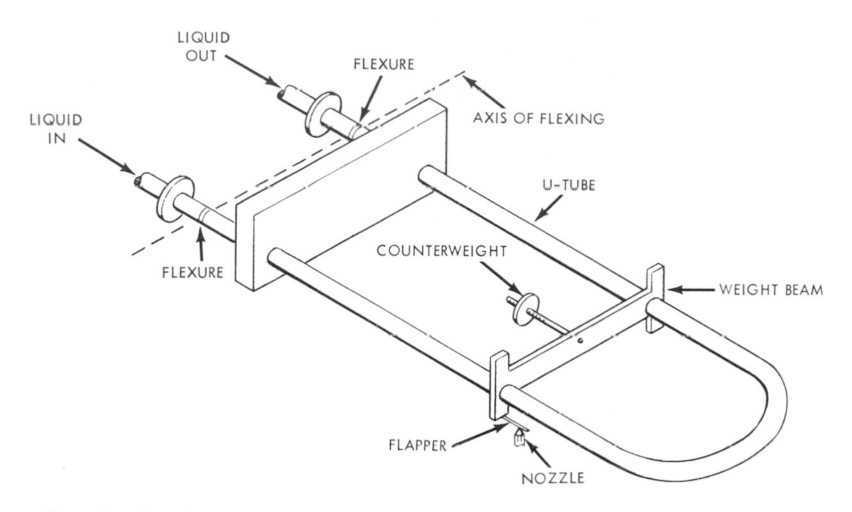

Fig. 13-12. A hairpin-shaped tube used for liquid density measurement.

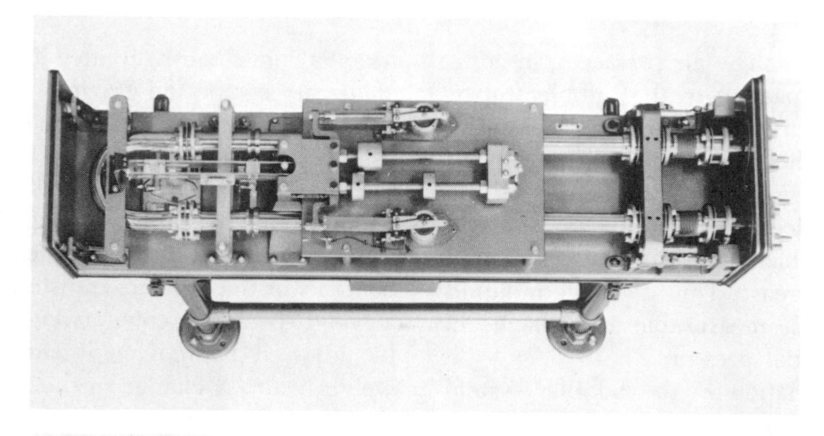

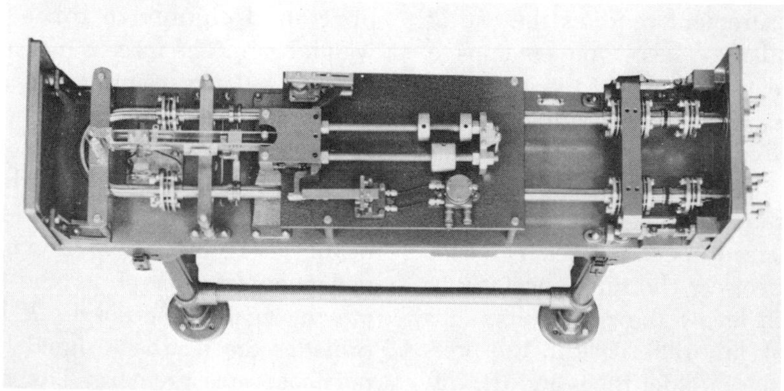

Fig. 13-13. Systems for measuring liquid density using a pneumatic output signal (top) and an electrical output (bottom). (Totco Div., Baker Oil Tools, Inc.)

In the electric system, shown in Fig. 13-13, *bottom*, the deflection bar acts on the capacitance transducer.

Radioactive Sensing Cell. The density of a liquid can also be measured using the radiation from a suitable radioactive isotope. A source of radioactivity, such as radium or cobalt 60, is placed at one side of a vessel. Its rays are directed across the vessel. A radiation detector is placed on the opposite side of the vessel or tube so that it receives the amount of radioactive energy remaining after passing through the vessel walls and the liquid through the vessel walls and the liquid. See Figs. 13-14A and 13-14B. The amount of radiation absorbed by any material varies directly with its density. Such a method is best for liquids that do not permit a sensing element to be immersed because of corrosion, abrasion, or other limitation.

Gas Density. Measuring the density of gases is important in the gas transmission and petrochemical industries. This measurement can be accomplished using a densitometer with a probe. See Fig. 13-15. The probe contains a vane, which is symmetrically positioned across the supporting cylinder. When the unit is to be put in operation, the probe is installed in the pipe line, which contains the flowing process fluid. The vane oscillates in a simple, harmonic motion. This causes an acceleration of the flowing process fluid. The frequency of the oscillation varies with the density of the fluid. As the density increases, the oscillation frequency decreases. The relationship is expressed:

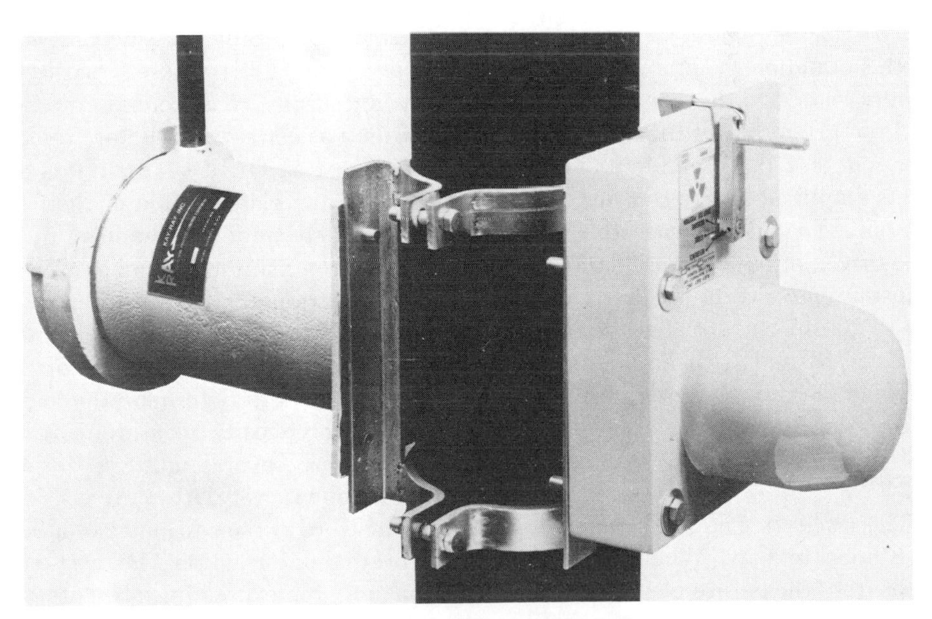

Fig. 13-14A. Measurement head for determining the density of process material with the use of a radiation source. (Kay-Ray, Inc.)

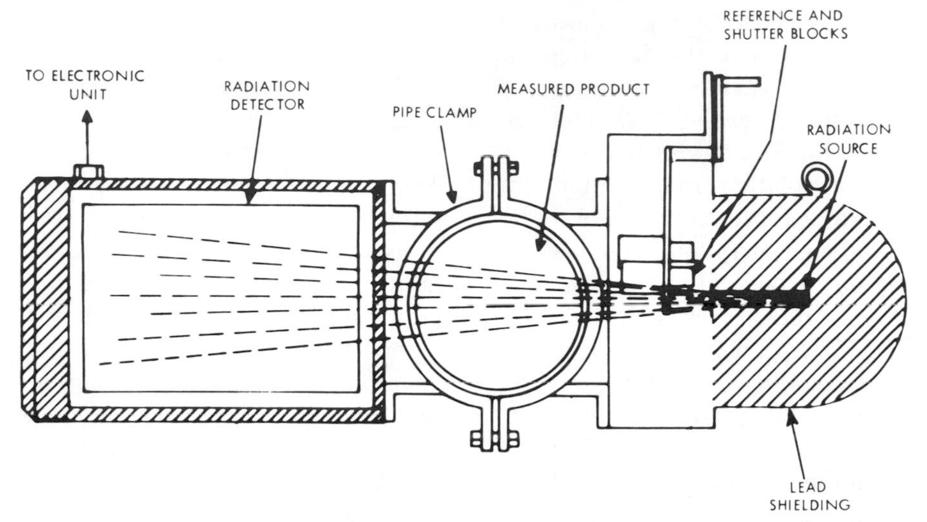

Fig. 13-14B. Interior view of the measurement head shown in Fig. 13-14A. Rays from the radiation source pass through the process material (measured product) and are sensed by the radiation detector. (Kay-Ray, Inc.)

$$\text{Density} = \frac{A}{f^2} - \frac{B}{f} + C$$

In this equation, *A, B,* and *C* are constants, related to the size of the probe and the properties of the fluid. *f* is the frequency of the oscillation. The signal is amplified by the transmitter and energizes a driver with the probe. The driver sustains the oscillation. The transmitter converts the frequency to a 4 to 20 mA current. This densitometer is also used with liquid. It is capable of measuring density, even if the fluid is not flowing.

Viscosity

The *viscosity* of a fluid expresses its resistance to flow. When viscosity is measured, the nature of the fluid must be known. The viscosity of some liquids varies with temperature, or changes considerably after the fluid has been shaken or is flowing. Such fluids are generally classified as *non-Newtonian.* Fluids whose viscosity is constant at any given temperature are classified as *Newtonian.* Before a measurement of fluid viscosity is attempted, the classification of the fluid must be determined. Newtonian fluids permit viscosity measurement by any of the instruments that are described in this section. Non-Newtonian fluids must be tested for their particular characteristics to determine their suitability for viscosity measurement.

There are several unit systems for expressing viscosity. Absolute viscosity is a measure of the resistance of a fluid to internal deformation. The most common unit for expressing absolute viscosity is the *centipoise.* Kinematic viscosity is the ratio of the absolute

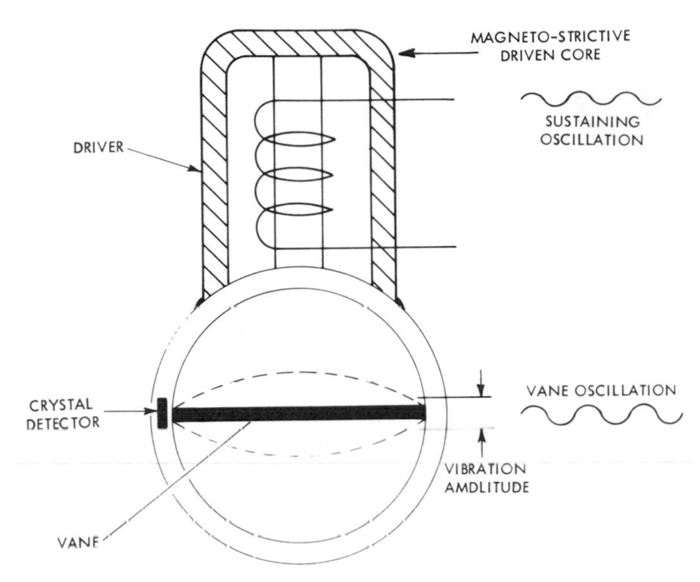

Fig. 13-15. A densitometer probe used for measuring the density of gas.

viscosity to the mass density. The common unit of kinematic viscosity is the centistoke.

$$\text{Centistokes} = \frac{\text{Centipoises}}{\text{Grams per cubic cm}}$$

The instruments that have been developed for continuous measurement of viscosity are based upon several laboratory-type devices:

1. Falling ball or piston
2. Rotating spindle
3. Measured flow through orifice.

Falling Ball or Piston. In the falling ball device, shown in Fig. 13-16, the time required for a ball to fall a certain distance through the liquid is measured. The time is proportional to the viscosity. The continuous viscosity measuring apparatus based upon this principle employs a rotameter. See Fig. 13-17. A metering pump draws some of the liquid from the main stream and pumps it through the rotameter at a fixed rate. The differential pressure across the rotameter is constant, therefore any movement of the

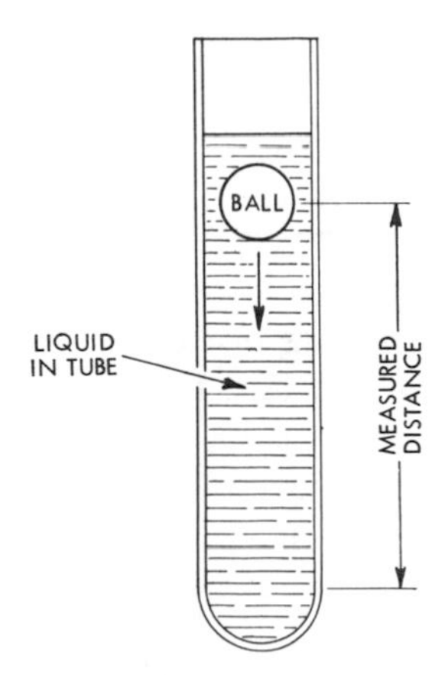

Fig. 13-16. Falling ball device used for measuring viscosity.

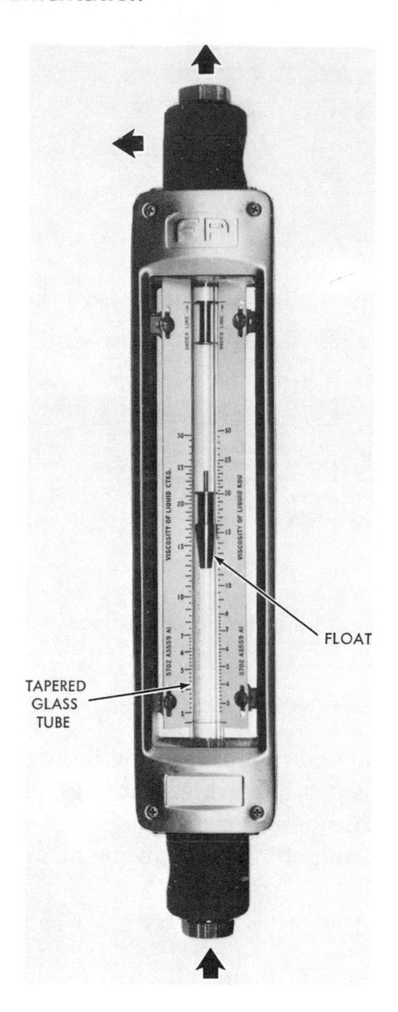

Fig. 13-17. Variable area device with rotameter used for measurnig viscosity. (Fischer & Porter Co.)

float is due to a change in viscosity. The rotameter can be calibrated in viscosity units.

In the falling piston laboratory device, shown in Fig. 13-18, the time required for a piston to fall a certain distance in a cylinder containing the sample liquid is measured. The time required varies with changes in viscosity.

The continuous measuring device employing this principle adds mechanical equipment, which automatically raises the piston after each fall. At the same time a new sample is drawn into the cylinder. See Figs. 13-19 and 13-20. Although this type is not really continuous, the operation can be accelerated sufficiently to make it satisfactory for continuous measurement. Such a device can measure viscosity over a range of .1 to 10^6 centipoise.

Rotating Spindle. The rotating spindle apparatus for measuring viscosity

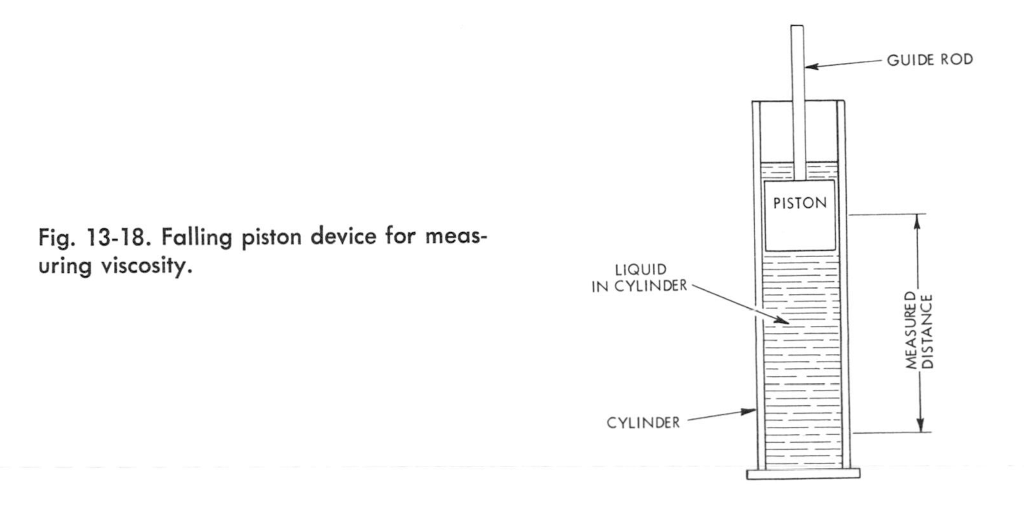

Fig. 13-18. Falling piston device for measuring viscosity.

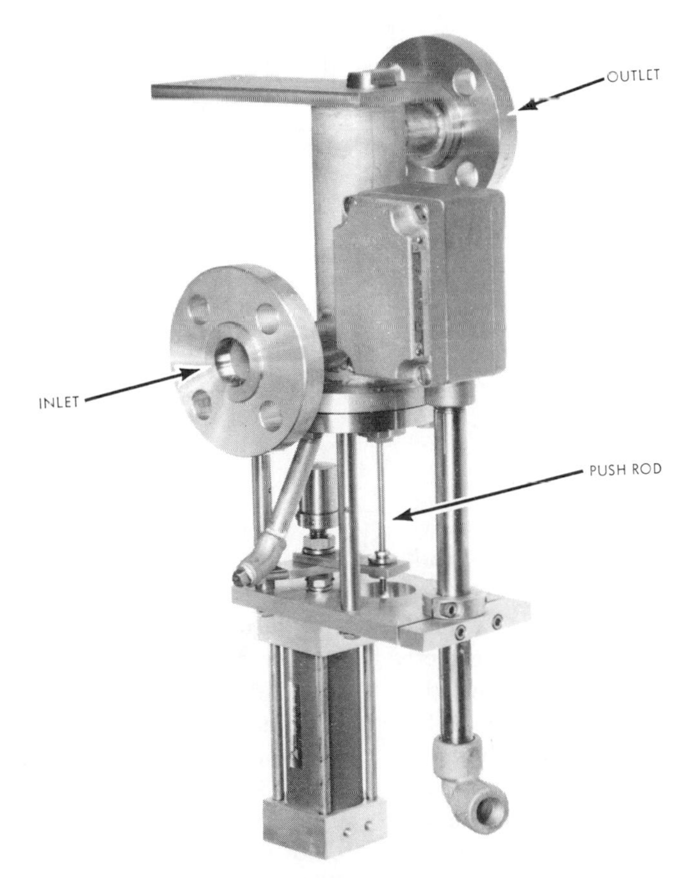

Fig. 13-19. An in-line viscometer. This mechanical device is used in viscosity measurement to automatically raise the piston. (Norcross Corp.)

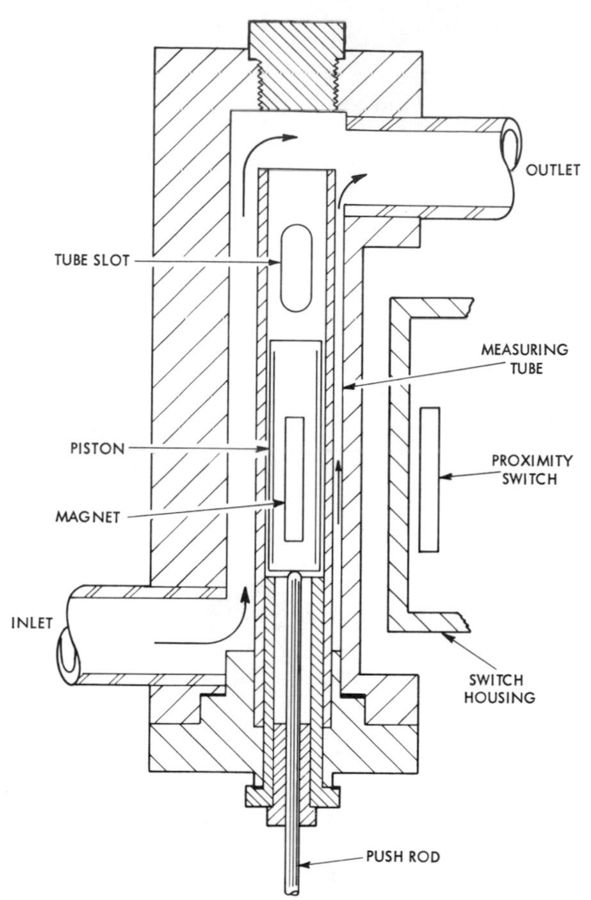

Fig. 13-20. Line drawing of the viscometer in Fig. 13-19 showing an interior view of the components. (Norcross Corp.)

in the laboratory consists of a spindle rotated in a container of the sample liquid. See Fig. 13-21. The principle involved is that the viscosity is directly proportional to the torque required to drive the spindle. Some units of this type rotate the container rather than the spindle. The principle, however, remains the same. The continuous viscosity measuring unit based on this principle resembles the laboratory apparatus, with the addition of a means of allowing the liquid to flow into and out of the container. See Fig. 13-22.

Measured Flow Through Orifice. A simple apparatus for measuring the kinematic viscosity of oils and other viscous liquids in the laboratory is the Saybolt universal viscosimeter. See Fig. 13-23. This apparatus consists of a temperature controlled vessel with an orifice in the bottom. The orifice is first plugged, and the liquid to be tested is poured into the vessel. The

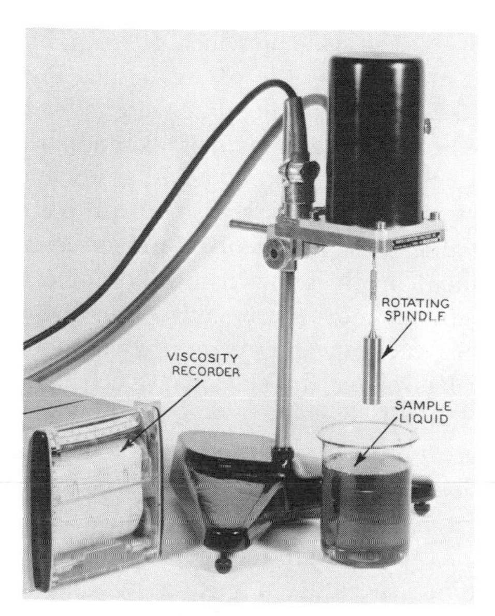

Fig. 13-21. Rotating spindle device for measuring viscosity. (Brookfield Engineering Laboratories, Inc.)

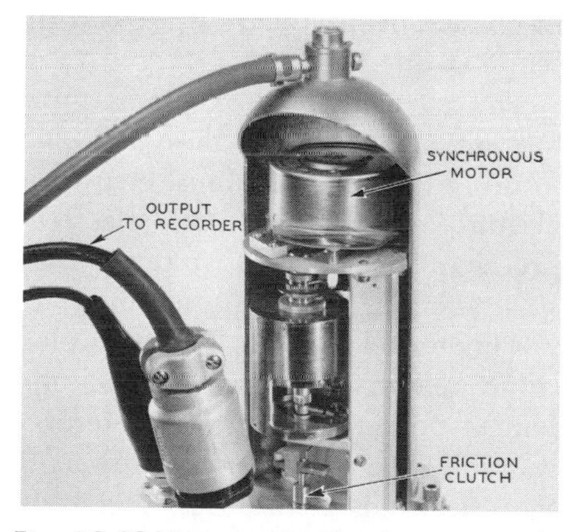

Fig. 13-22. Viscometran for in-process viscosity measurement. (Brookfield Engineering Laboratories, Inc.)

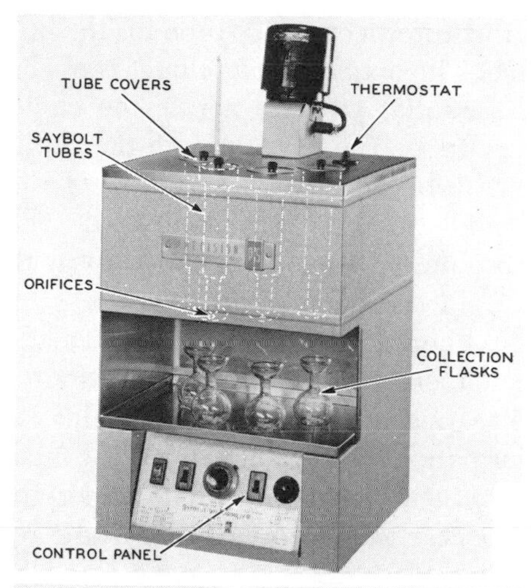

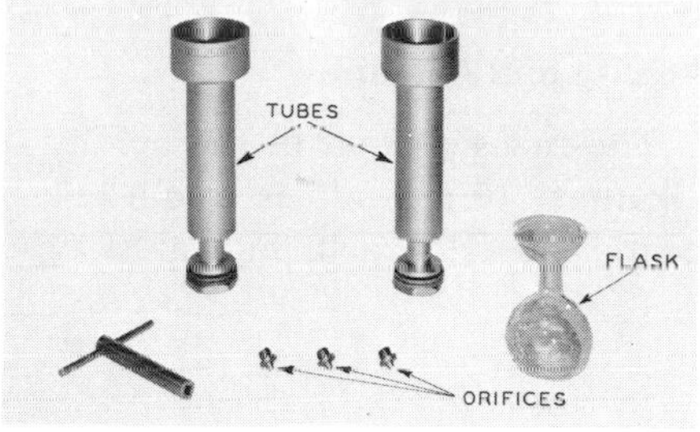

Fig. 13-23. Saybolt universal viscosimeter. (Precision Scientific Co.)

temperature of the liquid is regulated to the desired value. The plug is then quickly removed. At the same time, a stopwatch is started. When the liquid reaches the desired volume, the stopwatch is stopped. The units of viscosity measurement with this apparatus are Saybolt universal seconds.

Saybolt universal seconds = centistokes × 4.635.

For more viscous liquids the orifice is changed, and the resultant measurement is expressed as Saybolt furol seconds.

Saybolt furol seconds = centistokes × .470

An adaptation of the **Saybolt viscosimeter** for continuous measurement substitutes a capillary tube for the orifice. The actual measurement is of the differential pressure across the capillary tube. The flow through the capillary tube and the temperature of the liquid are accurately controlled. The absolute viscosity varies directly with the differential pressure.

Although the measurement made can merely result in an indication of the apparent viscosity values, this information is sufficient to make the measurement worth while. The important aspect is the reproducibility of the results.

Acidity and Alkalinity

The measurement of the acidity or alkalinity of a liquid is frequently of prime importance in industrial processes. The scale on which this variable is measured is the pH scale. See Fig. 13-24. On this scale, pure water, which is neutral (neither acid or alkaline), has a value of 7. A strong acid has a value of +1 and a strong base (alkali) has a value of 14. The theory of pH measurement involves the dissociation of certain chemical compounds, often in the presence of water. As these compounds (electrolytes) break down, electrically charged particles are formed. Some are positively charged, and some are negatively charged. In the case of water, hydrogen ions (H+) and hydroxyl ions (OH−) are formed. In acid solutions, there are a greater number of hydrogen ions. In alkaline solutions, there are a greater number of hydroxyl ions. The pH scale is based on the concentration of hydrogen ions in a certain volume of solution. A more complete discussion of pH theory is beyond the scope of this book.

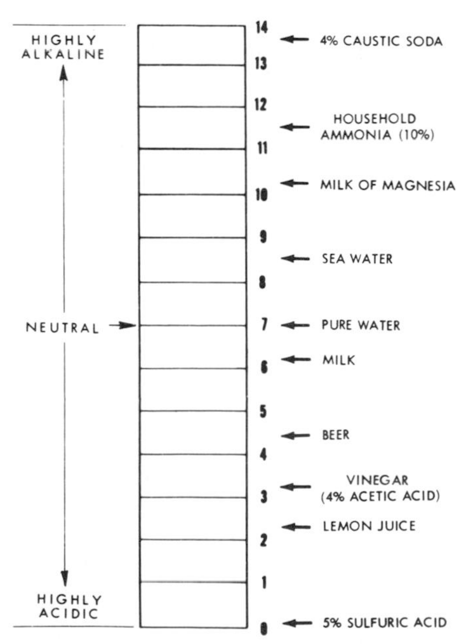

Fig. 13-24. pH scale showing the acidity or alkalinity of common liquids and substances.

The measurement of pH requires the use of specifically designed electrodes. Two electrodes must be used to obtain a measurement. One electrode produces a change in voltage (emf), since the pH of the solution in which it is immersed changes. The other electrode maintains a constant voltage (emf) when immersed in the reference solution. The most common pH-sensitive electrode is the glass electrode, shown in Fig. 13-25. The most common reference electrode is the calomel electrode. See Fig. 13-26.

Together these electrodes form an electrolytic cell whose output equals the sum of the voltage produced by the two electrodes. See Fig. 13-27. This voltage is applied as the input to a null balance millivolt potentiometer,

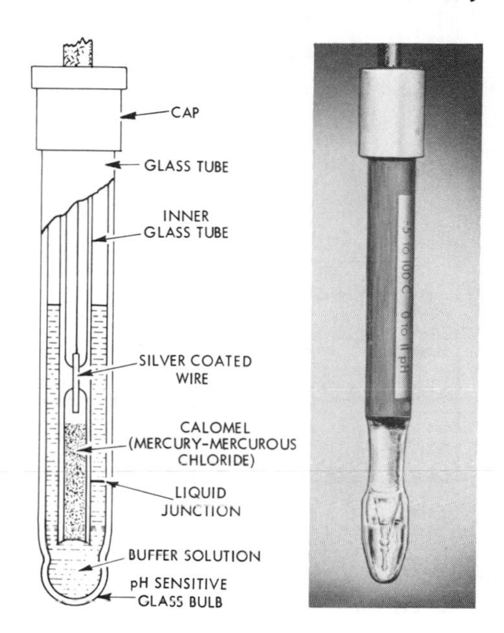

Fig. 13-25. The cross section at the left shows elements of the glass electrode depicted on the right. (Honeywell Process Control Div./Fort Washington, Pa.)

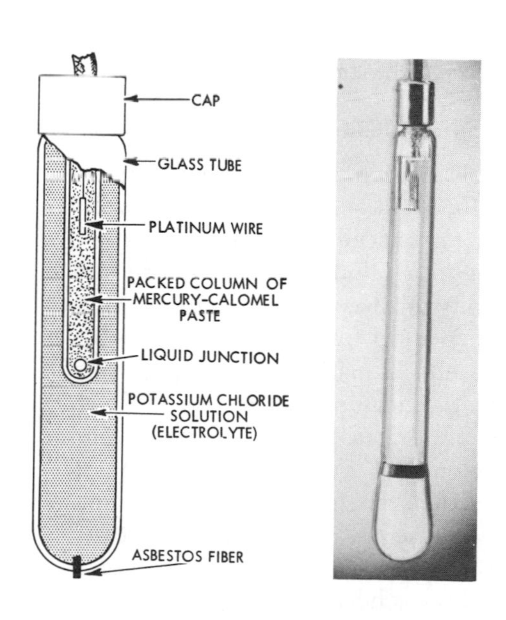

Fig. 13-26. The cross section at the left shows the elements of the reference electrode depicted on the right. (Honeywell Process Control Div./Fort Washington, Pa.)

similar to that used with a thermocouple. A temperature-compensating resistor, which is immersed in the solution, is frequently included in the circuit. Its resistance changes with the temperature of the solution, so that the

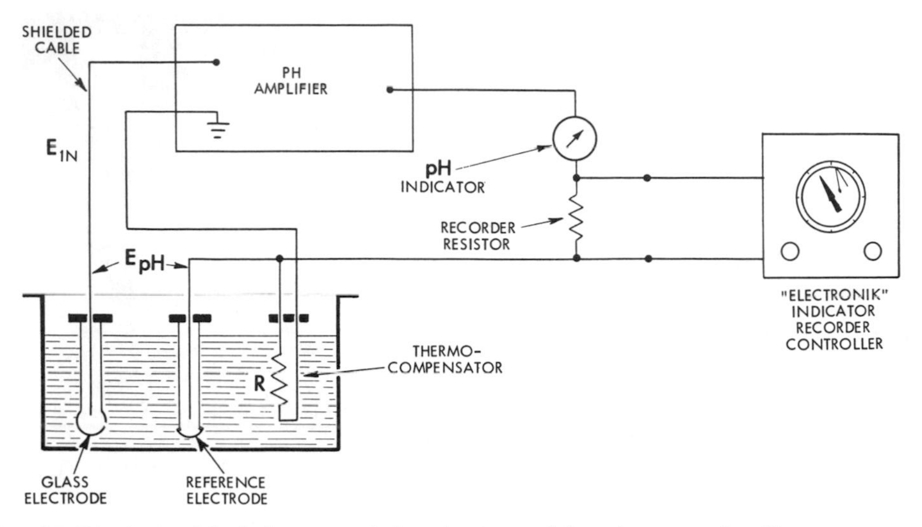

Fig. 13-27. A simplified diagram of the circuit used for electrometric pH measurement.

pH measurement is correct at the operating temperature.

Electrical Conductivity

pH measurement, as previously described, is concerned with the hydrogen ion concentration in a solution. All ions in a solution, however, affect its ability to pass an electric current. Measurement of this ability (conductivity) is useful in many industrial processes. Such a measurement can be made by immersing a pair of electrodes of known area a certain distance apart, and then measuring the resistance between them. Conductivity is expressed in mhos measured between two electrodes, each having an area of 1 square centimeter and placed 1 centimeter apart. See Fig. 13-28. A mho equals $\frac{1}{\text{Ohm}}$ and it represents the ability of one cubic centimeter of a substance to pass one ampere of current at one volt of potential. The conductivity electrodes used in industry often have areas greater or smaller than 1 square centimeter, and are placed closer or farther

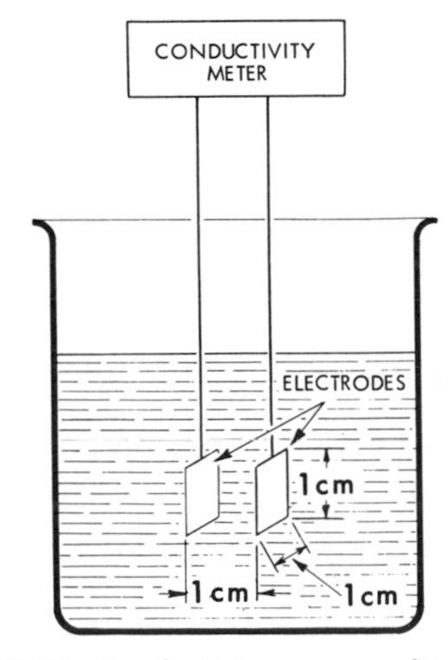

Fig. 13-28. Conductivity expressed in mhos.

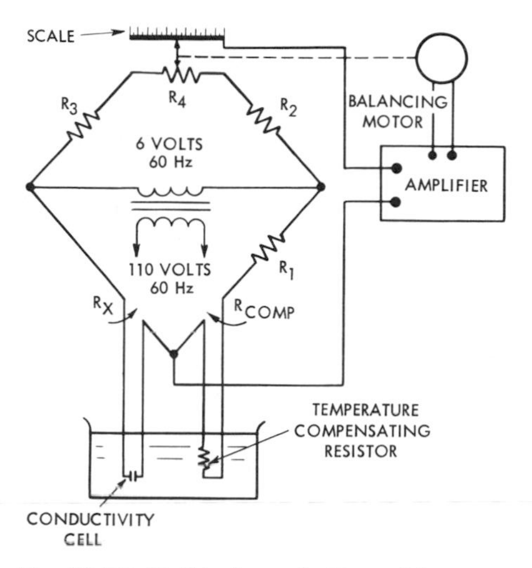

Fig. 13-29. Null balance bridge with a conductivity cell.

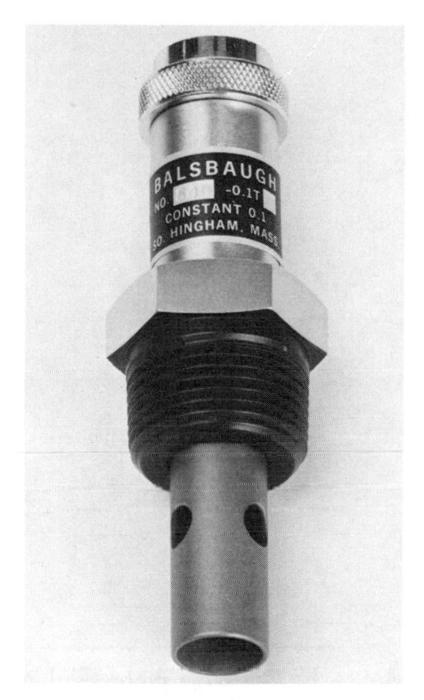

Fig. 13-30. Conductivity cell in probe form. (Balsbaugh Laboratories)

apart than 1 centimeter. The dimensional relationship between the actual electrodes and those of the standard conductivity cell is termed the *cell constant*. These constants range from .01 to 100.

The conductivity cell can be used with an ac null balance bridge, as shown in Fig. 13-29. As with the pH measuring system, the circuit uses a temperature compensating resistor to correct variations in the temperature of the solution.

Conductivity cells are available in probe form. See Fig. 13-30. They contain titanium-palladium or graphite electrodes with plastic insulation. A thermistor is enclosed in the center electrode for temperature compensation. The thermistor is part of a series-parallel network in the measuring circuit.

A conductivity measuring system without electrode is also available. See Fig. 13-31. In this system, two toroidally wound coils are encapsulated in close proximity to a flow-through probe immersed in the liquid solution. The coils surround the bore of the probe so that a loop of the solution couples them. An ac signal produced by an oscillator is applied to one coil which generates a current in the loop of the solution. The current is proportional to the conductivity of the solution. The current generates a current in the second coil which is the output of the probe. See Fig. 13-32. This output is converted by solid state electronic circuitry to either a standard current signal (4 to 20 mA) or voltage signal (0 to 10V). As in the electrode

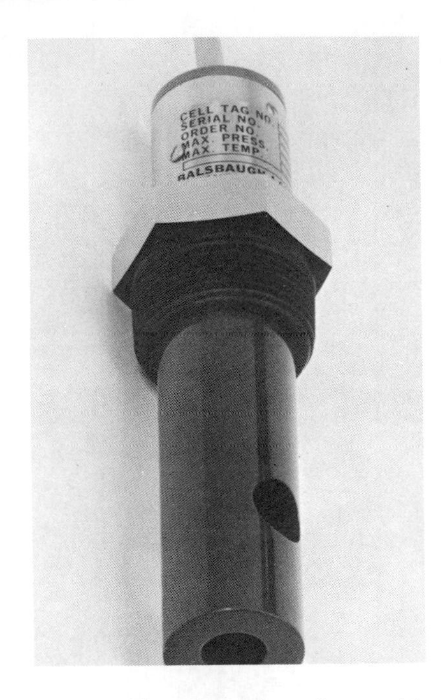

Fig. 13-31. Flow-through probe used in a conductivity measuring system. (Balsbaugh Laboratories)

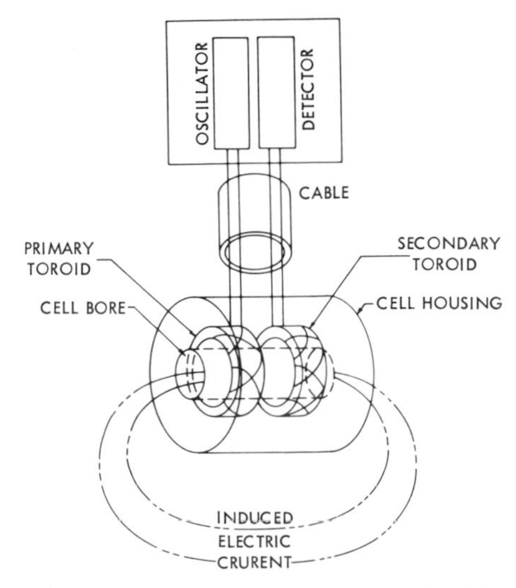

Fig. 13-32. Operation of the conductivity system without an electrode. (Balsbaugh Laboratories)

type, a thermistor enclosed in the probe provides temperature compensation.

Since the measurement is an indication of the total ions present in a solution, the scale of the instrument can be calibrated in percent concentration of electrolyte. This is frequently an important consideration in determining the purity of water, or the completeness of a chemical reaction.

Thermal Conductivity

Gases differ in their ability to conduct heat. The thermal conductivity of air at 32°F is established as 1.0. The relative thermal conductivities of other gases are:

CO_2	.585
CO	.958
Helium	6.08
Hydrogen	7.35
Nitrogen	1.015
Oxygen	1.007

The measurement of the thermal conductivity of a gas mixture provides knowledge about the proportions of its constituents. The most common method of making this measurement involves the use of a hot wire thermal conductivity gas-analysis cell. Such a cell, as shown in Fig. 13-33, consists of two chambers, each containing a wire filament. One chamber permits the sample gas to flow through it, while the other is sealed and contains a reference gas (such as air).

The temperature of the filament in the sampling chamber rises as the thermal conductivity decreases, because the heat of the filament is unable to pass out of the chamber through its walls. The preferred circuit used with

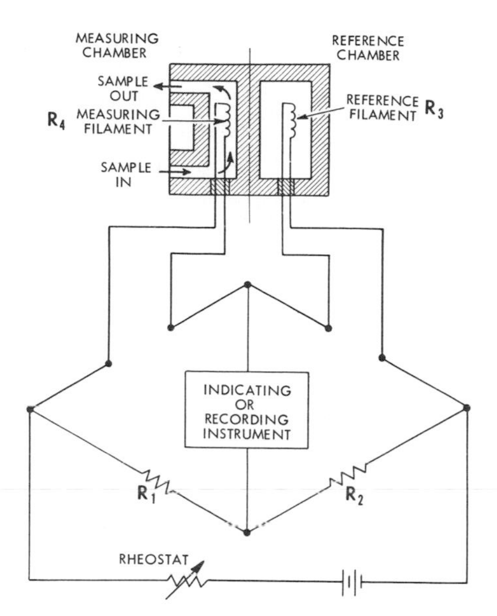

Fig. 13-33. Detail of a thermal conductivity bridge and cell.

thermal conductivity cells is the Wheatstone bridge, using two cells (four chambers), as shown in Fig. 13-34.

Although it is a simple matter to determine the proportion of two gases in a mixture, it is considerably more difficult to analyze a complex gas. Special techniques have been developed to accomplish this. The sample is passed through one chamber, then through an apparatus which removes one constituent of the sample, and finally through another chamber. See Fig. 13-35. The difference in the thermal conductivity in each of the chambers is, therefore, due to the constituent removed. Hence, the percentage of this constituent can be determined.

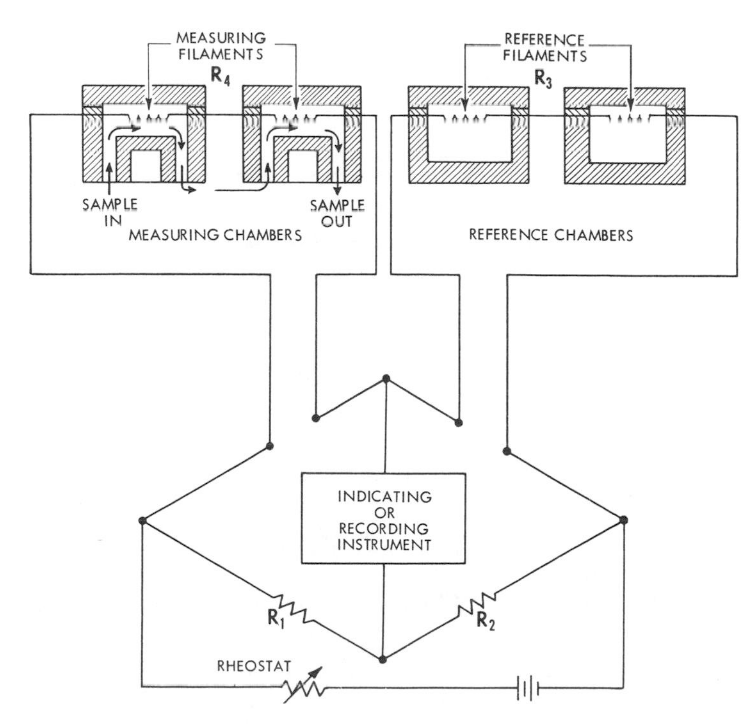

Fig. 13-34. Two cell (four chamber) thermal conductivity bridge.

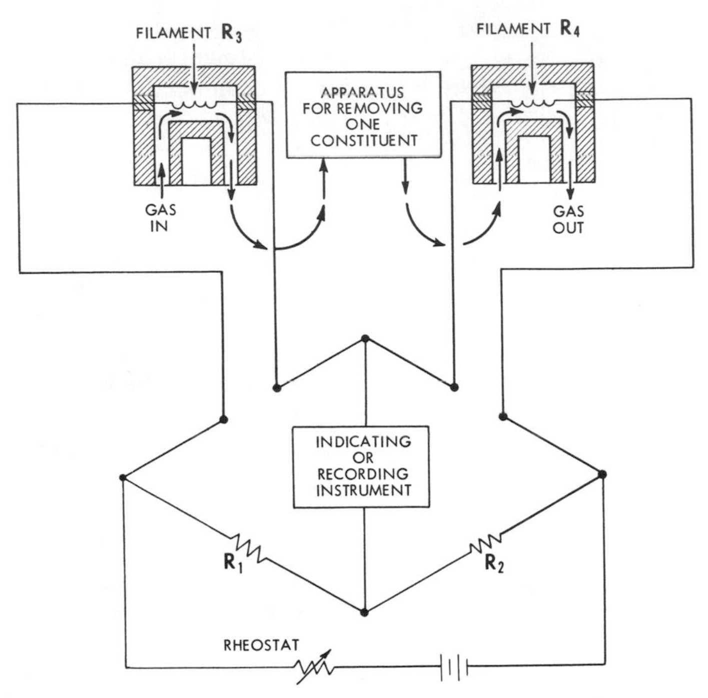

Fig. 13-35. Thermal conductivity bridge for double gas analysis. The gas moves through the system twice.

Thermal conductivity gas analyzers can be used to determine hydrogen in air, helium in air, hydrochloric acid vapor in air and ammonia in air.

Combustibility

Combustion is a process to which analysis instrumentation has been applied for many years. The amount of heat developed when a fuel is burned depends upon the completeness of combustion. The products of complete combustion, in addition to heat, are carbon dioxide and water vapor. When sulfur is present, sulfur dioxide is a combustion product. When carbon monoxide is given off, it is an indication that combustion is not complete. An instrument that measures the amount of carbon monoxide in fuel gas is able to be used to determine combustion efficiency. A thermal conductivity gas analyzer is suitable for this application. See Fig. 13-36.

The flue gas is passed through one thermal conductivity chamber and then into an apparatus that converts the carbon monoxide into carbon dioxide. The resultant gas is then passed through a second thermal conductivity chamber. The difference in the output of the two chambers is due to the carbon monoxide present in the flue gas. Air or carbon dioxide can be used in the reference chambers of the two cells.

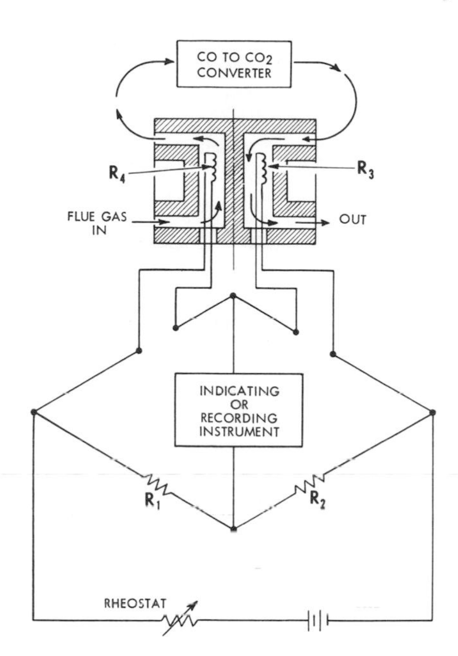

Fig. 13-36. Thermal conductivity bridge for flue gas analysis.

The combustion process presents another opportunity for the application of analyzers—the continuous analysis of fuel gas to determine its combustibility. A thermal conductivity analyzer is used for this measurement to provide information about the percentage of combustible constituent in the fuel gas.

Chromatography

Chromatography is the name given to a method of analysis that permits the continuous measurement of the amounts of each constituent in a complex vapor or gas mixture. There can be many constituents in the mixture. The method involves combining the sample gas with a carrier gas and passing the combination of gases through a column, which is made of metal tubing and filled with an adsorbent such as activated alumina, silica gel, or activated carbon. See Fig. 13-37.

The effect of moving the gases through the column is the separation of the constituents of the sample gas. Each constituent travels through the column at a different rate, because each is retained for a different period of time by the column adsorbent. The carrier gas, which forces the sample gas through the column, emerges from the column continuously, so that the constituents of the sample gas actually leave the column in combination with the carrier gas. The most common carrier gases are helium, nitrogen, air, or hydrogen. A typical sample gas might contain such constituents as ethane, propane, acetylene, butane, pentane, among others.

Fig. 13-38 shows a composition control system with a process gas chromatograph. The control section is at the bottom. A high speed strip chart recorder is also shown.

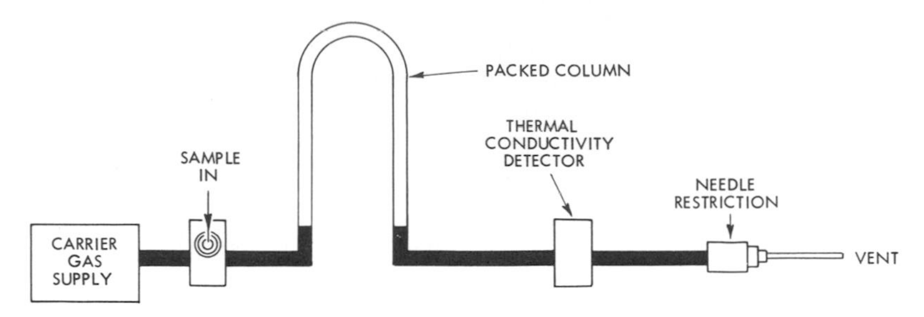

Fig. 13-37. Gas analysis by packed column method.

Fig. 13-38. Composition control system with a process gas chromatograph (Honeywell Process Control Div./Fort Washington, Pa.)

Thermal Conductivity Cell. Thermal conductivity cells are generally used as the detecting elements. During operation, a metered amount of sample gas is injected into the carrier gas stream, the combination passing to the measuring chamber of the cell. At the same time, the carrier gas is passed separately through the reference chamber. The cell, therefore, measures the difference in the thermal conductivity of the carrier gas and of the sample gas plus each constituent as it leaves the column. When the last constituent has emerged from the column, a new sample is injected into the carrier gas stream, and the analysis is repeated. The output of the thermal conductivity cell is applied to a bridge circuit. The output of the bridge circuit is amplified and becomes the input to a strip chart recording instrument. The record made by the instrument is a series of peaks and valleys. By comparing the height of each peak with that produced as the result of the passage of a calibrated sample, the percentage by volume of each constituent can be determined.

This has been a very brief description of a complicated process of analysis. For best results, the following variables must be controlled and recorded:

Flow rate
Pressure
Temperature

In addition, the column material must be carefully selected, and the sample injection apparatus must be capable of excellent reproducibility.

Words to Know

density
specific gravity
hydrometer
load cell
viscosity
acidity
alkalinity
mho
electrical conductivity
thermal conductivity
combustibility
chromatography

Review Questions

1. Define *density* and *specific gravity*.
2. List the four methods used for measuring liquid density or specific gravity.
3. How is a U-tube manometer used for measuring liquid specific gravity?
4. What are the units used to measure viscosity?
5. List the three devices for measuring viscosity in the laboratory and in industry.
6. What is means by the *pH of a solution?*
7. Why is a reference electrode necessary in the measurement of pH?
8. What is the relationship between electrical conductivity and resistance?
9. What is the operating principle of the thermal conductivity cell?
10. How is a thermal conductivity cell used to measure combustion efficiency?
11. What is the function of the column used in chromatography?
12. How is the actual measurement of each ingredient made in a chromatograph?

Chapter

14

Control

In order to achieve proper automatic control of a process, it is necessary to know the characteristics of the process material itself, as well as the devices used for process control. Frequently, there are more than one variable to be controlled, and several methods available to control them. A given process might require the control of temperature, pressure, density, and level. Also, regulating one variable can affect other variables. For example, regulating temperature in a process might affect the pressure and the density. Regulating the flow rate of material to a process also affects the level of material in storage vessels. It is important, therefore, to select a controlling device which most accurately affects the variable to be controlled.

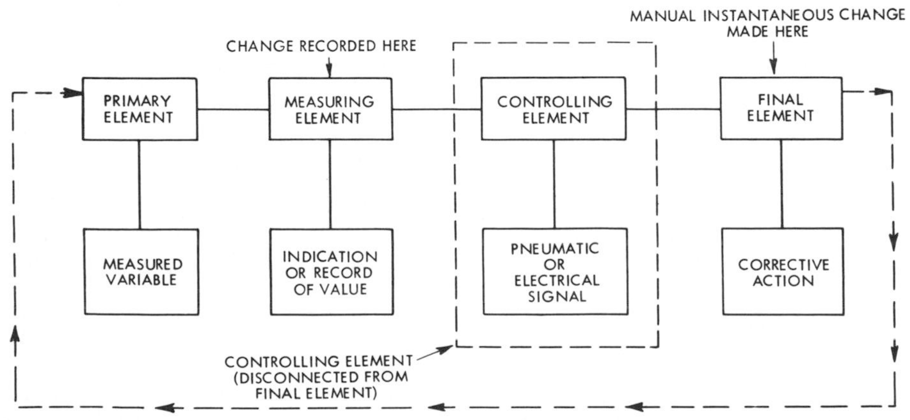

Fig. 14-1. The step function response method of studying the effects of controller changes on the measured variable.

Once the proper controller is selected, it is necessary to study the effects of controller changes on the variable. The simplest method of studying this effect is called the *step function response method.* See Fig. 14-1. Only the equipment to be applied to the operating process is required. A rapid stepwise change of the known amount in the control means is made. The effect of the change on the measured variable is then recorded graphically. The controller action is eliminated during this operation.

In the step function response study, the final element position is manually changed a small amount by regulating the air pressure to the final element. The resultant effect on the measured variable, which is sensed by the primary element, is recorded by the measuring element. This record is called the *process reaction curve.* Fig. 14-2 shows typical process reaction curves resulting from a step change.

Curve *A,* shown in Fig. 14-2, illustrates a simple process in which the measured variable begins to change as soon as the step change in the final element position is made. This means that there is little or no resistance to an energy change in the process. At the start, the rate of change is very rapid. As the process continues, however, the rate of change slows down. This means that the process is storing up some of the energy change. This characteristic of a process is called *capacity.* The rate of change is termed the *process reaction rate.*

Curve B, shown in Fig. 14-2, illustrates a more complicated process in which there is resistance, more than one capacity, and another process characteristic called *dead time.* Notice that the value of the measured variable at

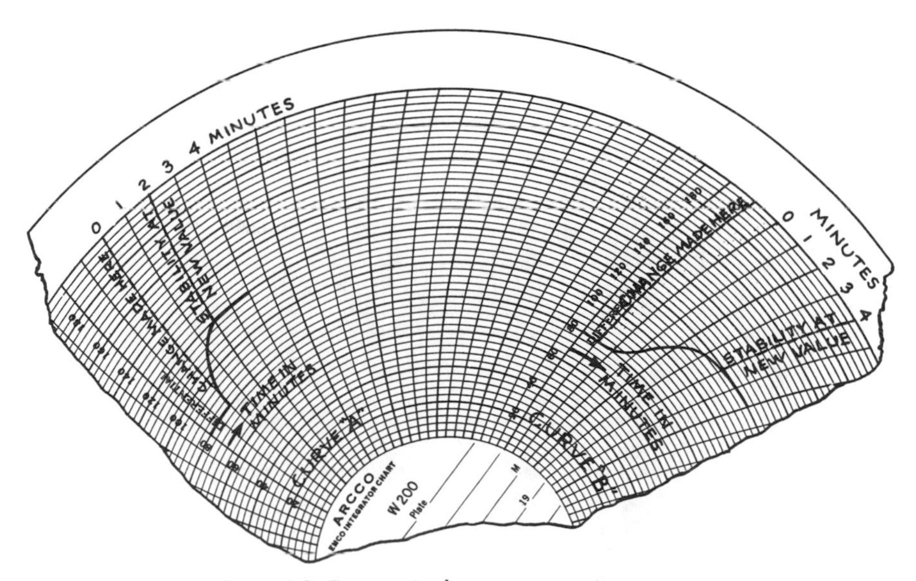

Fig. 14-2. Two typical process reaction curves.

the start, and for a short time thereafter, does not change. This is due to the effect of dead time. When the variable does begin to change, it changes slowly. Then it increases speed until it approaches the final value. At this point it slows down again. This change in reaction rate is caused by a combination of capacity and resistance and particularly by the passage of energy from one capacity through a resistance into another capacity. This characteristic is called *transfer lag*.

A process which produces a reaction curve similar to curve A may be controlled by the simplest form of controller.

A process which produces a reaction curve similar to curve *B* presents quite a different problem. Selecting a controller for this process demands an understanding of the control actions available and of their ability to handle the process characteristics.

On-Off Control

On-Off or two-position control, when applied to a process, produces a control pattern, as shown in Fig. 14-3.

The following characteristics should be noted. Fig. 14-4 shows that a faster reaction rate decreases the period. A greater differential between the *On* and *Off* positions increases the period and the amplitude. See Fig. 14-5. Dead time causes the value of the measured variable to go beyond the limits set by the differential, since dead time means a delay in the corrective action of the controller. A longer dead time increases the amplitude and the period. See Fig. 14-6. Transfer lag effect differs from dead time effect in that it is not a delay in response but, rather, a slowing down of the response. Transfer lag in a two position control system produces a pattern as shown in Fig. 14-7. With transfer lag, the measured variable exceeds

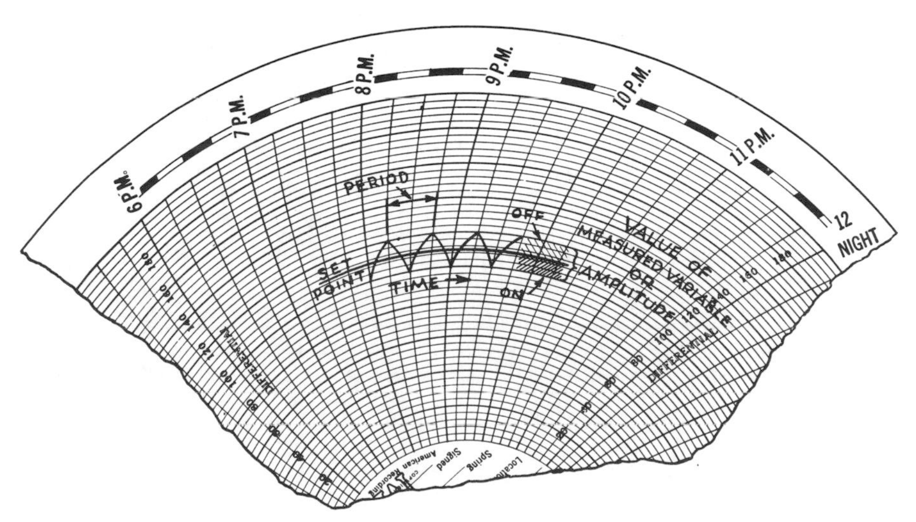

Fig. 14-3. Control pattern produced by an On-Off, or two position, control.

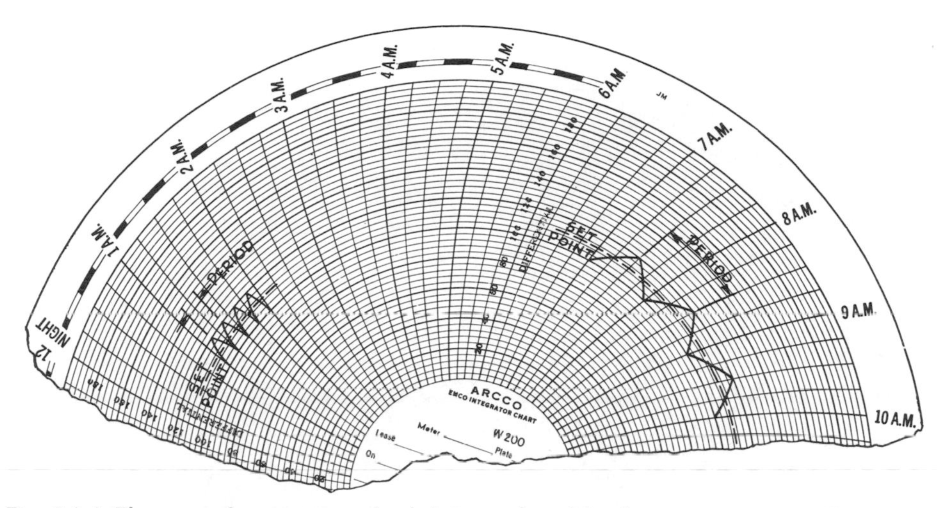

Fig. 14-4. The control pattern on the left is produced by fast reaction rate. The pattern on the right is produced by slow reaction rate.

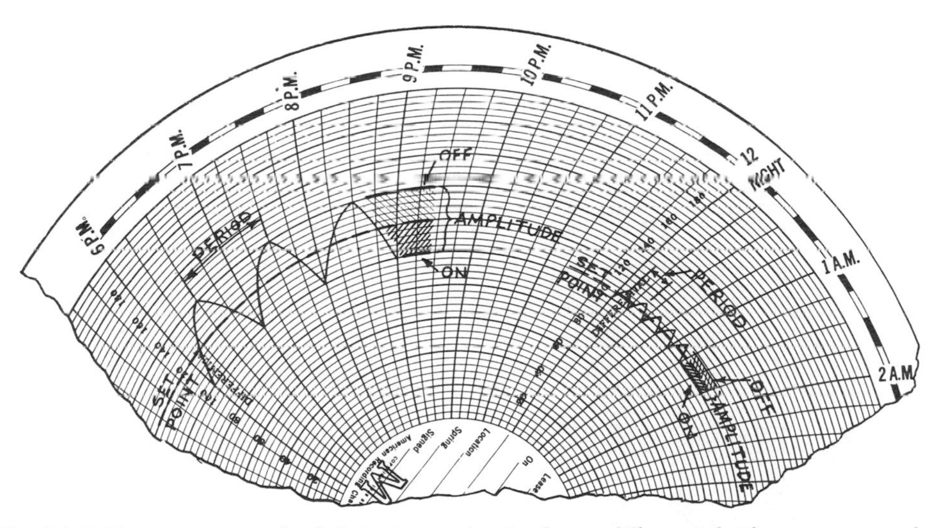

Fig. 14-5. The pattern on the left is the result of a large differential. The pattern on the right is the result of a small differential.

the controller differential values, as it does when there is dead time. Instead of the sharp peaks which appear when there is no transfer lag, these are rounded off. With transfer lag, all the changes are more gradual.

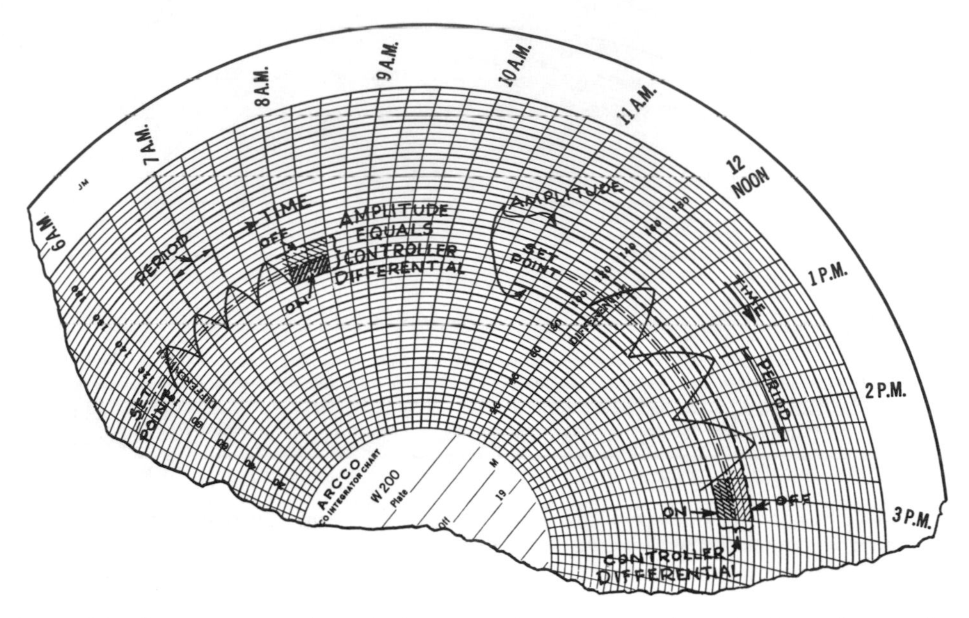

Fig. 14-6. The pattern on the left shows no dead time, and amplitude and controller differential are equal. The pattern on the right shows no dead time, and amplitude and period are increased.

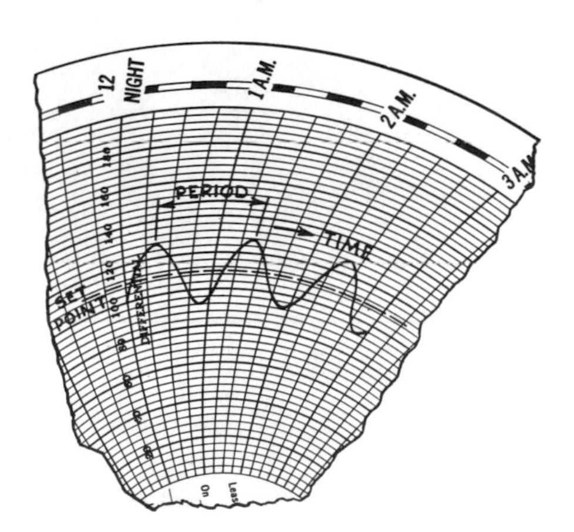

Fig. 14-7. Pattern showing presence of transfer lag.

A two position control system performs best when the reaction rate is slow and there is little or no dead time or transfer lag.

Proportional Control

Proportional control, when applied to a process that is stable, produces a control pattern like the one shown

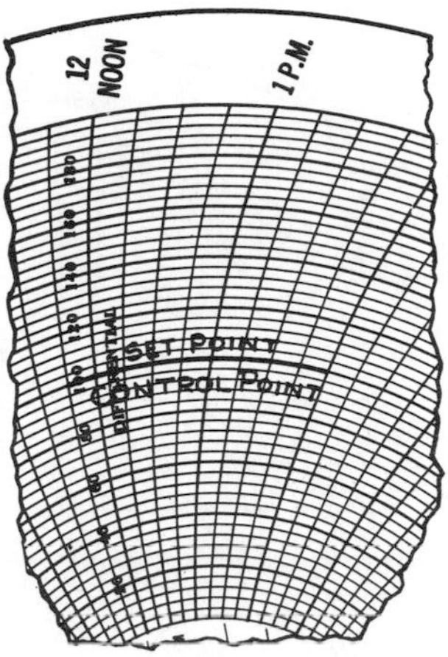

Fig. 14-8. Pattern produced by proportional control when process is stable.

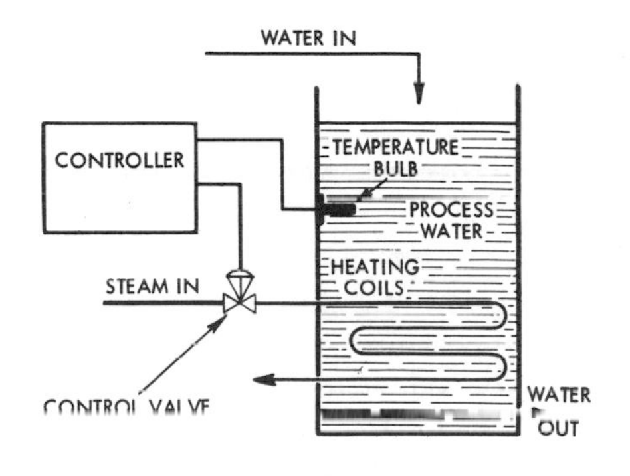

Fig. 14-9. A simple temperature control system.

in Fig. 14-8. Unfortunately, however, an industrial process is rarely stable. Fig. 14-9 shows a typical process in which steam is used to heat water to a particular temperature. The control system should maintain the temperature of the water at set point.

Ideally, the water enters the process at a fairly constant temperature and enters and exits at a constant rate. The steam pressure and, therefore, the temperature also remain constant. If any of these conditions change, the control system must make the necessary corrections to maintain the exit water at constant temperature. The changes are called *load changes*. Fig. 14-10 graphically shows the desired result response of the control system to a change in input conditions, load change.

If the reaction curve produced by a step function response study of the process shows that dead time and transfer lag are present, proportional control alone is not satisfactory.

With proportional action, there is a fixed relationship between the position of the final element and the amount of deviation of the measured variable from the set point. In the temperature

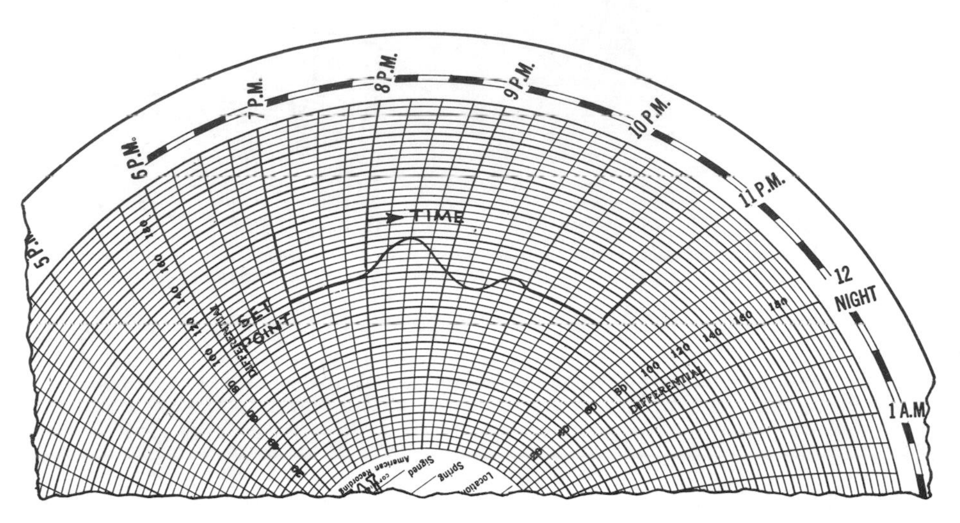

Fig. 14-10 Ideal response curve produced by a load change.

control process shown above, this means that the steam valve has a particular position for each temperature value above or below the set point of the temperature controller. The system is thus dependent upon the valve passing the necessary amount of steam at each position to bring the temperature back to the control point after a departure. If the valve position called for by the controller does not allow sufficient steam to pass, the system will not be able to restore control at set point. This is precisely what happens when a load change is large or lasts for a long time. The system will settle down after such a load change, but not at the set point. Fig. 14-11 graphically shows the departure from set point that results. It is called *offset*.

A fast reaction rate of the process gives poor results. The process settles down fast but the offset becomes too large. See Fig. 14-12.

If the reaction curve, shown in Fig. 14-13, indicates that there is dead time or transfer lag, the results are worse. In addition to a large offset, the time required for stability is considerably longer.

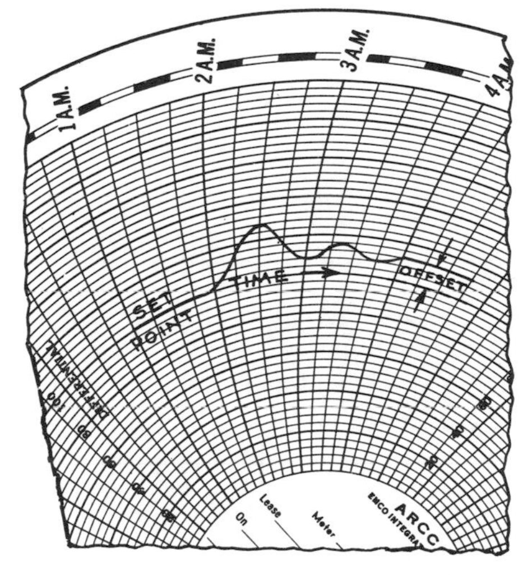

Fig. 14-11. Pattern illustrating effect of offset.

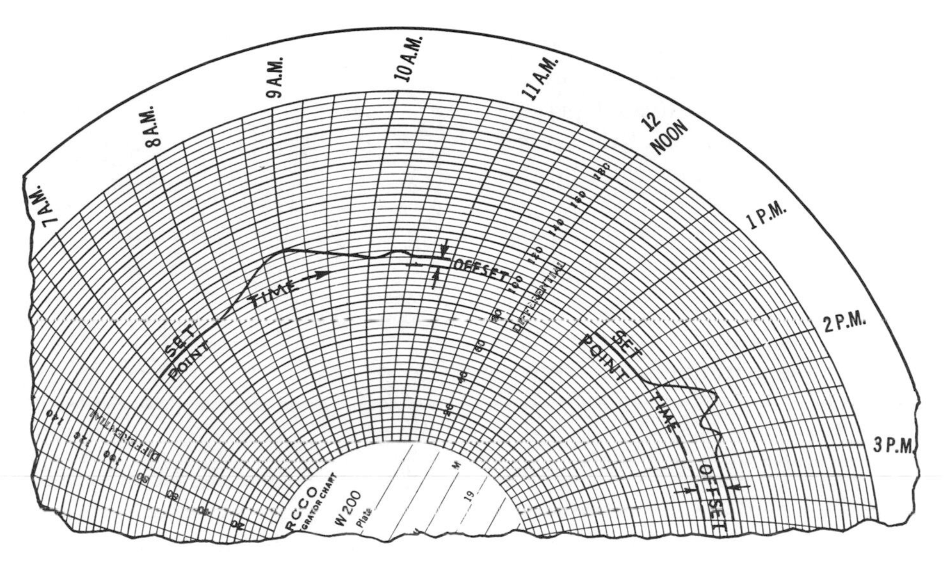

Fig. 14-12. The pattern on the left shows the result of a slow reaction rate. The pattern on the right shows the result of a fast reaction rate. The fast reaction rate increases the offset.

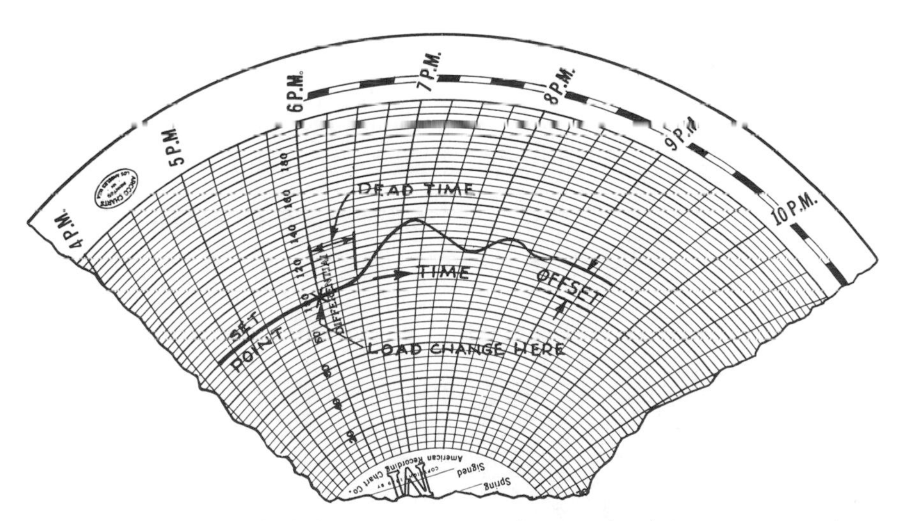

Fig. 14-13. A reaction curve indicating dead time.

We can conclude, therefore, that proportional control is not satisfactory under various circumstances. The process reaction rate is fast. There is considerable dead time or transfer lag. An examination of the process reaction curve provides this information. The load changes which occur are rapid, large, or for extended periods of time. An examination of the record of the control system provides this information.

With proportional control systems, three modifications are available to improve the system performance: proportional band adjustment, adjustable reset response, and adjustable derivative action.

Proportional Band Adjustment

Proportional band adjustment is a sensitivity, or gain, adjustment. In proportional control, there is a specific position of the final element for each unit of departure of the measured variable from set point. With a narrow proportional band, the final element moves to its limit when the measured variable only departs a small amount from set point. The narrower the proportional band, the closer the result is

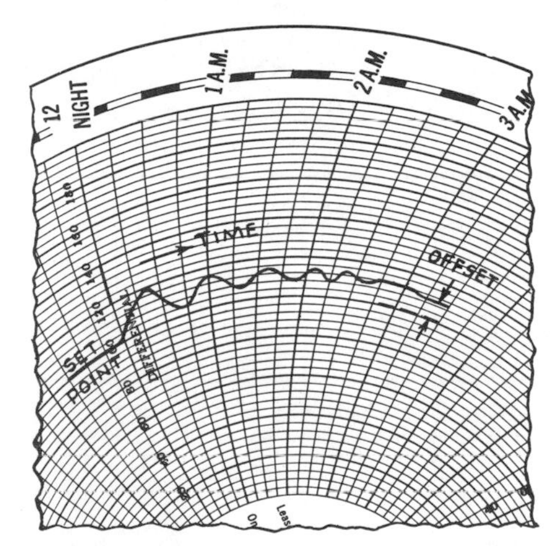

Fig. 14-14. With a narrow proportional band, stability time is long and offset is small.

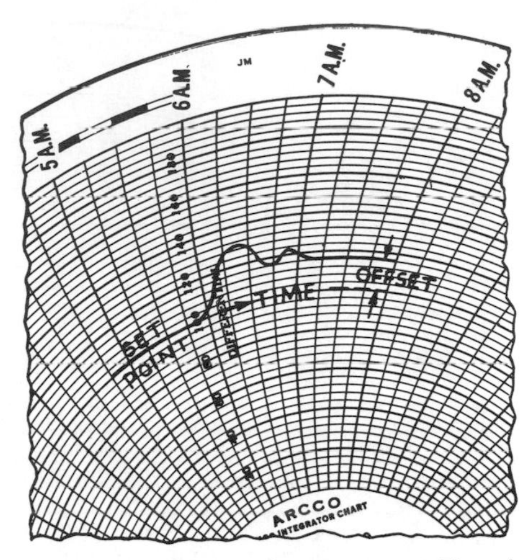

Fig. 14-15. With a wide proportional band, stability time is long and offset is large.

to two position control. See Fig. 14-14. The proportional band adjustment affects the response of the system to a load change, as shown in Fig. 14-15.

When the reaction rate is fast, and the dead time or lag is great, the proportional band must be wide.

Adjustable Reset Response

Reset response, when added to proportional control, acts to eliminate the offset produced by a load change. This happens because reset response continues to change the signal to the final element as long as the measured variable is not at set point. In a proportional plus reset control, the rate of motion of the final element changes with the amount of deviation from set point.

In the temperature control process above, change in steam pressure results in less heat being supplied to the process for a particular valve opening. This

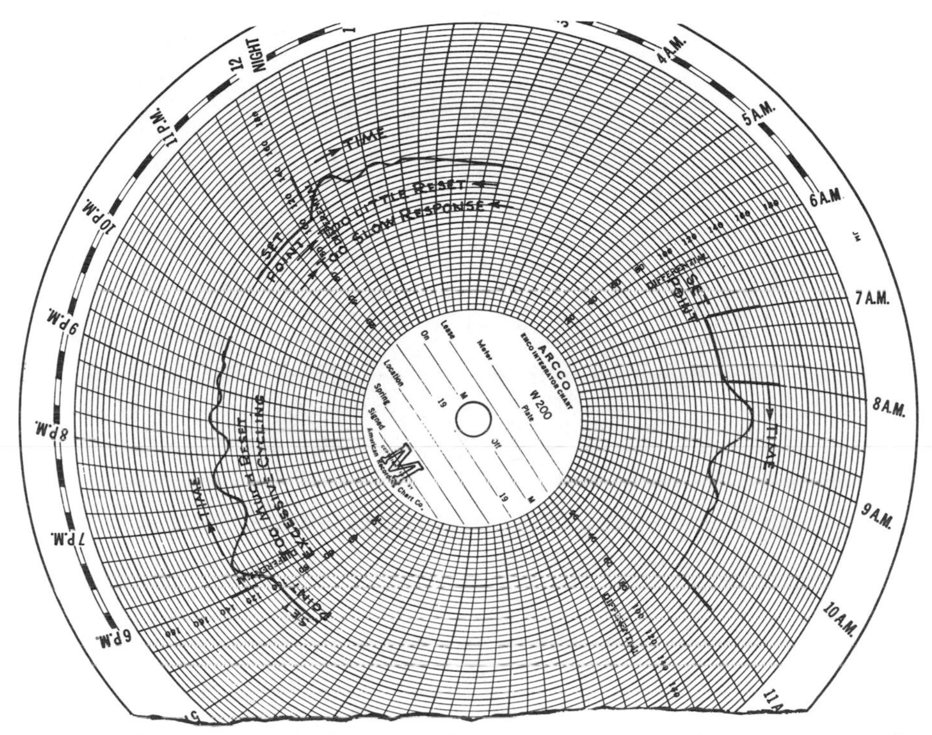

Fig. 14-16. The pattern on the left shows that too much reset results in excessive cycling. The pattern in the center shows insufficient reset results in an excessively slow response. The pattern on the right shows the correct reset results in good response.

is a load change. With proportional control alone, the temperature of the exit water is altered as a result of this change. The valve position signaled for by the controller is not able to supply the amount of heat required to maintain set point. The addition of reset response allows for the change of the valve position. The amount of reset response is adjustable. The adjustment affects the length of time necessary to return to set point. See Fig. 14-16.

Notice that the offset is eliminated in all cases but that the amount of reset determines the length of time required to do this. It is possible to have an excess of reset so that the controller overcorrects and, instead of becoming stable, the system becomes completely unstable as shown in Fig. 14-17. Proportional band and reset response are both adjustable. It should be noted that the adjustment of either one affects the other. The effect of reset adjustment has already been illustrated. The addition of reset response to a proportional controller requires a widening of the proportional band setting. See Fig. 14-18.

We can conclude from this discussion that when properly adjusted proportional plus reset response will over-

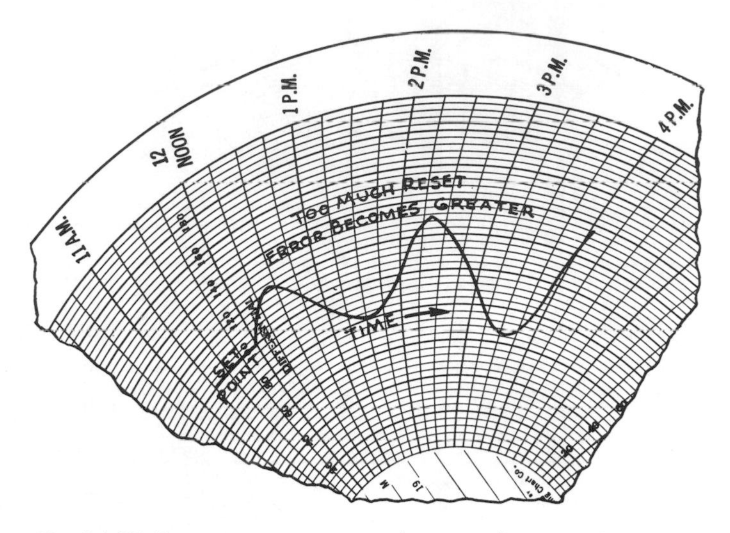

Fig. 14-17. Excessive reset causes the control error to increase.

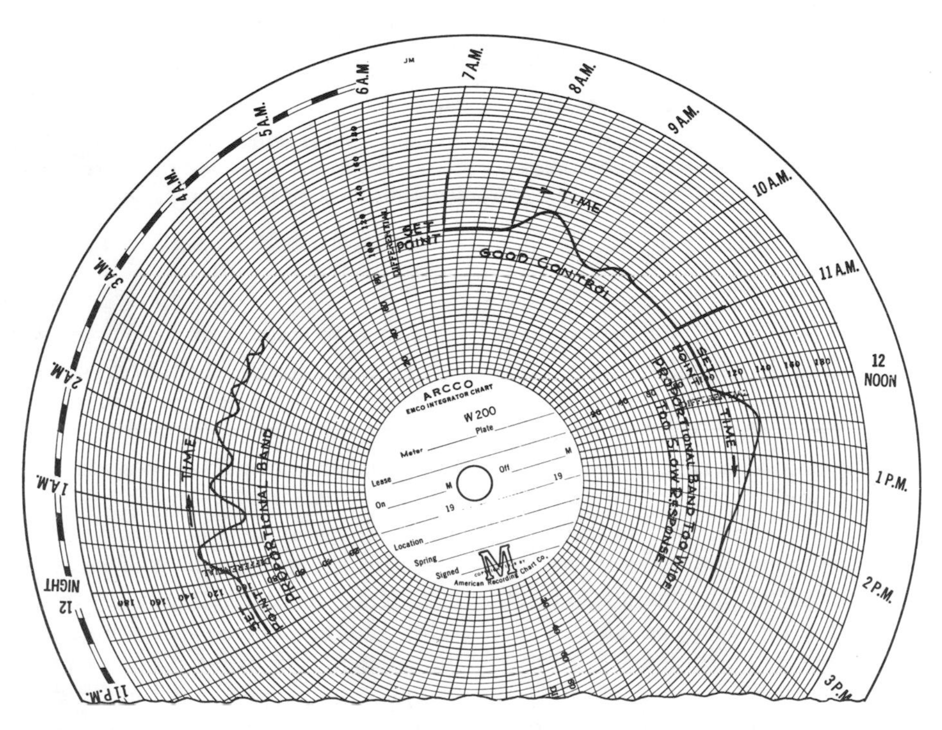

Fig. 14-18. The pattern on the left indicates that the proportional band is too narrow. The pattern in the center indicates that the proportional band is providing good control. The pattern on the right indicates the proportional band is too wide, which results in slow reset response.

come most of the limitations of proportional control alone. Excessive dead time or lag and large rapid load changes are two conditions which cannot be completely satisfied by proportional plus reset control.

Adjustable Derivative Action

Derivative action, when added to a proportional plus reset controller, acts to overcome the disturbance caused by dead time or lag and large, rapid load changes. This is possible because derivative action makes its contribution to control in accordance with the rate, not the amount, of departure of the measured variable from set point.

Without derivative action, a proportional plus reset controller, used to control a process having considerable dead time and lag, would produce the result shown in Fig. 14-19, when subjected to a load change.

We have already seen that the addition of reset to a proportional controller requires a change in proportional band setting. The addition of derivative action to a proportional plus reset controller also requires a change in proportional band and reset settings. The proportional band should be made narrower and the reset action increased.

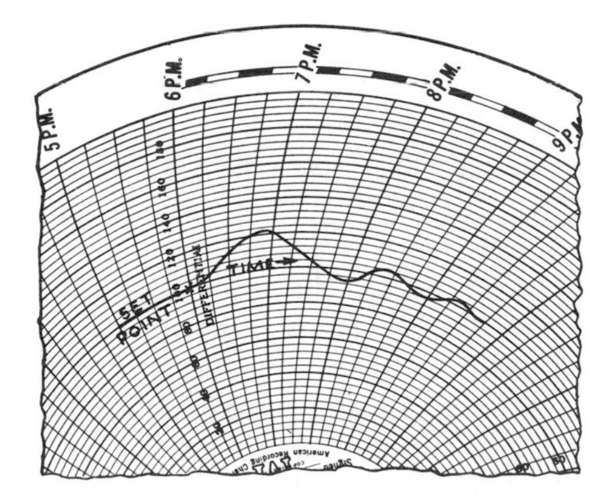

Fig. 14-19. This pattern indicates that the pressure has considerable dead time and lag.

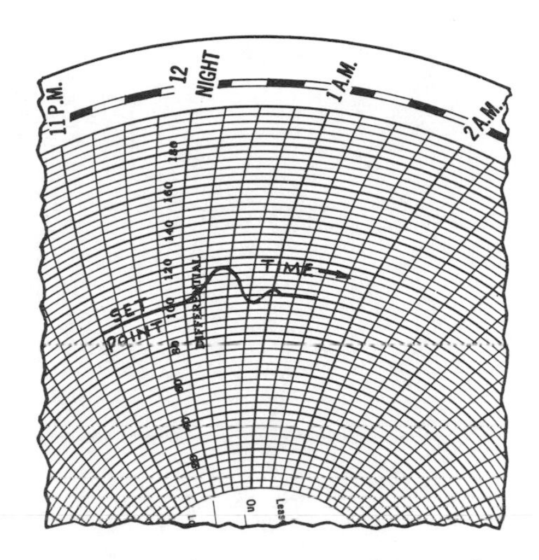

Fig. 14-20. This pattern indicates that proper corrections have been made to the proportional band and the reset response.

The result of adding derivative action to a proportional plus reset controller is shown in Fig. 14-20. This, of course, assumes that the proper corrections to the proportional band and reset response have been made. Both the amount of deviation and the stability time have been considerably reduced. Since derivative action itself is adjustable, the effect of such adjustment, shown in Fig. 14-21, should be noted.

When the proper adjustments have been made in a controller with proportional plus reset plus derivative action, the system can maintain control at set point. It maintains control even

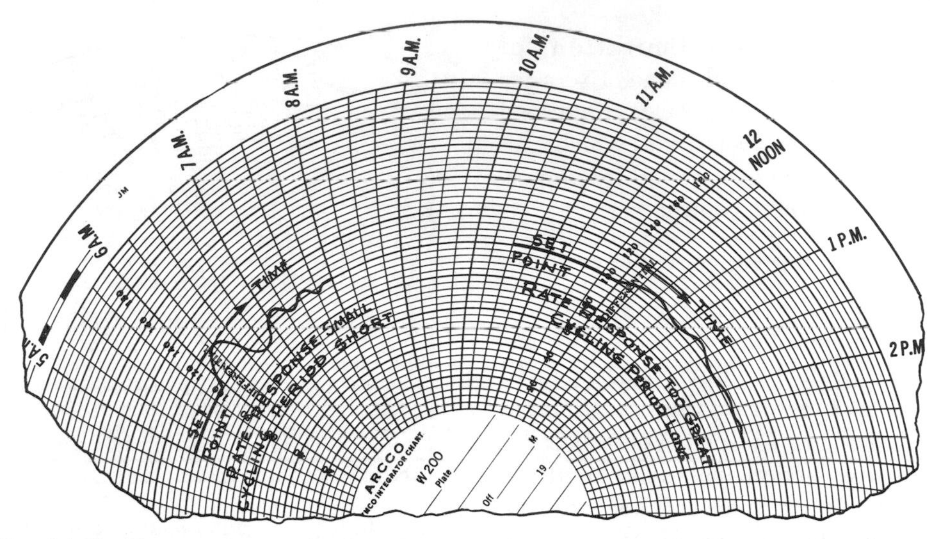

Fig. 14-21. The pattern on the left shows that an insufficient adjustment to derivative (rate) response results in a short cycling period. The pattern on the right shows that an excessive adjustment results in a long cycling period.

if the process reaction rate is fast or slow, dead time or transfer lag are present, and load changes are large or small, fast or slow.

When a controller is applied to a process, the order in which the adjustments are made is important. The proportional band adjustment should be made first. If reset or derivative action adjustments are available, they should be set at 0 during the adjustment of proportional band. The derivative action adjustment should be made before the reset adjustment. When adjusting derivative action, reset should be at 0. Reset response adjustment should be made last, since its only function is to eliminate the offset resulting from the adjustment of the other two responses.

The control responses described above are all available in pneumatic, electric, and hydraulic controllers.

Pneumatic Controllers

There are two types of pneumatic controllers available. One type, the flapper-nozzle type, has been previously described. (See Chapter 8.)

The other type, the force balance type, operates entirely on a balance of air pressures. The value of the measured variable is converted to an air pressure by a pneumatic transmitter. The set point is converted to an air pressure by a valve which is adjusted manually.

The schematic, shown in Fig. 14-22, of the force balance pneumatic control unit is shown in two sections, the unbalance detector section and the amplifier section.

The unbalance detector section consists of a stack of pressure chambers, including a set point chamber, meas-

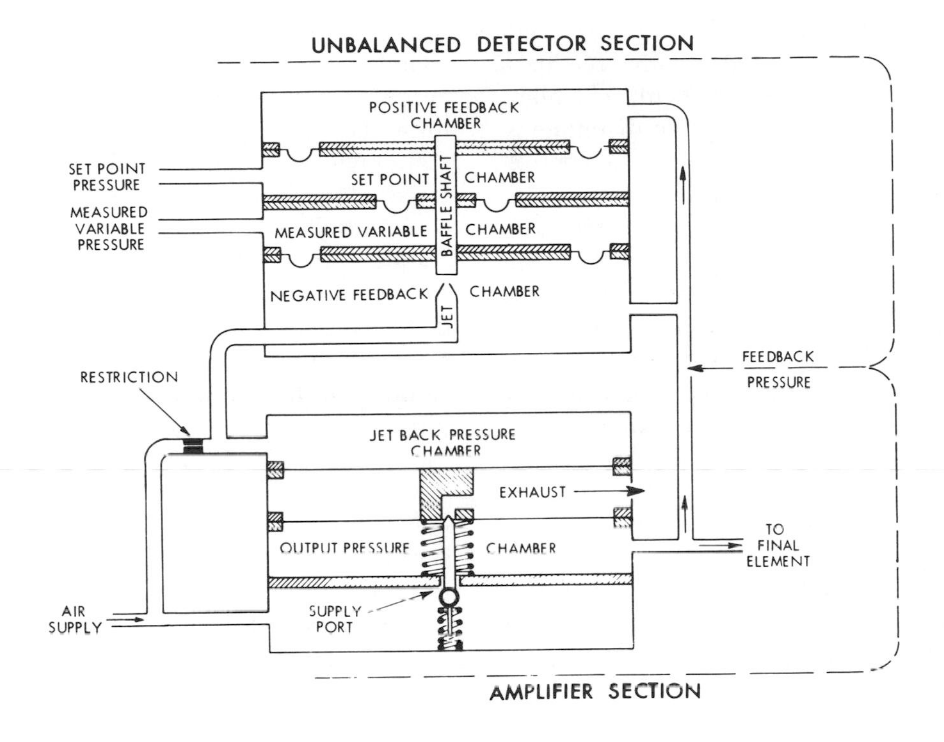

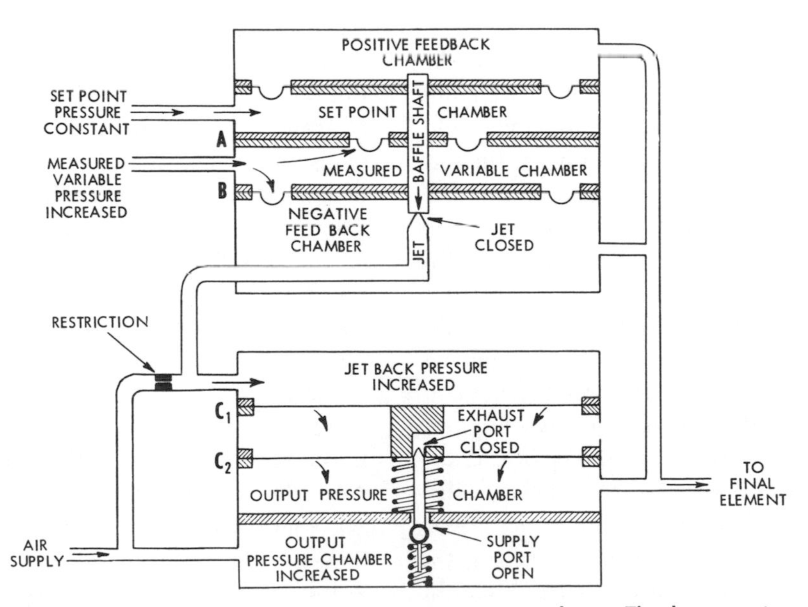

Fig. 14-22. Schematics of a force balance pneumatic control unit. The bottom view shows the measured variable pressure in the unit increases as the value of the measured variable changes.

ured variable chamber, positive feedback chamber, and negative feedback chamber. These pressure chambers are separated by flexible diaphragms. This section of the pneumatic controller compares various pairs of pressures. Set point pressure is compared with measured variable pressure, and positive feedback pressure is compared with negative feedback pressure. The controller produces a single output signal in the form of motion of the baffle shaft. Since the baffle shaft is attached to all the diaphragms, its movement results from the total unbalance of both pairs of pressures acting upon the diaphragms.

The amplifier section consists of a pair of pressure chambers, a jet back pressure chamber and an output pressure chamber. These pressure chambers are separated by flexible diaphragms. This section of the controller produces the pneumatic output to the final element.

The measured variable pressure increases slightly as a result of a change in the value of the measured variable. This increase pushes upward on diaphragm A and downward on diaphragm B. Since diaphragm B has twice the area of diaphragm A, the effect of the increase is to create a greater downward force. The baffle shaft, therefore, is forced downward. This causes a restriction in the passage of air from the jet, causing the pressure in the jet chamber of the amplifier section to increase. Double diaphragm C_1, C_2, therefore, moves downward against the spring, carrying the valve stem downward. This closes the exhaust port and opens the air supply port, allowing air to pass through the supply port to the output pressure chamber and to the final element. The output pressure continues to increase until the upward force acting upon diaphragm C_2 equals the downward force acting upon C_1. The force on C_2 is the result of both the air pressure and the spring. When the upward pressure on diaphragm C_2 equals the downward pressure on C_1, the valve stem is returned to its original position, shutting off the air supply to the output pressure chamber. The pressure to the final element reaches a new value.

With On-Off action, a change of only 2% in the measured variable pressure moves the baffle shaft sufficiently to produce a full range change in the output pressure to the final element.

Proportional Action. In the proportional action unit, shown in Fig. 14-23, the set point pressure enters the set point chamber. It then passes through a fixed restriction to the positive feedback chamber. An adjustable restriction is placed in the positive feedback line to the set point chamber. The pressure in the positive feedback chamber affects the amount of motion of the baffle shaft for each change in the measured variable pressure. This is the result of the above arrangement. For each unit the measured variable pressure changes, there is a new position of the baffle shaft and a new controller output pressure.

The two restrictions in this pneumatic circuit form an adjustable pressure divider circuit and are shown schematically in Fig. 14-24.

With this adjustable pressure divider circuit, it is possible to establish different ratios between the positive feedback chamber output pressure and

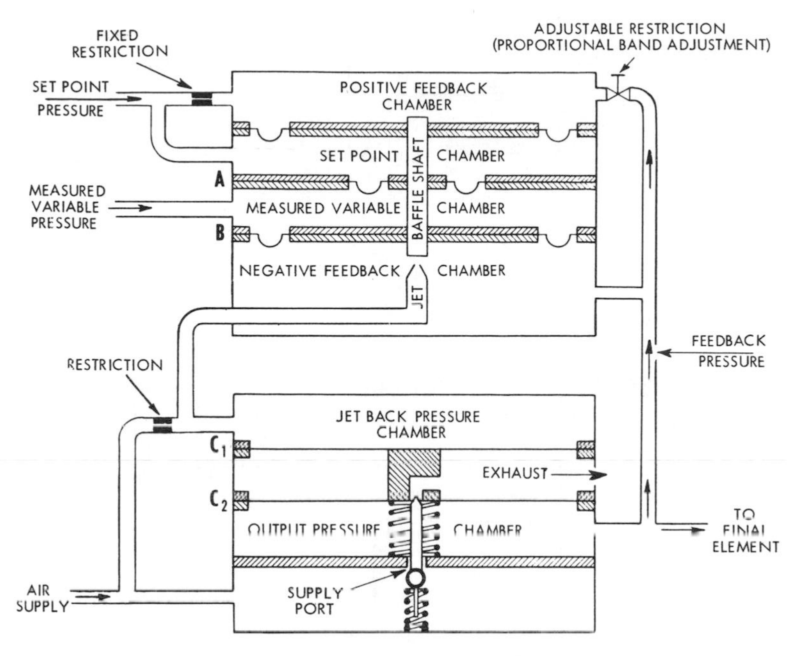

Fig. 14-23. Proportional action unit.

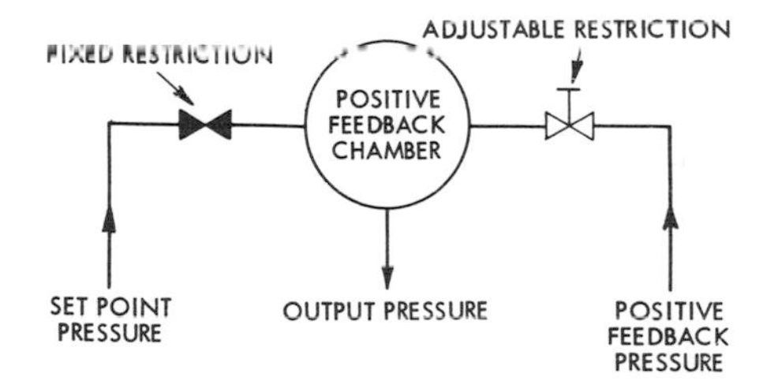

Fig. 14-24. Schematic of an adjustable pressure divider circuit.

the positive feedback pressure. This is done by adjusting the amount of the restriction in the positive feedback line. The set point pressure can be considered as a constant. Since its restriction is fixed, it contributes a constant reduced pressure to the positive feedback chamber. The adjustment of the restriction in the positive feedback line is a proportional band adjustment.

When the adjustable restriction is fully closed, as shown in Fig. 14-25, there is no positive feedback. A negative feedback pressure opposes any change in the measured variable pressure. This means that the measured variable must change considerably to to produce a full range change in the output pressure to the final element. Therefore, completely closing the needle valve provides the widest proportional band possible.

With the adjustable restriction fully open, there is very little difference between the positive and negative feedback pressures. This cancels out most of the feedback effect. See Fig.

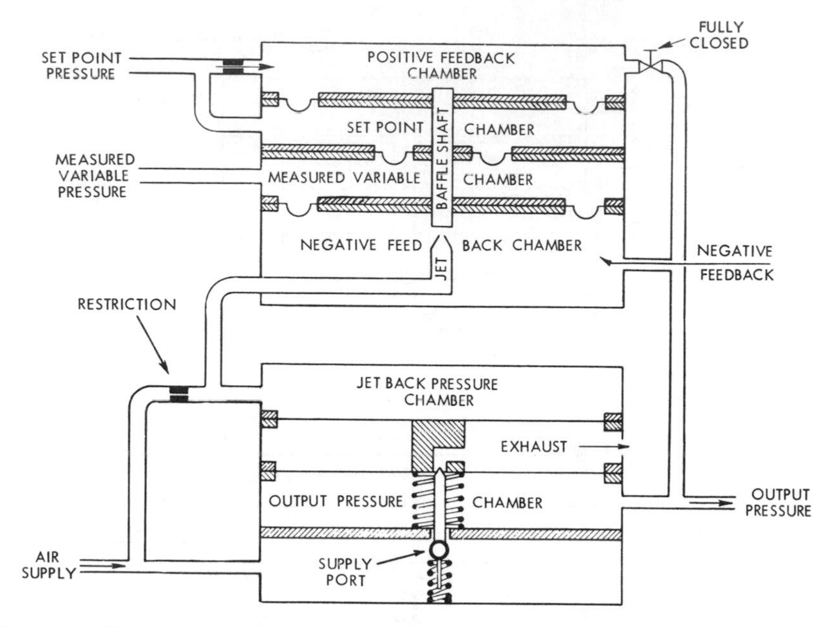

Fig. 14-25. Fully closed restriction (proportional band adjustment) in a proportional action controller.

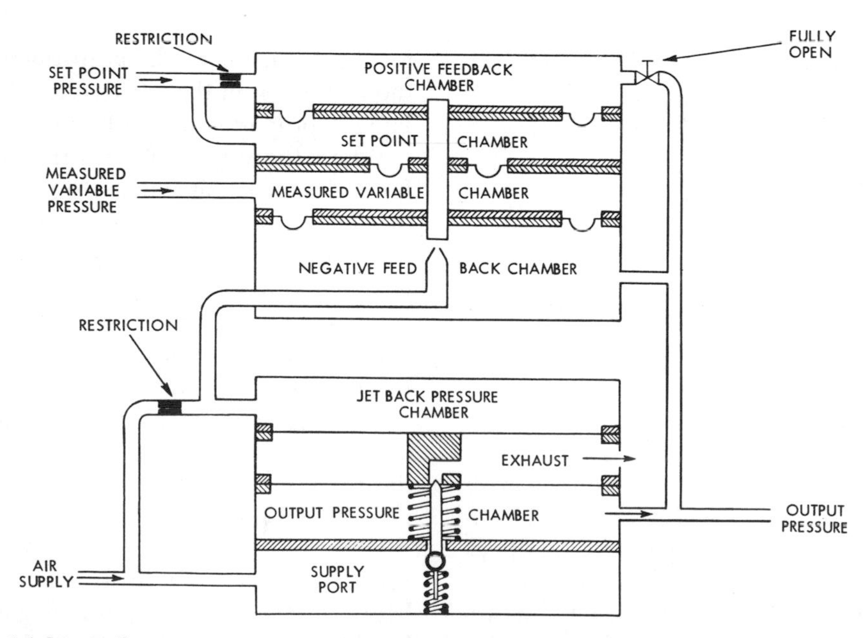

Fig. 14-26. Fully opened restriction (proportional band adjustment) in a proportional action controller.

14-26. As in the On-Off type, the motion of the baffle shaft depends almost entirely on the difference between the set point and the pressures of the measured variable. Therefore, completely opening the needle valve provides the narrowest proportional band possible.

Proportional plus Reset Action. Fig. 14-27 shows a proportional plus reset controller. The proportional controller remains unchanged except that the set point pressure does not pass through a restriction into the positive feedback chamber. A reset section is added. It consists of a reset chamber and a reset reference chamber, which are separated by a diaphragm. The diaphragm acts to close or open an exhaust port connected to the reset chamber.

When the measured variable is at set point, the pressures in the feedback chambers and the pressures in the set point and measured variable chambers are equal. The output pressure of the controller remains constant. See Fig. 14-28.

When the measured variable deviates from set point, the pressures in the set point and the chambers of the measured variable become unbal-

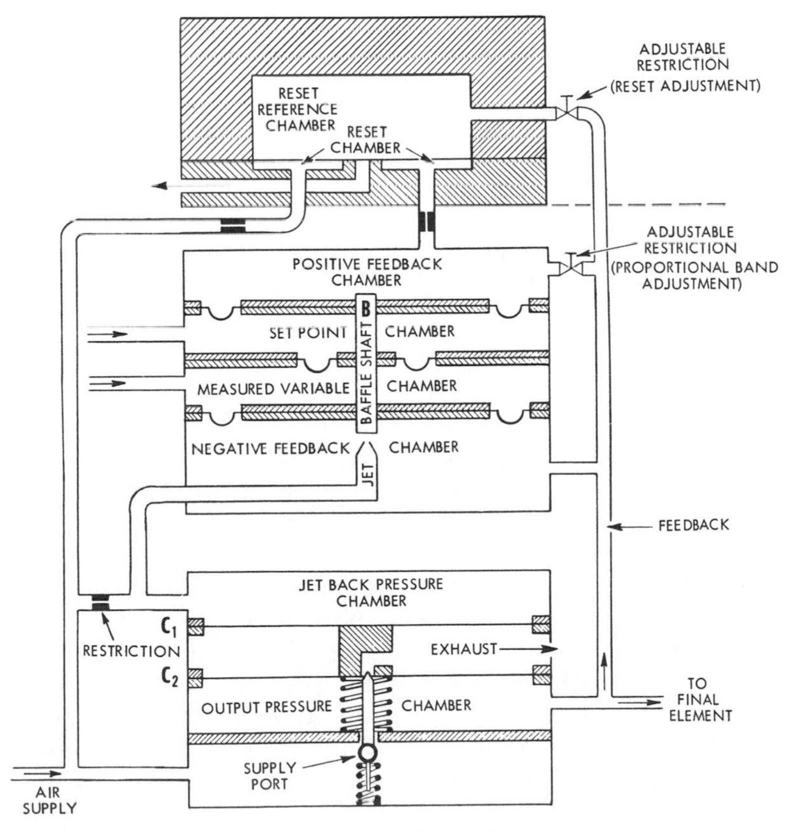

Fig. 14-27. Force balance pneumatic controller with proportional plus reset action.

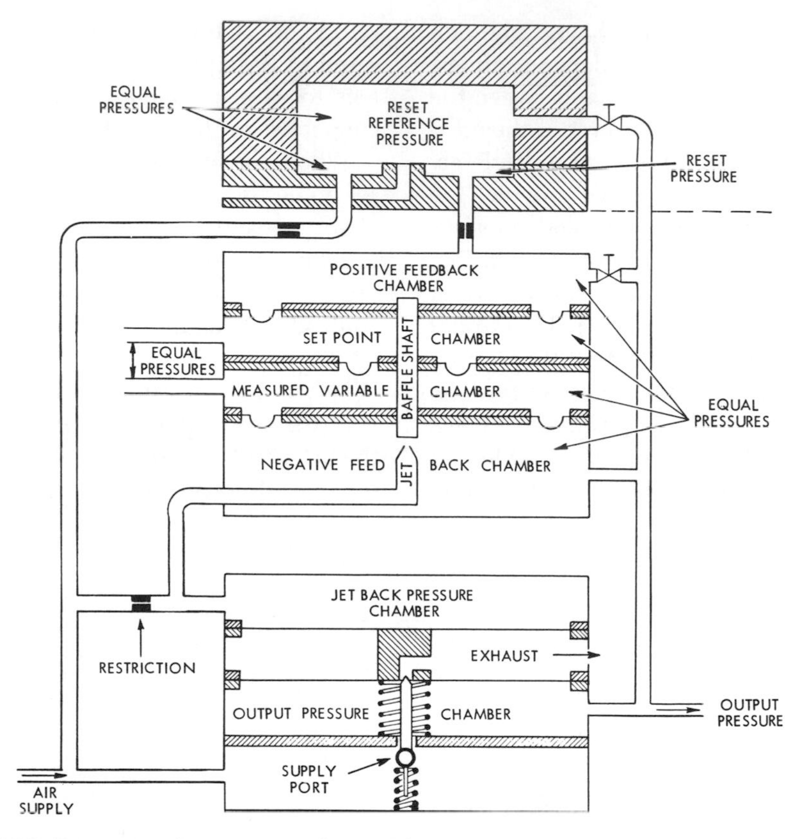

Fig. 14-28. Pressures when measured variable is at setpoint. There is no change in the output pressure of the controller.

anced. Hence, the baffle shaft changes position. See Fig. 14-29. If we assume that the measured variable pressure exceeds the set point pressure, the baffle shaft moves up, allowing air to enter the negative feedback chamber from the jet. At the same time, the double diaphragm in the amplifier section moves upward, reducing the pressure in the output chamber.

The output pressure change passes to four separate locations simultaneously. First, it passes to the final element, which moves to a new position as the output pressure changes. Then it moves to the negative feedback chamber, where the lowering of the pressure tends to allow the baffle shaft to move downward. This direction is opposite the original movement. Then the output pressure change passes through an adjustable restriction to the positive feedback chamber. Here the lowering pressure tends to allow the baffle shaft to move upward. This is the same direction as the original movement. Lastly, the output pressure change passes through an adjustable

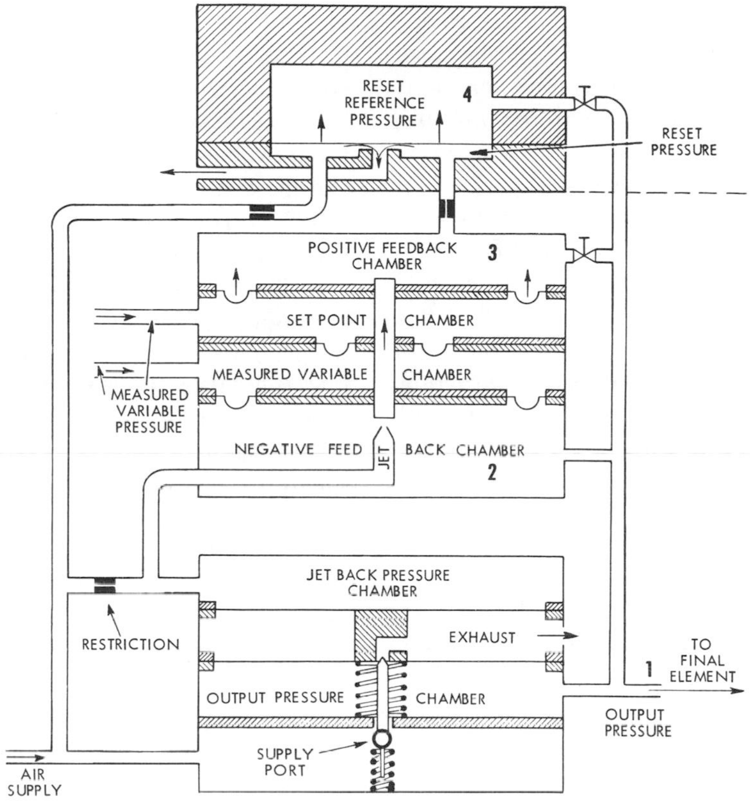

Fig. 14-29. Increase in the measured variable signal pressure has the following effects: (1) output pressure is changed, (2) pressure in negative feedback chamber is lowered, (3) pressure in the positive feedback is lowered, and (4) pressure in the reset reference chamber is lowered.

restriction to the reset reference chamber. The lowering of the pressure causes the diaphragm in the reset unit to move upward, allowing air to escape from the reset chamber. The reset chamber receives pressure from the air supply line after it passes through a fixed restriction.

Reset action is achieved by slowly changing the reset pressure to the divider circuit. The rate at which the reset pressure is varied is determined by the setting of the reset adjustment. When the reset reference chamber pressure is greater than the reset chamber pressure, the diaphragm stops the escape of reset pressure. This pressure then builds up until it is greater than the reference pressure. At the same time, the positive feedback pressure is increased, causing the output pressure to increase. This in turn increases the reset reference pressure. Then the cycle begins all over again. As long as the positive and negative feedback chamber pressures are unbalanced because of the difference between set point and the pressure of the measured variable reset action continues. The output pressure continues to change.

Thus the requirement of automatic reset response has been satisfied. The final element is continuously repositioned as long as there is an offset between set point and control point.

Derivative Action. Derivative action in a pneumatic controller is achieved by interposing a derivative amplifier between the proportional and the power amplifiers. The derivative amplifier is a complete force balance unit. See Fig. 14-30.

If there is a change in the measured variable pressure, there is a greater change in the measured variable plus derivative action pressure. The baffle shaft in the proportional amplifier moves more than if there were a change in the measured variable pressure alone, that is, without derivative action. See Fig. 14-31. The gain in the movement of the baffle shaft depends on the setting of the adjustable restriction in the derivative amplifier. This is a derivative setting. The gain in the movement also depends on the rate change of the measured variable.

The effect of derivative action, therefore, is to cause the proportional amplifier to produce an output greater in magnitude than would be produced by proportional action alone and at a rate related to the rate change of the measured variable.

Pneumatic controllers with diaphragms are commonly called the *stacked diaphragm type.*

Fig. 14-32 is a schematic of a pneumatic force-balance controller using bellows units to effect the balance. The set point pressure is introduced by a remote pressure regulator. The measurement input is supplied by a pneumatic transmitter. The set point and measured variable pressures each enter one of the four bellows which act upon opposite sides of a force-balance floating disc. Any unbalance between

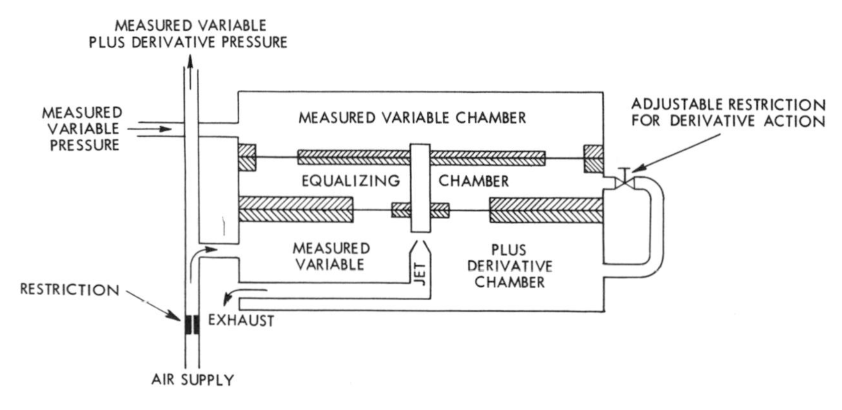

Fig. 14-30. Derivative action amplifier.

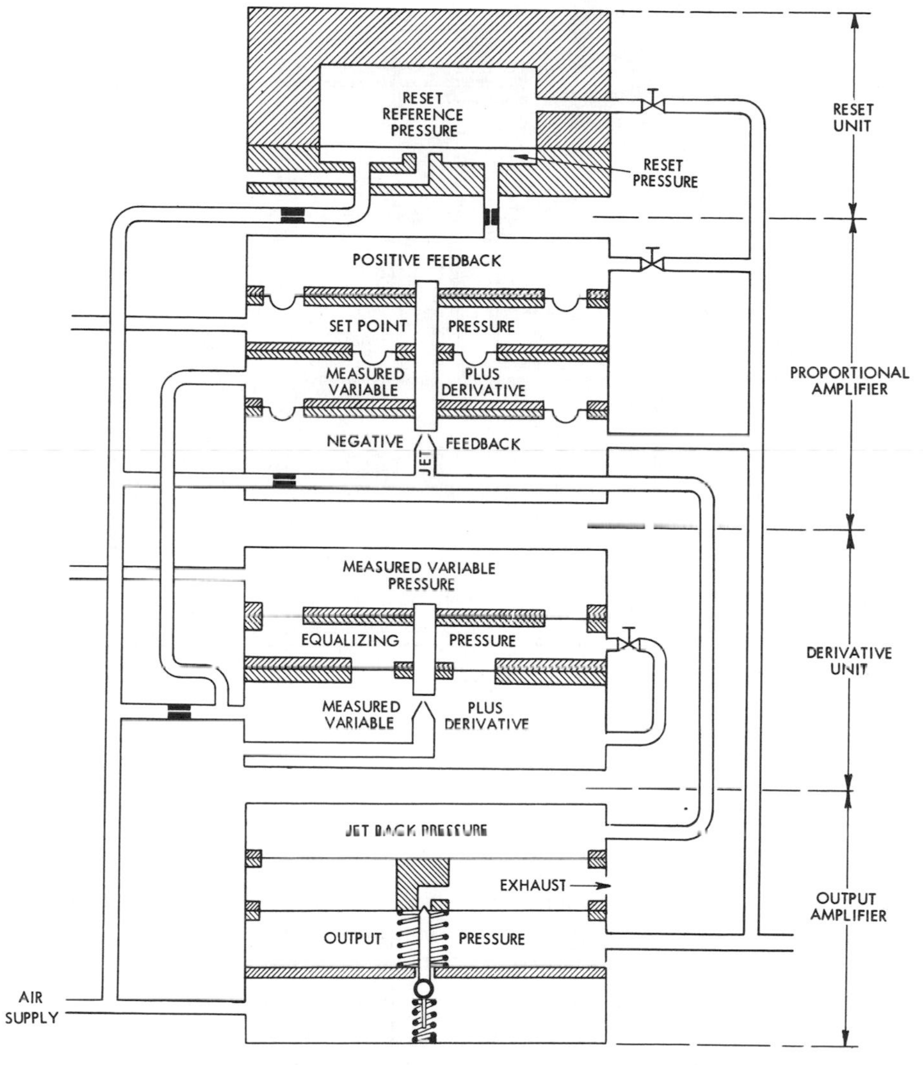

Fig. 14-31. Pneumatic force balance controller with proportional plus reset plus derivative action.

the set point and measurement pressures causes this disc to move. The plate acts as a flapper. The flapper presses against the nozzle which is mounted on another plate immediately above the floating disc. The nozzle is separated from the floating disc by an adjusting arm and fulcrum bar which provide proportional band adjustment. Proportional band adjustment is accomplished by rotating the nozzle to different positions from the fulcrum of the lever, changing the magnitude of floating disc motion necessary to pro-

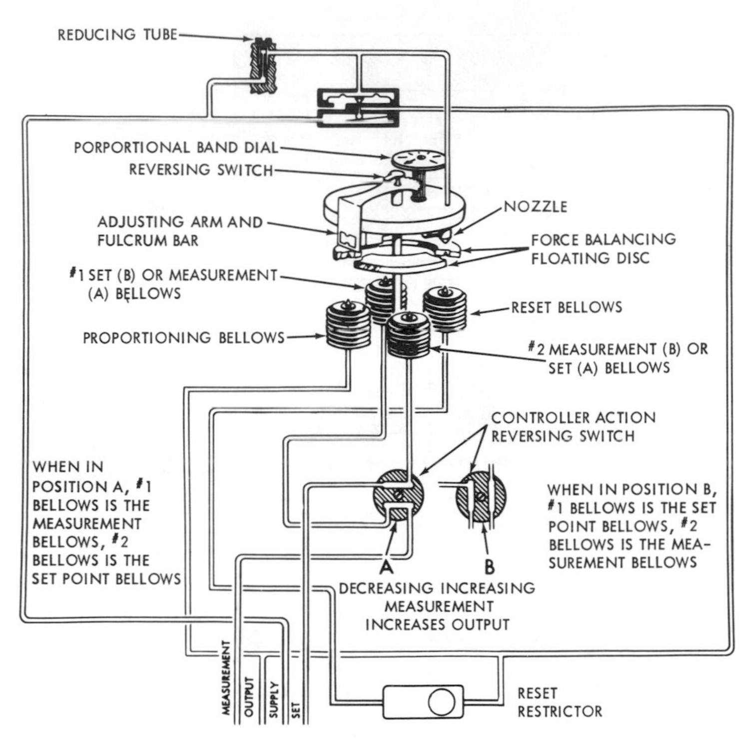

Fig. 14-32. The moment arms of the four bellows are fixed by the position of the proportional band adjusting lever. (The Foxboro Co.)

duce the full change in controller output. Two other bellows act upon the floating disc: the reset bellows and the proportional bellows. These receive as their input the feedback pressure from the air relay. The reset bellows acts upon the side of the floating disc opposite to the proportional bellows so that its effect on the nozzle-flapper position is counter to that of the proportional bellows. The adjustable restriction and capacity tank in the line to the reset bellows permit adjustment of the reset time.

If the fulcrum bar, shown in Fig. 14-32, is positioned directly over the proportional feedback bellows and the reset bellows then the controller has simple On-Off control action. Fig. 14-33 shows a topview and a sideview of such an arrangement.

If the fulcrum bar is adjusted so that b is 4 times distance a, as shown in Fig. 14-34, then the controller has a 25 percent proportional band. When the value of the measured variable increases, the output of the relay increases four times the measured variable increment. Fig. 14-34 provides a topview and a sideview of this arrangement.

If the fulcrum bar is adjusted so that distance a is four times distance b, as shown in Fig. 14-35, the result is a 400

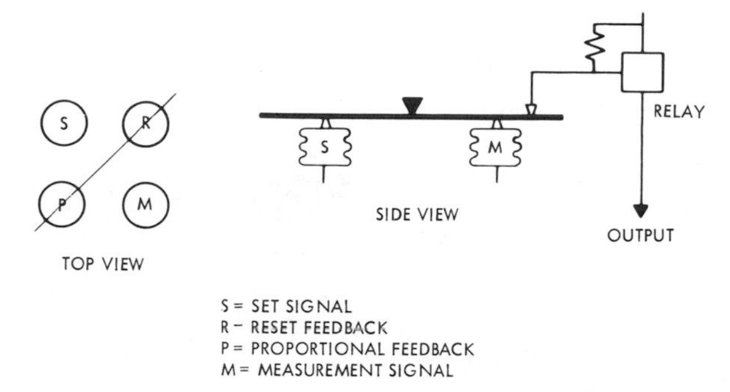

Fig. 14-33. Because the fulcrum bar is positioned over the proportional feedback bellows and the reset bellows, the controller has simple On-Off action. (The Foxboro Co.)

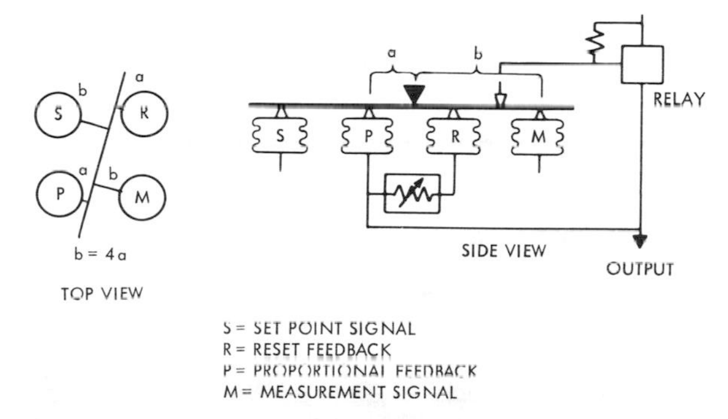

Fig. 14-34. Because of the positioning of the fulcrum bar, the controller has a 25 percent proportional band. (The Foxboro Co.)

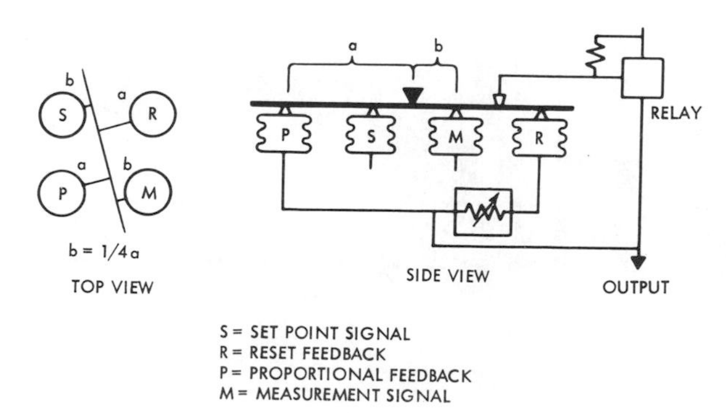

Fig. 14-35. The positioning of the fulcrum bar provides a 400 percent proportional band. (The Foxboro Co.)

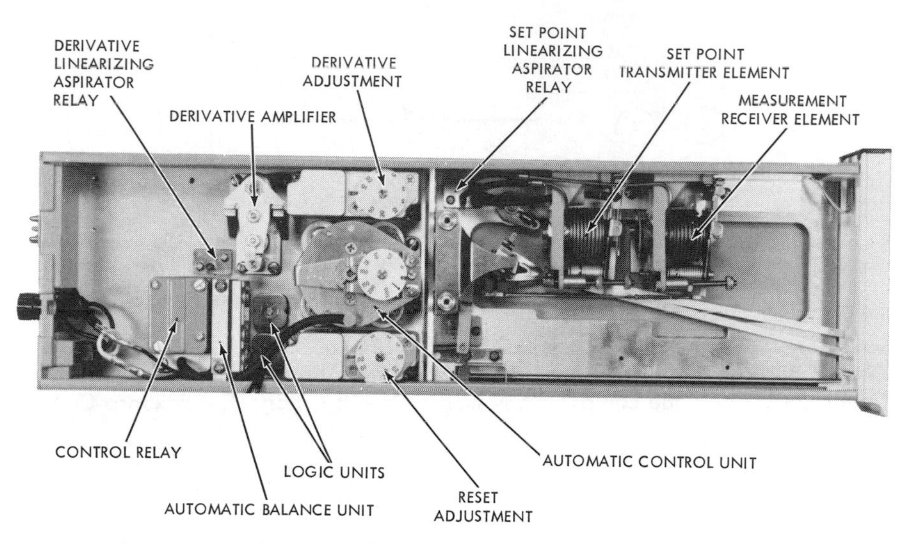

Fig. 14-36. The interior of a controller unit, showing the various components. (The Foxboro Co.)

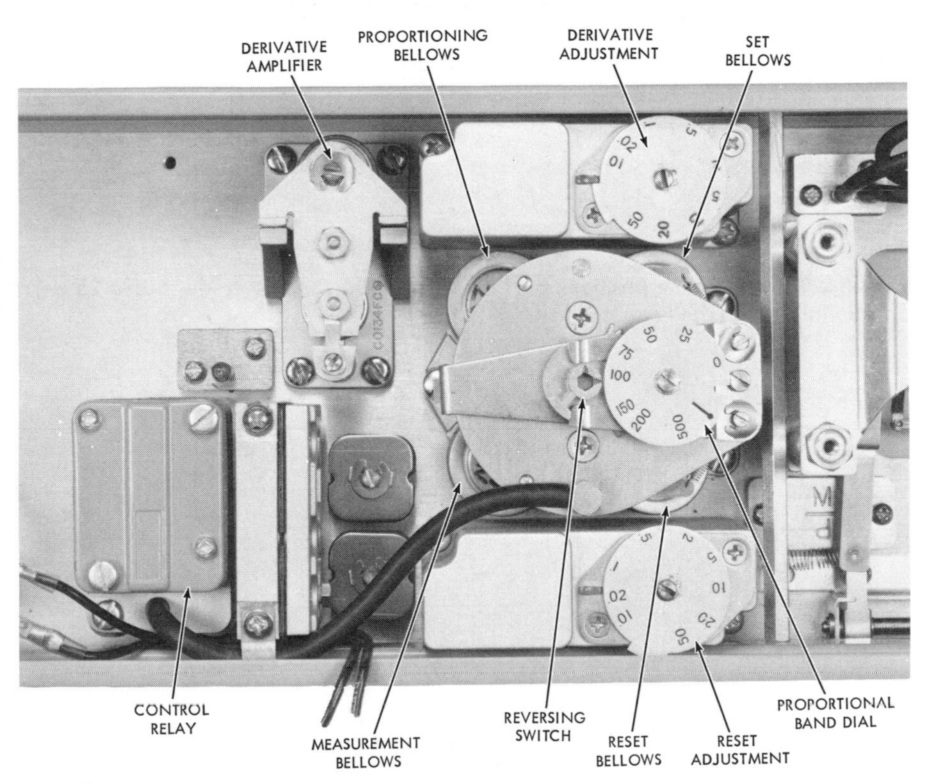

Fig. 14-37. Close-up of one portion of the controller unit, showing the bellows and the reset and derivative adjustments. (The Foxboro Co.)

percent proportional band. When the value of the measured variable increases, the output of the relay only increases one fourth this amount to restore the balance of forces. Fig. 14-35 depicts both a topview and a sideview of the fulcrum bar in relation to the bellows.

Fig. 14-36 shows the interior of an actual controller unit. The unit shown schematically in Figs. 14-32 through 14-35 is identified as the *automatic control unit.* Fig. 14-37 shows a close-up of this portion of the controller.

Electric Controllers

Electric controllers for proportional, proportional plus reset, and proportional plus reset plus derivative actions can be divided into two types:

1. The null-balance type in which there is an electrical feedback signal to the controller from the final element
2. The direct type in which there is no electrical feedback signal. This is also called *feed-forward control.*

Fig. 14-38 is a diagram of an electrical null-balance controller. As with the pneumatic controller, an electrical null-balance controller provides the various control actions by modifying the feedback signal. This is done by

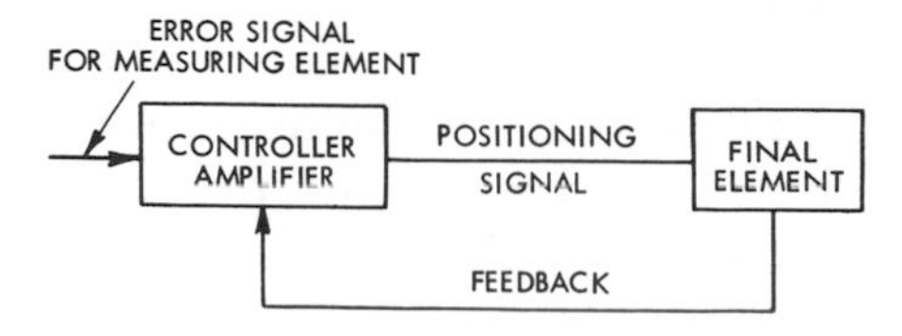

Fig. 14-38. Schematic of an electrical null balance controller.

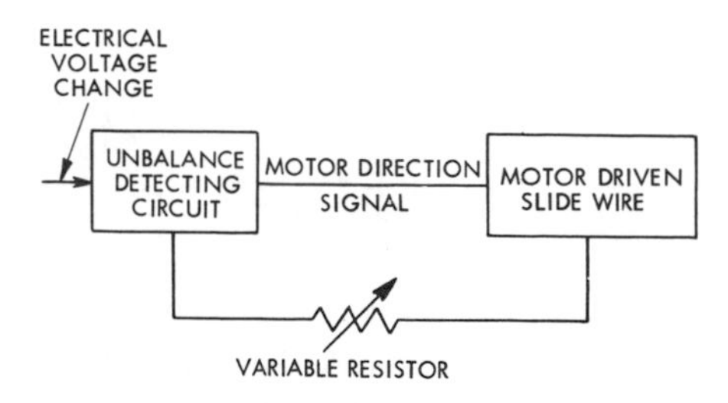

Fig. 14-39. Proportional action is provided by addition of a variable resistor, which allows adjustment of the proportional band.

adding properly combined electrical resistances and capacitances to the feedback circuit, just as restrictions and chambers were added in the pneumatic circuits.

A sensitivity, or gain, adjustment is the only addition to the controller required to achieve proportional action. This permits the adjustment of the proportional band. The sensitivity adjustment is made by inserting a variable resistor in the feedback line. This provides for regulating the magnitude of the feedback signal. The feedback signal depends on the position of the final element. The amount the final element must move is established by the setting of the variable resistor. This movement produces electrical balance in the controller. See Fig. 14-39.

The feedback signal is modified by the addition of a resistor-capacitor arrangement to provide proportional plus reset action. See Fig. 14-40.

Any change in the feedback voltage from the final element slidewire causes a current to flow into the capacitor C. This is where it is stored as a voltage. The capacitor is then said to be charged. This results in a voltage drop

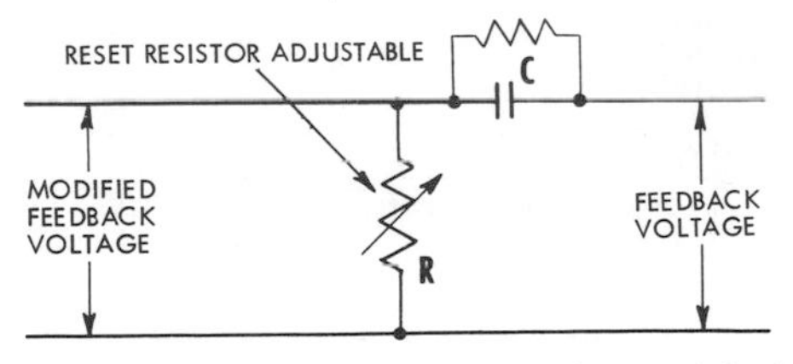

Fig. 14-40. The feedback signal is modified to provide proportional plus reset action.

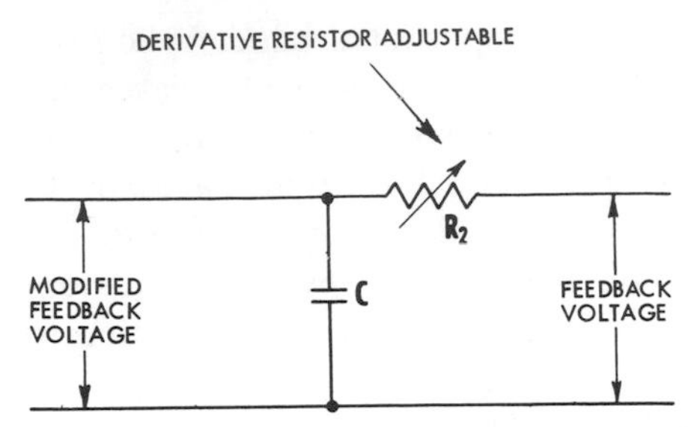

Fig. 14-41. The feedback signal is modified to provide proportional plus derivative action.

across resistor R. For a current to continue to flow through R, the capacitor must remain charged. Hence, it must continue to receive current. It receives current by a continuing change of the feedback voltage. The slidewire of the final element and the final element itself must, therefore, continue to change position to produce the necessary voltage change. The voltage continues to change as long as there is an unbalance signal from the controller. The reset action ceases when the error signal is eliminated and the value of the measured variable is at set point. The setting of the resistor R determines the rate at which reset action proceeds.

The feedback signal is modified with an additional resistor-capacitor network to provide proportional plus derivative action. See Fig. 14-41. With this arrangement, any change in the feedback voltage causes the capacitor C_2 to draw either more or less current, whichever is required to delay the effect of the change in feedback voltage. The final element can move more than it would if only proportional action were used.

The setting of the resistor R_2 determines the rate at which the capacitor is charged. Therefore, the modified feedback voltage varies with the rate the final element slidewire changes the feedback voltage. The final element slidewire position changes as the value of the measured variable changes. The modified feedback voltage changes with the rate the measured variable changes. Derivative action is thus accomplished.

Fig. 14-42 shows a null balance electric controller with proportional plus derivative plus reset action. The controller includes a control bridge, a feedback bridge, a detector-amplified relay, and a power motor. The rate and reset networks are also shown.

The power motor positions the sliders of two different slidewires. One slidewire is in the control bridge, and the other is in the feedback bridge. In some electric controllers the power motor can also be used to position the final element.

The input signal of the detector is the difference in voltage between the control bridge and the feedback bridge. This signal is sufficiently amplified so that the relay is actuated. The relay causes the power motor to run in a direction which repositions the

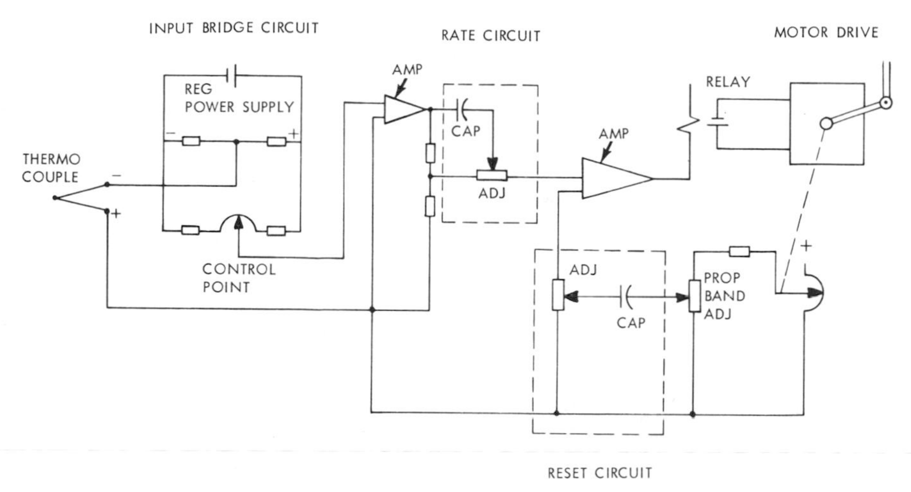

Fig. 14 42. An electronic temperature controller with proportional plus derivative (rate) plus reset action.

sliders. This reduces the input signal to zero.

The control bridge contains the proportional band adjustment. This is the sensitivity adjustment. The adjustment determines the relationship between the input signal to the control bridge and the resultant output signal of the control bridge.

The feed-forward electric controller operates on a different principle. There is no feedback signal from the final element. Fig. 14-43 is a schematic of a typical feed-forward electronic controller with proportional plus reset plus derivative action. This controller operates from a 0 to 10 volt dc input signal. If the primary element does not provide an input, the conversion module provides the operating voltage.

The percentage of total input voltage is shown on the process indicator. The matching set point voltage of 0 to 10 volts is introduced by a precision potentiometer. Any difference between the input signal voltage and the set point voltage enters the proportional plus reset module. The percentage of total input voltage is also shown on an error indicating meter.

Proportional band adjustment is accomplished by a precision potentiometer which regulates the output of the amplifier in the proportional plus reset module. The regulated output of the amplifier is further changed by a precision potentiometer in the reset RC network to provide reset action. For rate response the input signal voltage is modified by an RC network before entering the proportional plus reset module to provide derivative action. The output of the derivative module is then proportional to the change rate of the input signal voltage. The output meter indicates the percentage of total output voltage being produced.

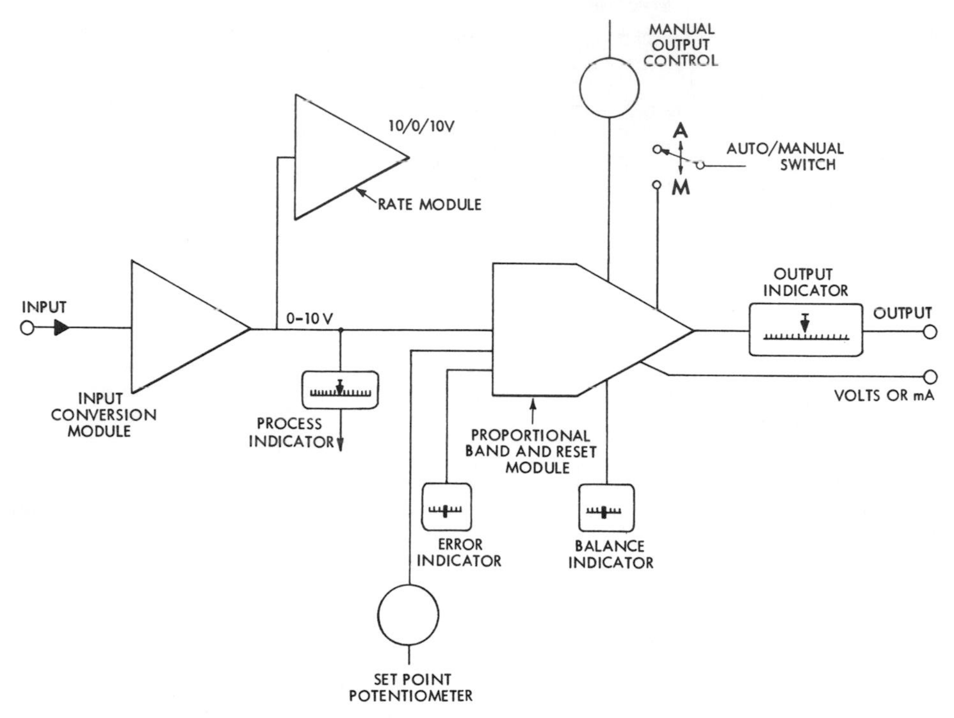

Fig. 14-43. A direct-type electric controller using proportional plus derivative plus reset action.

A balance-indicating meter is provided to facilitate switching to manual control. The manual adjuster must be positioned so that the meter is at the null position when it is switched from automatic to manual control. A capacitor in the instrument circuit eliminates the need for manual adjustment when the meter is switched from manual to automatic. This feature of a meter is called *automatic bumpless transfer*.

Hydraulic Controllers

Hydraulic controllers are available which provide the three control responses. Essentially, the hydraulic controller resembles the pneumatic controller except that the system must remain completely closed. Jet pipes and pistons, or four-way valves, are used in place of the flapper-nozzle and air relay used in pneumatic controllers.

The jet pipe resembles the pneumatic nozzle. It directs a fluid stream into either of two receiving chambers of a double-acting cylinder. When the jet pipe is moved to the left by the measured variable, as shown in Fig. 14-44, more fluid enters Chamber 1 than enters Chamber 2. This causes the pressure in Chamber 1 to increase, moving the piston to the right. The piston movement can be used to position a final element.

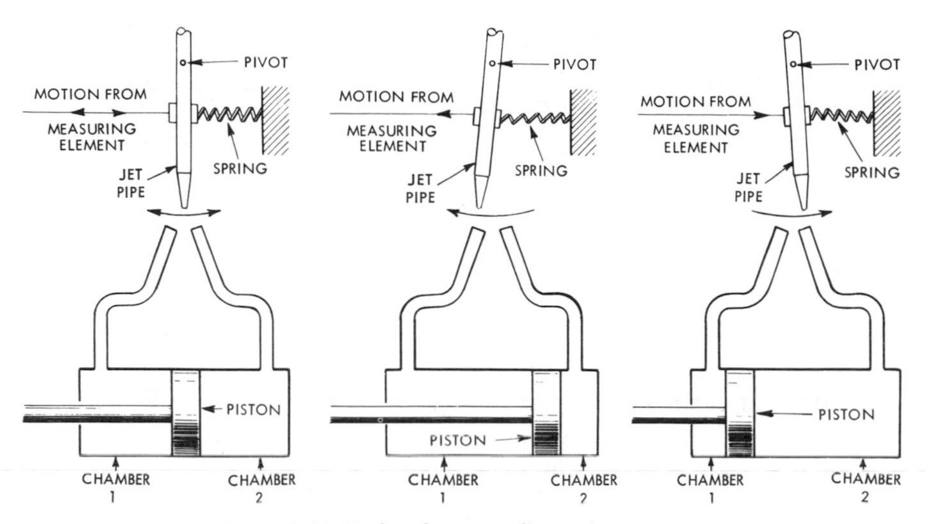

Fig. 14-44. Hydraulic controller with jet pipe.

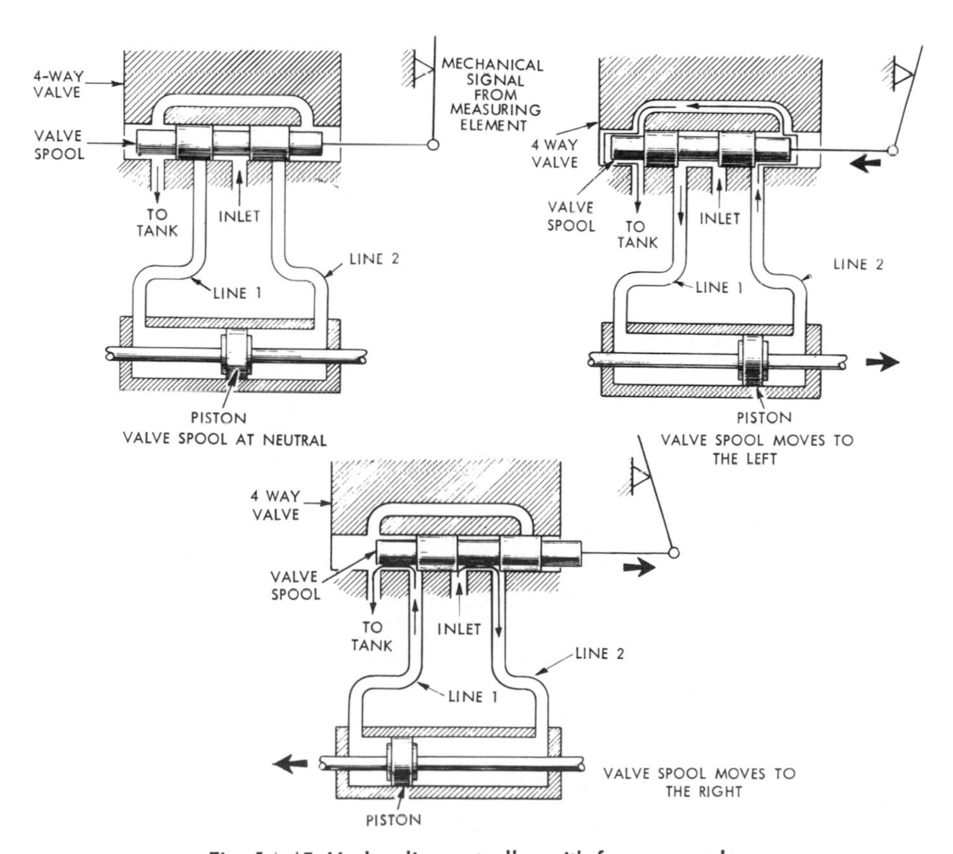

Fig. 14-45. Hydraulic controller with four-way valve.

The four-way valve can be used in the same manner as a jet pipe. See Fig. 14-45. When the valve spool is moved to the left, the fluid path to Line 1 is opened and the path to Line 2 is closed. The piston moves to the right. Similarly, when the valve spool is moved to the right, the fluid path to Line 1 is closed, and the path to Line 2 is opened. The piston then moves to the left. The action of the four-way valve resembles the action of the jet pipe system.

On-Off Action is accomplished in a hydraulic controller by very rapid movement of the piston to either extreme position when the measured variable deviates a small amount from set point. The jet pipe or valve spool is positioned at neutral to establish the set point at a particular value of the measured variable.

Proportional Action requires a feedback signal from the piston to the jet pipe or spool. Proportional action is provided by the addition of a feedback linkage from the piston. When the jet pipe is moved to the right, as shown in Fig. 14-46, because of a change in the value of the measured variable, the piston moves to the right. The feedback linkage is attached to the piston. It brings the jet pipe back to neutral position which stops the movement of the piston. Thus there is a piston position for each value of the measured variable. Changing the location of the pivot provides proportional band adjustment. This regulates the amount of piston motion required to restore the jet pipe to its neutral position.

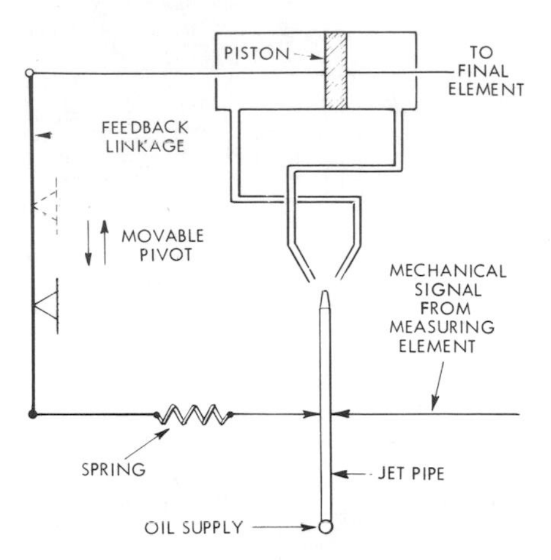

Fig. 14-46. Proportional action in a jet pipe hydraulic controller.

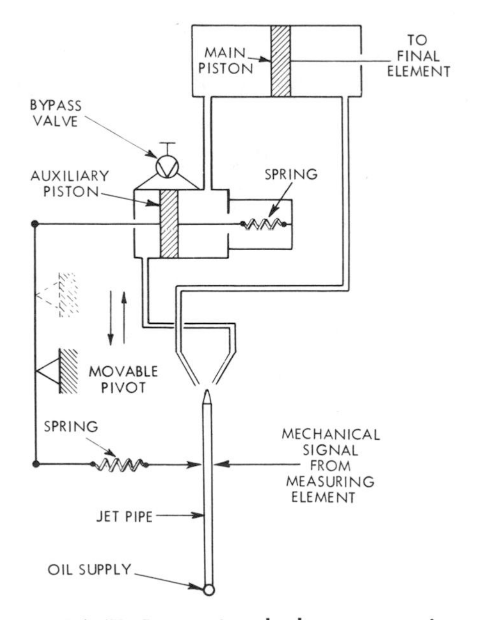

Fig. 14-47. Proportional plus reset action in a jet pipe hydraulic controller.

Automatic Reset Action can be added to the proportional controller by modifying the feedback signal. Fig. 14-47 illustrates a jet pipe hydraulic controller with proportional plus reset control action. The addition of an auxiliary piston with a bypass provides the reset action. When the jet pipe is moved to the right, both the main piston and the auxiliary piston move to the right. The feedback linkage is attached to the auxiliary piston. This linkage returns the jet pipe to its neutral position. Because of the bypass, the auxiliary piston returns to its mid position, moving the jet pipe to the right again. This causes motion of the main piston in the original direction.

Derivative Action can be added to a hydraulic proportional plus reset controller. In the schematic of a controller shown in Fig. 14-48, a four-way valve is used to provide the power to the reset and power cylinders. The proportional bellows provides the actuation of the four-way valve. There are two separate feedback systems, one for reset action and a second for rate action. The reset feedback signal is provided by the power cylinder. The jet pipe sends a signal to the proportional bellows when the value of the measured variable deviates from set point. Reset action allows the power piston to move as long as the proportional bellows receives this signal. The

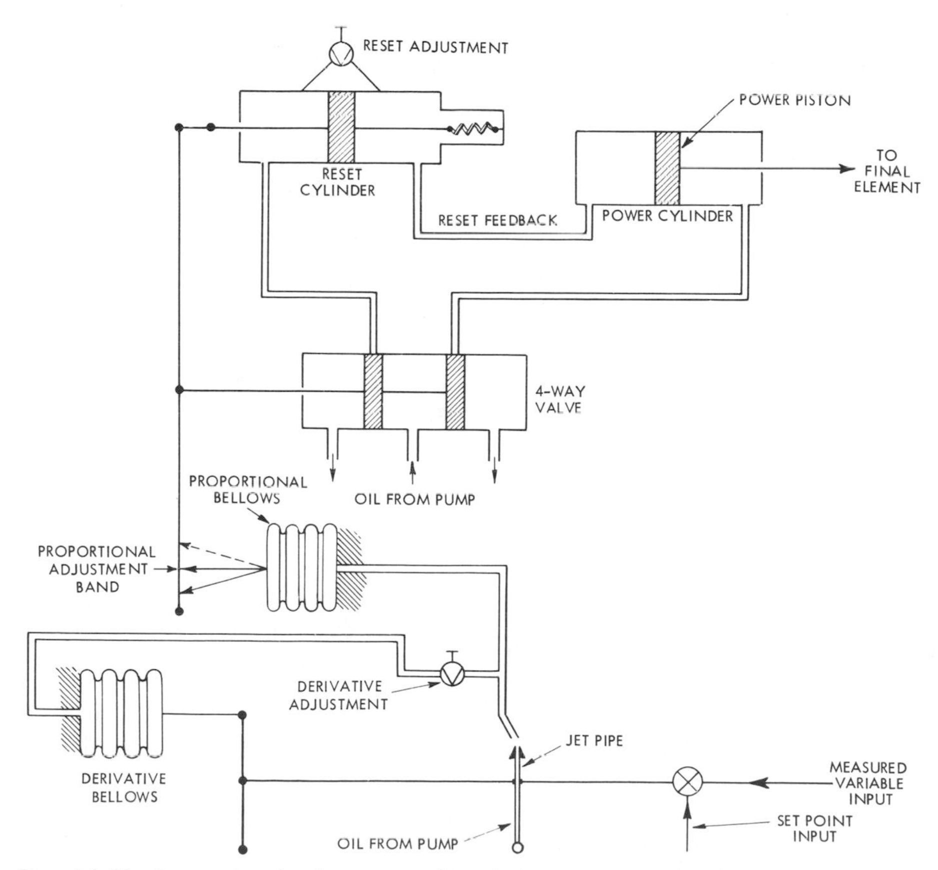

Fig. 14-48. Proportional plus reset plus derivative action in a hydraulic controller.

adjustable restriction in the bypass line of the reset cylinder determines the reset time.

The linkage from the derivative bellows provides the derivative feedback signal. The derivative bellows moves the jet pipe back to a neutral position. The adjustable restriction between the jet pipe and the derivative bellows determines the rate time. The derivative action delays the feedback to the jet pipe as long as necessary in the particular process. The proportional bellows is able to operate longer with derivative action in the controlling system. Proportional band adjustment is made at the point where the proportional bellows acts on the lever assembly which connects the four-way value spool to the reset piston.

Control Valves

The control valve is by far the most common and the most varied in type of the many final elements available.

A valve is essentially a variable orifice and behaves according to the following formula:

$$Q = Kd^2 \sqrt{\Delta p}$$

when

Q = rate of flow
K = flow coefficient constant—corrected for units
d = nominal valve size
Δp = pressure drop across valve

When the formula is applied to valves, it is usually modified by substituting a valve coefficient C_v for Kd^2. The equation then becomes:

$$Q = C_v \sqrt{\Delta p}.$$

For liquids with specific gravity,
$\Delta p = 1.0$.

Most control valve manufacturers provide C_v values which are based upon actual performance tests.

When a control valve is selected for a particular process application, this formula is used to determine the C_v required for maximum flow. It is very important that the valve be capable of passing the maximum amount of fluid. Before such a calculation can be made, however, the correct differential pressure must be determined. The differential pressure must, of course, be accurate if the calculation is to be reliable. Time spent in studying the pressure losses in the piping both upstream and downstream from the valve is well invested. If the differential pressure across a valve can vary, the lowest estimate of differential pressure should be given in the calculation of C_v. With compressible fluids, the flow rate does not increase in accordance with these formulas except when the pressure drop across the valve is less than one half the entering pressure.

The valve formula for various fluids using common units of measurement is:

$$Q_l = C_v \sqrt{\frac{\Delta p}{G}}$$

when

Q_l = flow rate of liquid in gpm at flowing temperature
G = specific gravity

When the flow rate is given in lbs per hour (W_l), the formula becomes

$$W_l = 500\, C_v \sqrt{\Delta p G}$$

This calculation, as with sizing of orifice plates, requires a viscosity correc-

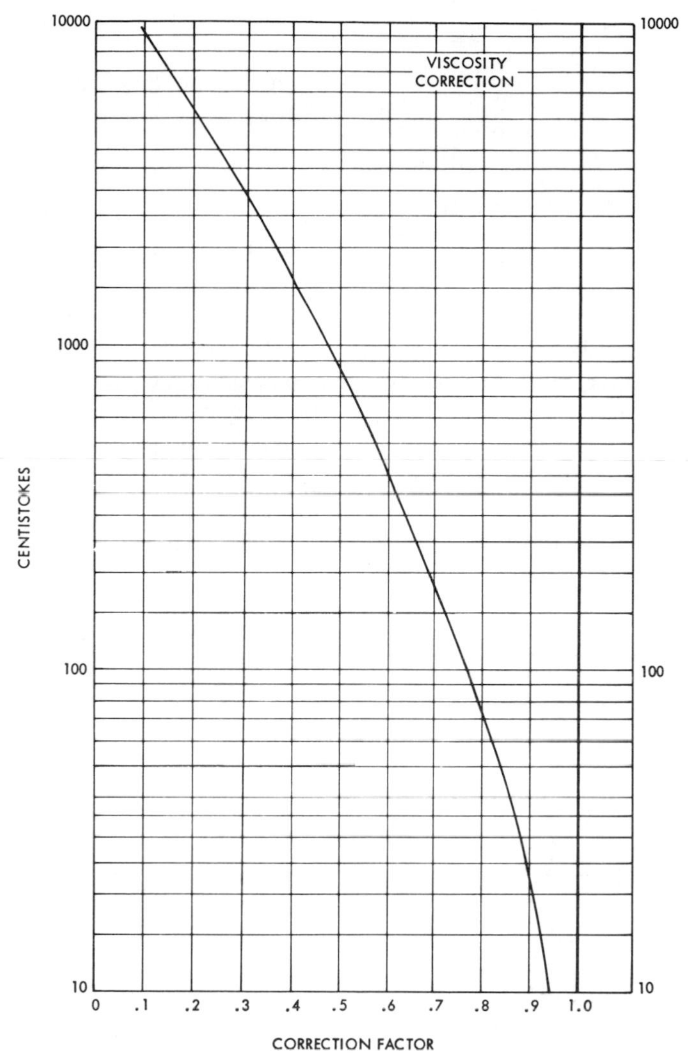

Fig. 14-49. Graph showing correction factor for viscosity.

tion, if the viscosity of the flowing liquid is greater than 10 centipoises. The graph in Fig. 14-49 shows the factor for a range of viscosities. To allow for viscosity, divide the C_v by the viscosity correction.

$$W_s = 3C_v \sqrt{P_2 \Delta p}$$

when

W_s = flow rate of dry saturated steam in lbs/hr.

P_2 = downstream pressure in psia.

(*psia* means pounds per square inch absolute.)

$$Q_g = 61\, C_v \sqrt{\frac{P_2\,(\Delta p)}{G}}$$

when

Q_g = flow rate of gas in cu. ft. per hour (at 14.7 psia and 60°F).

When P_2 is less than one half of P_1, the expression $\sqrt{P_2 \Delta P}$ becomes $\frac{P_1}{2}$.

This means, as mentioned earlier, that if the pressure drop is greater than 50% of the absolute inlet pressure, that portion of the pressure drop greater than 50% does not increase the flow. All compressible fluids follow this rule.

Correction factors must be applied for gas and steam flowing at elevated temperatures. Valve manufacturers have developed slide rules for sizing their valves. The slide rule includes a means to correct for specific gravity, steam super heat, and elevated gas temperatures.

Since the amount of fluid passing through a valve at any time depends on the opening between the plug and the valve seat, there is a relationship involving stem position, plug shape, and rate of flow. The stem position is determined by the operator, which receives its signal from the controller. There are many different types of valve bodies, each of which has its own flow characteristic. In addition, globe valves with specially shaped plugs to provide particular flow characteristics are available.

Valve characteristics are usually described graphically in terms of percent of flow vs. percent of lift (or travel).

Fig. 14-50 contains a graph and an illustration of a quick opening globe valve. The graph shows plug char-

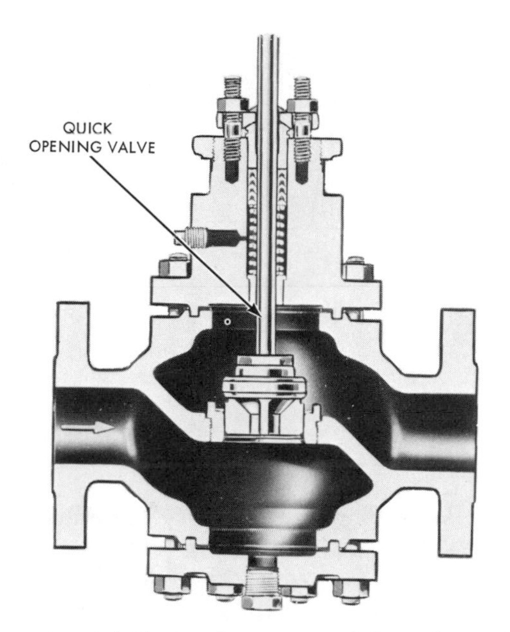

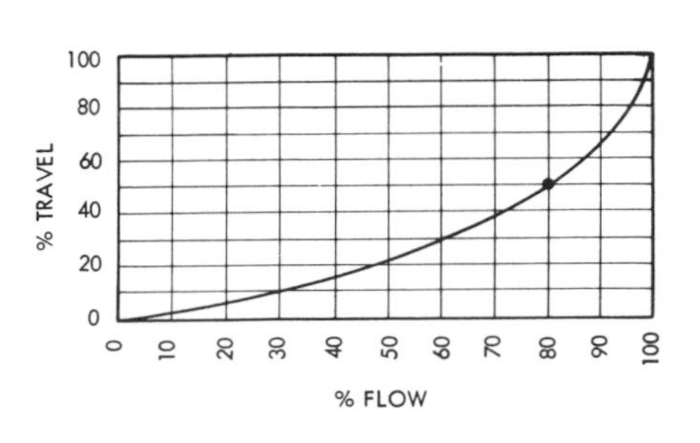

Fig. 14-50. Quick opening globe valve with graph showing plug characteristics. (Taylor Instrument)

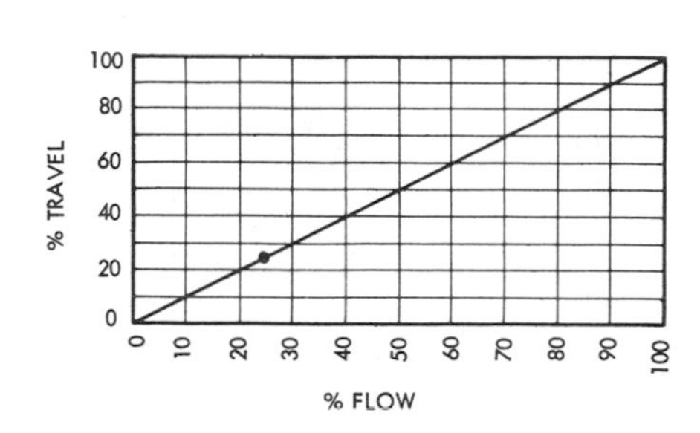

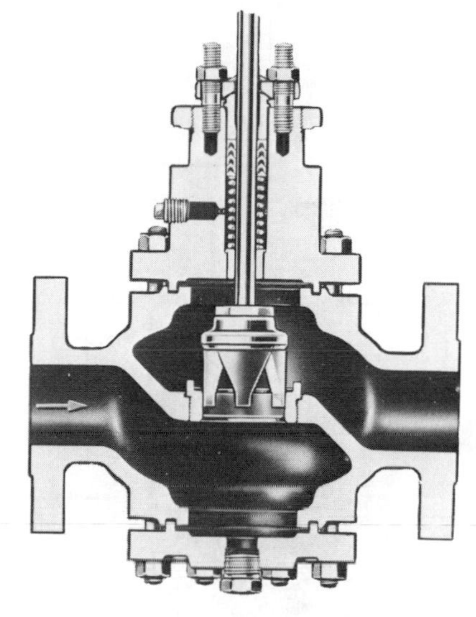

Fig. 14-51. Linear globe valve with graph showing plug characteristics. (Taylor Instrument)

acteristics. The quick opening globe valve, along with those which follow, is one of the basic valve types.

Fig. 14-51 shows a linear globe valve. The graph plots the plug characteristics. When the plug is 50% open, there is 50% flow. When the plug is 25% open, there is 25% flow, etc.

Fig. 14-52 depicts an equal percentage globe valve. The graph shows plug characteristics. It can be seen from the graph that there is almost a constant change of flow rate for changes in lift (travel).

Fig. 14-53 shows a Saunder-type valve. The graph shows that relationship between lift and flow is approximately linear up to the point where the plug is 50% open.

Fig. 14-54 contains an illustration of a butterfly valve. The graph demonstrates that the relationship between flow and lift is approximately equal, up to the point where the plug is 50% open. Beyond that range, the relationship is linear.

The selection of a control valve for a particular application should be made carefully. The selection should be based on a careful analysis of the process characteristics. If the reaction rate of the process is fast and the projected load changes are small, control is only slightly affected by the valve characteristics. If the process rate is slow and load changes are great, valve characteristics become important.

The nature of load changes should be examined and compared. Some load changes are linear, while others are not. If most of the load changes are linear, the valve should have a linear characteristic. Similarly, if the load

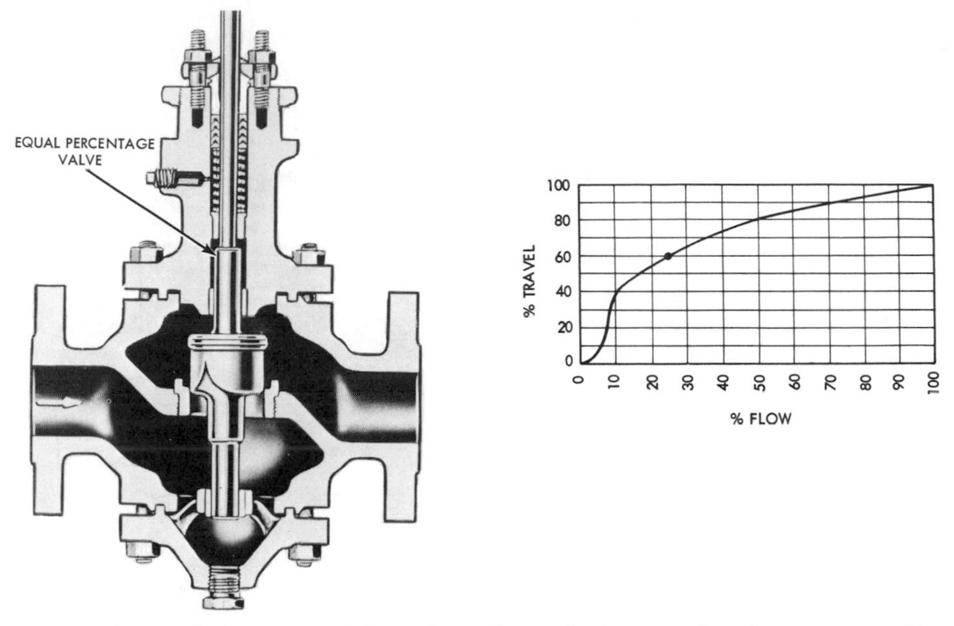

Fig. 14-52. Equal percentage globe valve with graph showing plug characteristics. (Taylor Instrument)

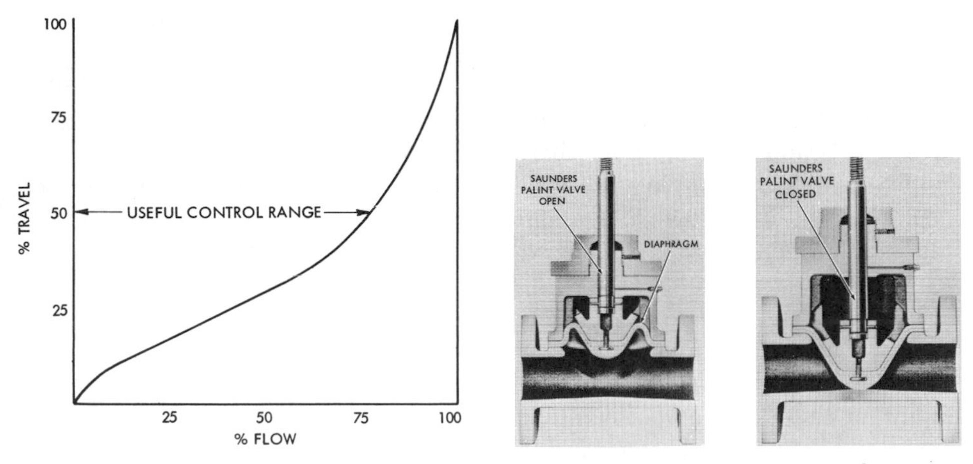

Fig. 14-53. Saunders patent valve with graph showing plug characteristics. (Taylor Instrument)

changes are non-linear, the valve should have a non-linear characteristic (equal percentage). Good control is difficult to attain if, for example, a non-linear valve is used in a process which has linear load changes.

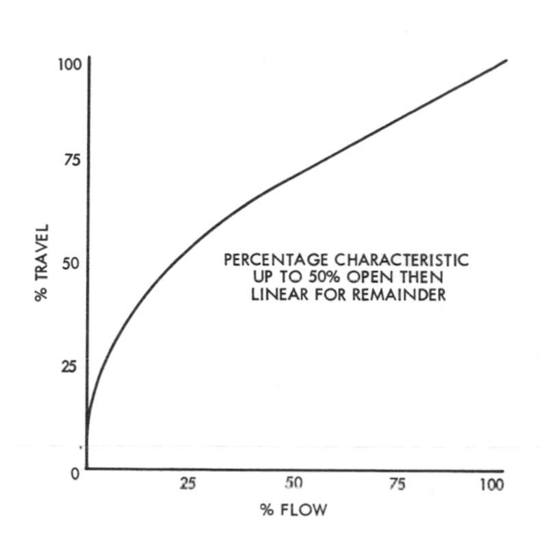

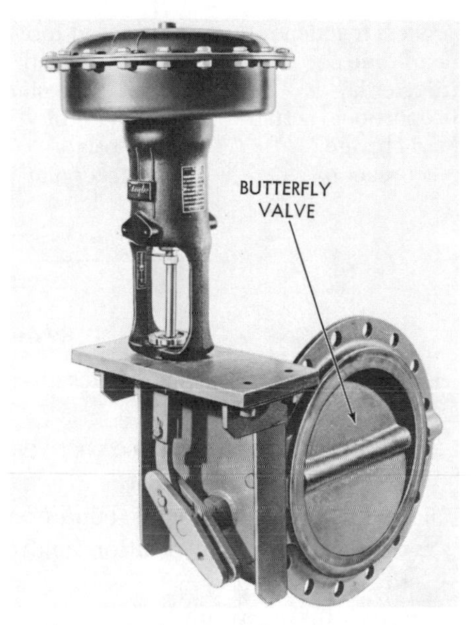

Fig. 14-54. Diaphragm-operated butterfly valve with graph showing plug characteristics. (Taylor Instrument)

In addition to the flow characteristics of a valve, there are other factors which influence the selection of a valve type. Valves are available which resist corrosion, operate with fluids of particularly high or low temperatures, high or low pressures, and high or low velocities. Fluids containing suspended solids require special consideration. Special valve types are available for such service. Some applications may require an extremely tight closing of the valve. The single-port globe valve, as an example, permits tighter closing than the double-port type. The quantity of fluid being processed can limit the selection. Globe valves are generally manufactured only in sizes up to 16 inches, so other types must be used when the size of the flow exceeds 16 inches.

The above information is only a partial review of control valves. Many of the fundamental considerations have been presented. It is suggested that the manufacturers be consulted when particular problems arise.

Words to Know

process reaction curve
capacity
nozzle
fulcrum bar
derivative action
diaphragm

process reaction rate
dead time
transfer lag
proportional action
load change
reset reaction
feed-forward
automatic bumpless transfer
viscosity
psia
set point
bellows
measured variable
final element
feedback
flapper

Review Questions

1. What process characteristics affect reaction rate?
2. What is the difference between dead time and lag?
3. List the three process conditions necessary for two position control to be satisfactory.
4. Give three examples of process load changes.
5. In proportional action control, there is a fixed relationship between the position of the final element and the amount the measured variable deviates from set point. Assuming a control valve is used, explain this statement in terms of temperature control.
6. Why does dead time and transfer lag cause poor control when only proportional action is used?
7. Is a wide proportional band adjustment more sensitive than a narrow one? Give the reason for your answer.
8. How does the addition of reset response to a proportional controller eliminate the disturbance caused by load change?
9. When is it necessary to have derivative action in a controller?
10. Describe the order in which the adjustments of a proportional plus reset plus derivative controller are made. Explain this procedure.

part three

PART THREE concerns itself with the application of instruments to actual processes. The previous sections have dealt with the operating principles of measuring and controlling instruments. The characteristics of a process, and the means by which control devices accommodate these characteristics have been explained. The variables to be measured and controlled are presented in the same order as in previous sections. To point up the "systems concept" more than one method of handling each application is described. It is expected the student will be able to develop other combinations. Part Three also includes appendixes which the student will find useful in his continued study of instrumentation.

Water management analyzers measure pH and ORP (oxidation-reduction potential) values of neutralized acids and chrome reductions at an industrial waste treatment facility at McClelland Air Force base near Sacramento, California. This facility treats from 1.2 to 2.4 mgd (million of gallons per day) of wastewater from the base's service and repair complex. To protect the environment, the water is cleaned before being discharged into a creek leading to the Sacramento River. (Honeywell Process Control Div./Fort Washington, Pa.)

Application

Chapter

15

The two types of industrial manufacturing processes are the *batch-type* process and the *continuous-type* process. In the batch type process, the product material generally does not flow from one section of the process to another. In the continuous type process, however, the product material undergoes various treatments as it flows through the system.

Some processes combine the features of the batch and continuous types. In such processes, several product materials are treated, as in the batch-type processes, then stored and drawn off in a continuous process.

The control of batch-type operations is easy to understand. Actually, the two types of systems exhibit similar features, except for the flow of product material in the continuous type process.

The following discussion of measuring and controlling variables in process systems is primarily concerned with batch-type operations. However, the principles to be learned also apply to continuous systems.

Temperature Control

Steam Heat

Fig. 15-1 illustrates a simple batch-type temperature control system. In this system, the product material is heated to the desired temperature, and maintained at that temperature for extended periods of time, by steam coils. The temperature recorder/controller has pneumatic control. The control is actuated by a pressure spring.

Food products, such as ketchup, and chemical products, such as plastic or rubber compounds, are typical materials treated with a simple batch-type

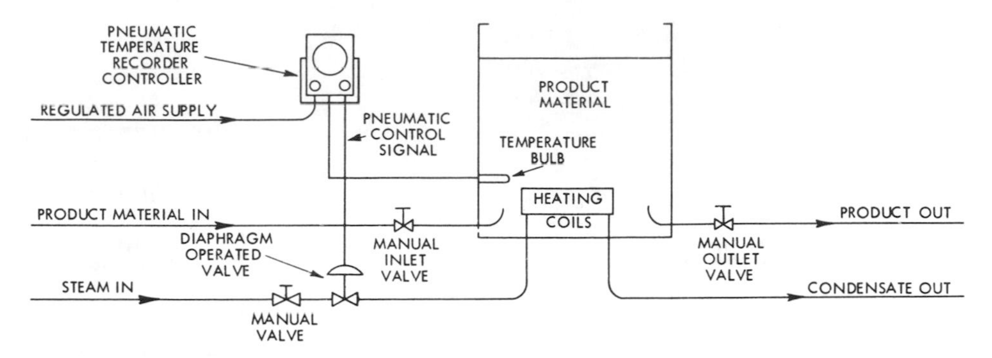

Fig. 15-1. Steam heat is used in this control system to provide the desired temperature.

system. The product material moves into a process storage vessel until the desired level is reached. The manual steam valve is opened so that steam can enter the heating coils.

A simple On-Off control can be satisfactory, if the temperature increases as soon as the manual steam valve is opened. With this type of control system, the control valve remains completely open until the temperature of the product material is above set point. Then the valve closes completely. When the valve closes, the temperature of the material begins to drop. As soon as the temperature drops below the set point, the control valve opens again. The temperature changes and the corresponding movements of the control valve are repeated over and over. It is important that the temperature be always held within allowable limits during this On-Off cycling. The temperature bulb must be very sensitive, and the controller and valve must respond quickly, making the entire control system fast acting.

If the rate of temperature increase is slow, but there is no dead time when the manual steam valve is opened, proportional control should be used. In such an application, the control valve remains completely open until the temperature is within the range of the proportional band. At that point, the valve begins to close. When the temperature reaches set point, the valve ceases to close and remains at a position which maintains the desired temperature. Usually, there is some initial valve overshoot (the valve closes too much and oscillates slightly) before it settles down to a fixed position.

The control valve, in a well-designed system, is near the mid vay position when the temperature is at set point. The valve opens or closes as much as necessary, which enables it to throttle the flow and maintain set point.

The proportional control system can correctly control the temperature as long as the temperature of the steam is constant and the heating requirements of the product material does not change. If either of these two conditions routinely changes during the process, however, then automatic reset action must be added to the proportional controller. The product material, for example, can require varying amounts of heat because of physical or chemical changes in its composition.

Automatic reset action enables the control valve to assume a different position for each deviation of the product material from the set point temperature, thus permitting a varying amount of steam, as required, to enter the heat exchanger. This compensates for any change in the temperature of the steam already in the heat exchanger or in the heat requirements of the product material.

If changes in the temperature of the product material are very slow, or if the action of the control valve is very slow, derivative action must be added to the controller. Derivative action allows the control valve to change settings in accord with the rate the product temperature changes.

The valve should be the air to open type for the sake of safety. If the air pressure fails, the valve closes. This prevents overheating of the material.

Gas or Oil Heat

Fig. 15-2 shows another simple batch-type temperature control system. The material is heated and maintained at the desired temperature for an extended period of time. A gas or oil burner is used in place of a steam coil to provide the heating source. A thermocouple is used as the temperature sensitive element. The electric controller signal is connected to a motor-operated valve. This type of installation can be used for melting metals or heating materials used in the manufacture of such products as varnish or glass.

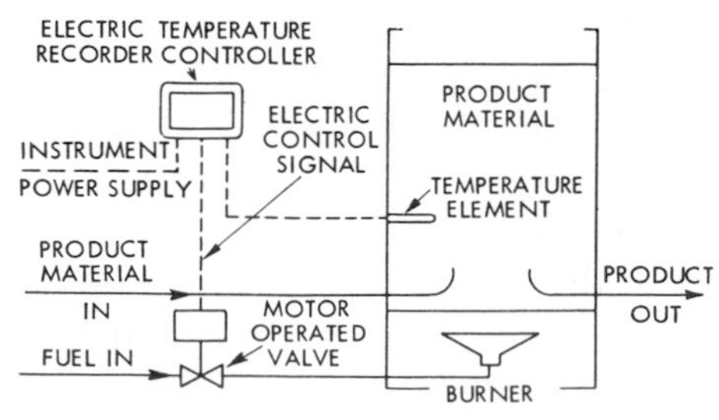

Fig. 15-2. Gas or oil can be used as the fuel in this control system.

Since the gas or oil burner should not be turned *On* and *Off* repeatedly, On-Off control cannot be used.

The proportional controller is the null-balance type, containing a slidewire. The control valve mechanism also contains a slidewire. These two slidewires are connected into the electronic control circuit. The slider on the instrument slidewire is positioned by a reversible motor which receives its signal from the measuring circuit. The slider on the control valve slidewire is positioned by the valve motor. Therefore, for each measured temperature there is a corresponding valve position. The proportional band adjustment determines the valve motion for each unit the measured variable deviates from set point.

If the heat required to maintain set point, or if the heat supplied by the burner (gas or oil) varies, automatic reset action must be added to the controller. Automatic reset action allows the valve position to change, if the position provided by proportional action does not result in sufficient heat. Automatic reset action is provided by adding a resistance-capacitance network between the slidewire of the instrument and the slidewire of the control valve.

If the system responds very slowly

to temperature changes, action must be added to the controller. This provides a derivative means for the valve to change its setting, depending on the rate the temperature changes. This is accomplished by adding a resistance-capacitance network between the instrument slidewire and the valve slidewire.

The resistance-capacitance networks for producing reset action and derivative action are described in Part Two. Derivative action is rarely added to a proportional controller without the addition of reset action.

The control valve mechanism should include safety features to insure valve closure should the fuel supply cease. The valve should also close automatically if there is a power failure. The control instrument should include a battery back-up system to maintain control in case the power fails. It should also include a backup for closing the valve in case the measuring element fails. The slider on the instrument then travels to the upper end of the scale, actuating the mechanism which causes the valve to close. This mechanism can also be used to actuate an audible alarm.

Bumpless transfer is another important feature available in controllers. Bumpless transfer comes into play when the control is switched from automatic to manual, or vice versa. It provides a continuous control signal while the transfer is being made.

Electrical Heat

Fig. 15-3 shows a batch-type control system in which the process heat is provided by electric heaters. Although the heaters in the drawing are shown at

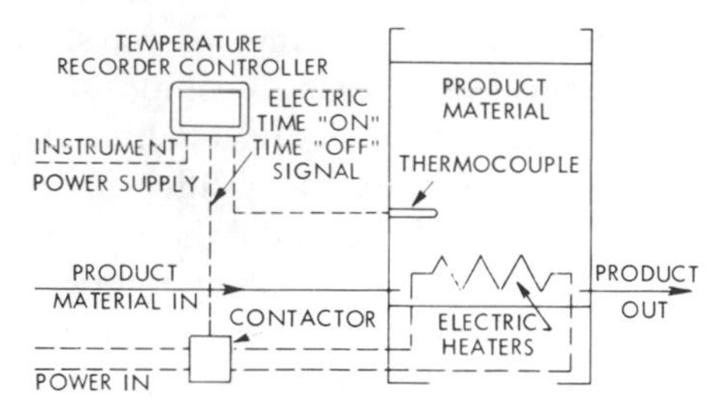

Fig. 15-3. Electricity is used to provide heat in this control system.

the bottom of the vessel, they can be contained in a jacket surrounding the vessel. Glue, wax, and various syrups are frequently processed using such heaters.

The heaters are operated by an electrical contactor which, in turn, is actuated by the controller. A thermocouple is used as the primary element. The controller is the electrical null-balance type with slidewire.

Proportional action is achieved in such a system by regulating the time-On and time-Off ratio. If the temperature falls below the set point, the heaters remained *On* longer. If the temperature rises above the set point, the heaters remain *Off* longer. There is a new time-On and time-Off ratio for every unit of rise or fall of the process temperature. If reset is needed, it can be added. Reset permits the ratio to change as long as the temperature remains above or below the set point. The controller should be equipped with battery backup, and thermocouple break protection.

If the electrical heaters are large, and turning them On and Off repeatedly is impractical, only part of the electric heat is controlled. This arrangement requires that the uncon-

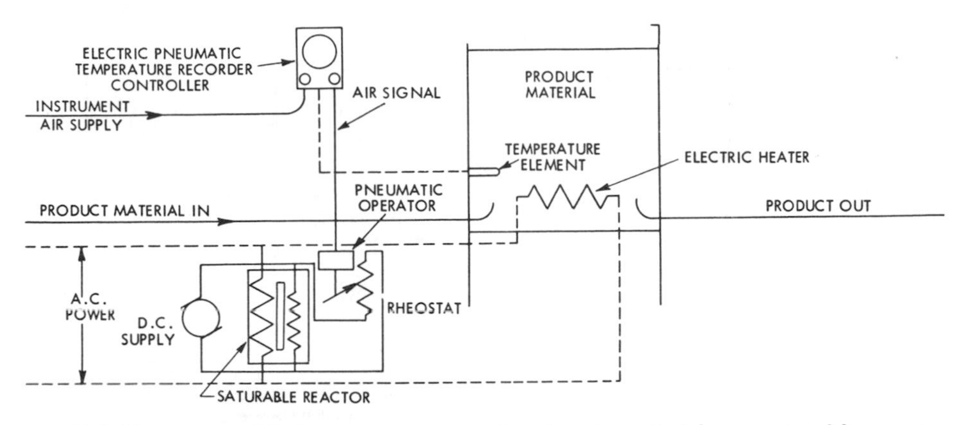

Fig. 15-4. Voltage to this temperature control system is varied by a saturable reactor.

trolled heaters should be capable of heating the product to set point. The controlled portion provides the additional heat required.

Another method of controlling electric heaters employs a device which varies the voltage to them. One such device is a variable transformer which can be motor driven. The control system resembles that used for the motor operated valve.

A saturable reactor, shown in Fig. 15-4, can also be used to vary the voltage to the heaters. A saturable reactor is a magnetic amplifier. It controls alternating current power by varying the direct current in the control circuit. This reactor is similar to a transformer with a dc primary and an ac secondary circuit. The controller contains an electrical measuring circuit, which uses the electrical primary element as the input device. The control circuit is pneumatic. It receives the error signal from an electronic amplifier, which actuates a pneumatic relay. The pneumatic signal is fed into a diaphragm-operated rheostat. The rheostat regulates the amount of dc voltage in the saturable reactor, which, in turn, regulates the amount of ac voltage going to the heaters. All the control actions described previously can be combined with this type of system.

Pressure Control

A frequent problem in pressure control is the regulation of fluid pressure. Water pressure must be regulated when it is used in heating systems and distribution systems. Gas or oil pressure must be regulated before entering a burner. Pressure regulators are commonly used to provide this service. The three classes of pressure regulators are direct-operated, pilot-operated, and instrument pilot-operated.

Direct-operated Regulators

Direct-operated regulators can be weight-loaded, pressure-loaded, or spring-loaded. See Fig. 15-5. Regard-

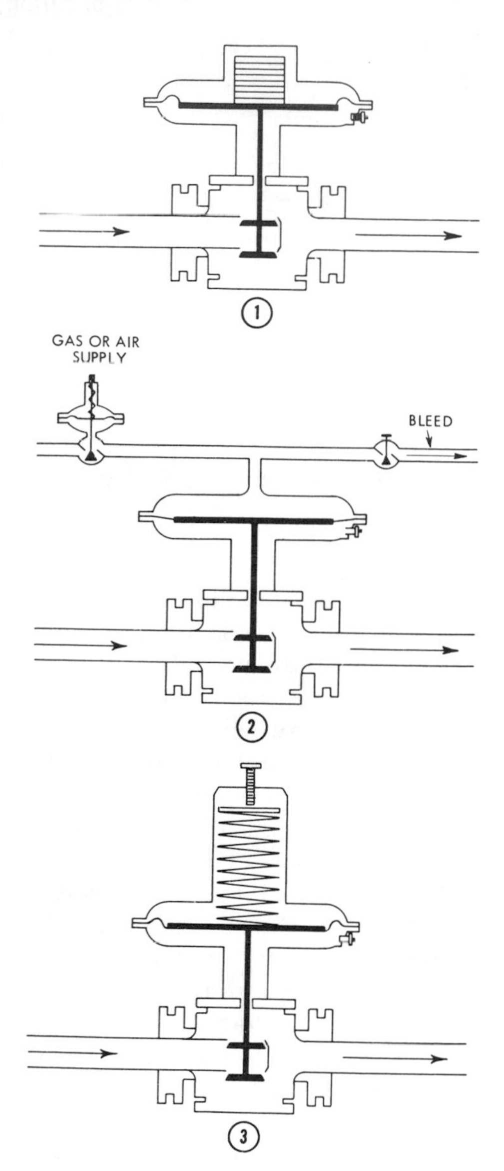

Fig. 15-5. Direct-operated regulators can be (1) weight-loaded, (2) pressure-loaded, or (3) spring-loaded.

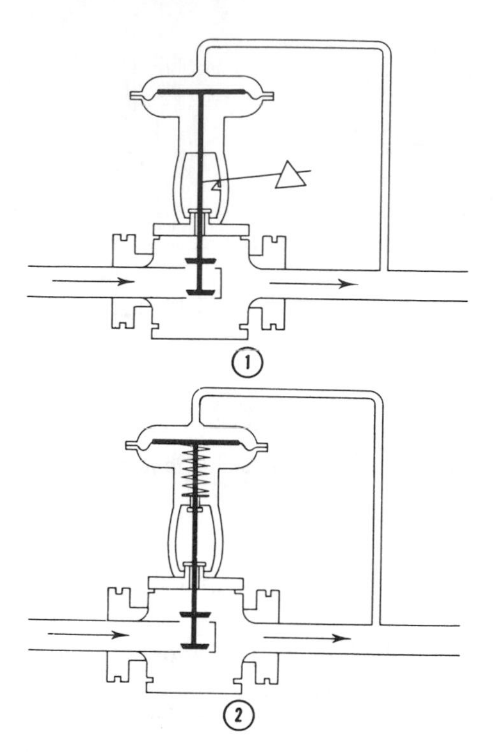

Fig. 15-6. Externally connected regulators can be (1) weight loaded or (2) spring-loaded.

less of which type is used, the pressure entering the regulator acts against the loading force. When this pressure rises, the loading force changes the valve opening so that the desired outlet pressure is maintained. Regulators can be self-contained. This means there is no external connection to the pressure pipe line. See Fig. 15-5. Regulators can also be connected externally, as shown in Fig. 15-6.

Pilot-operated Regulators

Pilot-operated regulators can have internal or external pilots. See Fig. 15-7. A pilot is a small regulator positioned between the pressure connection to the regulator, and the loading chamber. Controlled pressure is piped to the pilot, which varies the loading on the regulator. The addition of the pilot improves the control. Only a slight change in the controlled pressure is required to produce a full range change of the regulator.

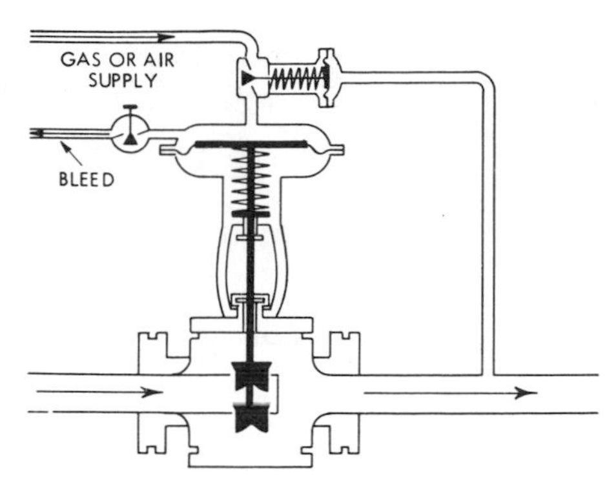

Fig. 15-7. Pressure regulator with a pilot-operated reducing valve.

oxygen, nitrogen, or hydrogen are stored in vessels in which the pressure must be regulated. Frequently a relief valve is all that is required. The relief valve opens in proportion to the magnitude of the overpressure. They are generally self-contained spring-loaded regulators.

When more precise pressure control is required, an instrument pilot and a diaphragm control valve are used. See Fig. 15-9. The instrument pilot can be a pneumatic pressure controller providing an air signal to a control valve. A variation of this system uses a pressure controller with an electrical output. The electrical signal is sent to an electro-pneumatic relay which provides the air signal required to operate the control valve. The electrical signal is introduced for faster control response, especially when the distance between the sensing point and the control point is great.

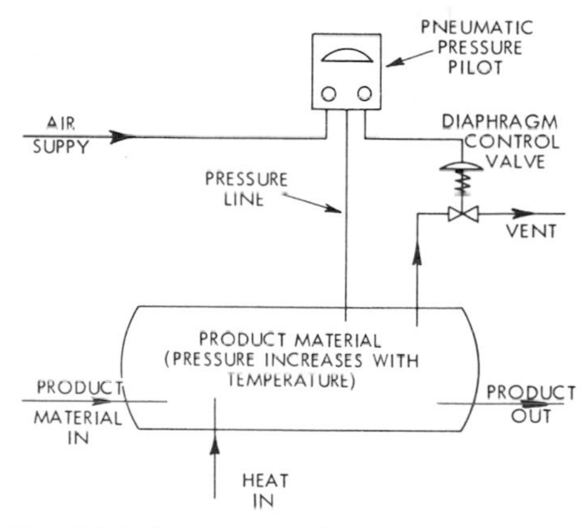

Fig. 15-9. In some applications, the pneumatic pressure pilot and a diaphragm control valve provide more precise pressure control.

Instrument Pilot-operated Regulators

Instrument pilot-operated regulators, shown in Fig. 15-8, provide more flexible control. In general, the pilot is a pneumatic controller with a bellows or pressure spring as the sensing element. The controller usually includes proportional band adjustment for varying the sensitivity. The regulation is actually performed by a control valve.

The regulation of pressure inside a closed vessel is a common control application. Gas holders (tanks for gas supply) require such control. Liquid

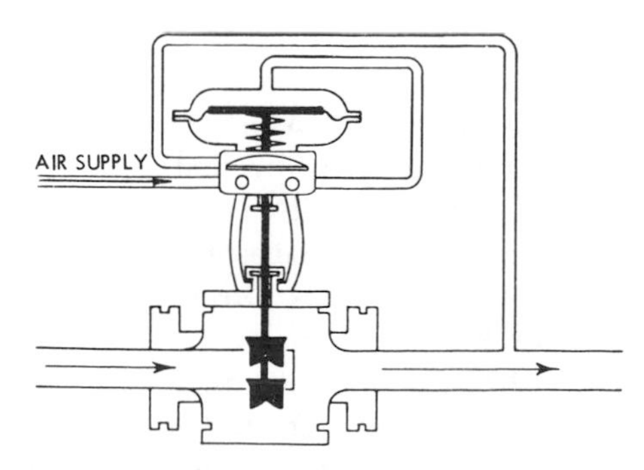

Fig. 15-8. This pressure regulator uses a reducing valve operated by an instrument pilot.

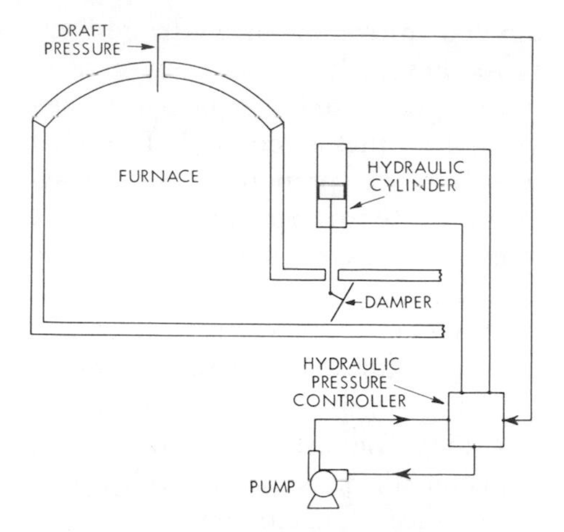

Fig. 15-10. Hydraulic system used for controlling industrial furnaces and boilers.

A hydraulic control system is often used for furnace pressure or draft control in industrial furnaces or boilers. This choice is made because hydraulic systems possess sufficient power to operate heavy dampers or large butterfly valves. Any of the elastic deformation pressure sensing elements may be used to vary the hydraulic signal. A hydraulic piston is used to position the final element. A typical hydraulic control system for controlling furnace draft is shown in Fig. 15-10. The draft pressure is sensed by a diaphragm which controls the hydraulic pressure to a double-acting cylinder which positions the outlet damper on the furnace.

Proportional, proportional plus reset, and proportional plus reset plus derivative actions may be included.

Level Control

The control of liquid level is necessary in many industrial processes. The control of fuel oil or water level in a combustion system is important for continuous operation. Controlling the level of the components of a mixture in a blending operation is a frequent requirement in the petroleum or chemical industries. The control of level in water or sewage treatment plants is essential. In general, all continuous processes require that the level of supply ingredients be controlled.

In the previous discussion of level measuring devices, we established two classes of instruments. *Direct-operated devices* are actuated by the varying amount of the process liquid. *Indirect-operated devices* are actuated by a variable which changes as level changes. The same classification can be applied to level control.

Direct-Operated Level Controllers

Ball floats, displacers, electrodes, and capacitance probes are some of the commonly used direct-operated liquid level controllers.

Ball Floats. The simplest ball float liquid level controller is the float-operated lever type shown in Fig. 15-11. The float is attached to one end of the lever system and rides on the surface of the liquid. The opposite end of the lever system operates a valve which regulates the flow of liquid into the storage vessel. This type of ball float is only suitable for open storage vessels.

Ball float-operated liquid level controllers are also available for use in

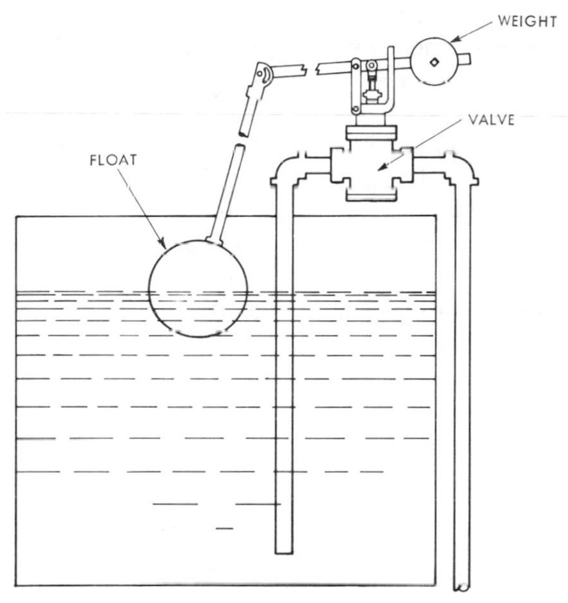

Fig. 15-11. Liquid level controller for use in open tanks. (Fisher Controls)

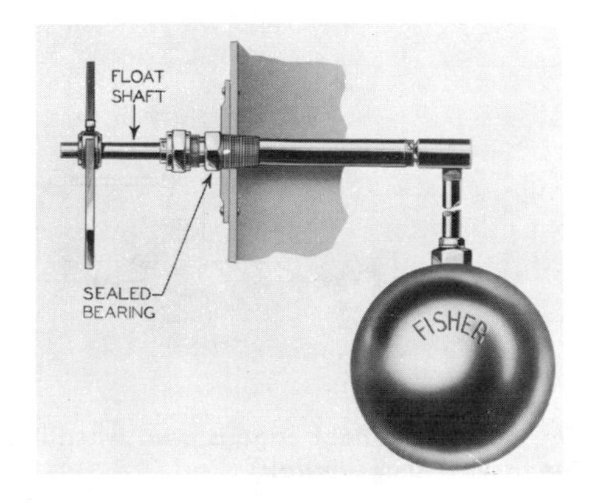

Fig. 15-12. Liquid level controller for use in closed tanks. (Fisher Controls)

closed tanks. This requires that the float shaft, which transmits the float motion to the lever system outside the tank, pass through a sealed bearing or stuffing box. See Fig. 15-12.

In some applications, the float cannot be situated in the vessel. When this is the case, a float cage is mounted on the side of the vessel with one pipe connection going to the top of the vessel and the other to the bottom. Any change in liquid level in the tank is matched by a change of level in the float cage. See Fig. 15-13.

When greater sensitivity and more versatile control are required, a pilot-operated float controller of the type shown in Fig. 15-14 is used. In this type of controller, the float lever system actuates a pneumatic or hydraulic relay. The control valve is then operated by pneumatic or hydraulic pressure. This arrangement makes it possible for the float mechanism to be located a considerable distance from the valve.

Ball floats are also used to actuate magnetic switches. See Fig. 15-15. The magnetic switch can be used to operate a solenoid valve or pump motor. In a typical magnetic-operated float switch, the float is enclosed in a chamber mounted on the side of the vessel. A magnetic piston is attached to the float. When the float rises and carries the piston to the level of the magnetic switch the electrical circuit is completed or broken.

Displacers. Displacer elements, when used for level control, provide greater sensitivity and permit the use of a proportional control system which

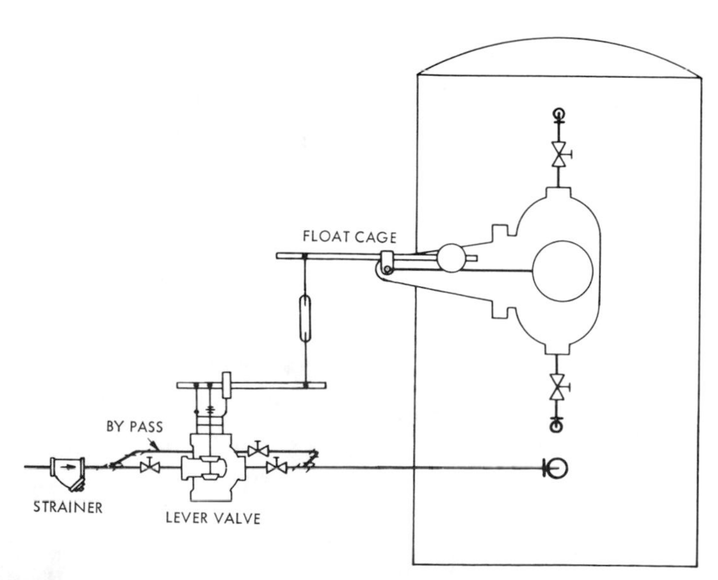

Fig. 15-13. The float cage is used when liquid level controller cannot be located inside the tank. (Fisher Controls)

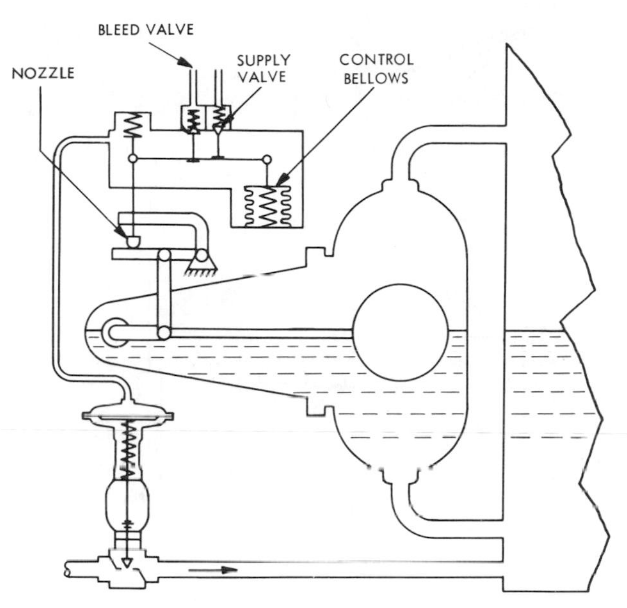

Fig. 15-14. A liquid level controller operated by pilot to provide greater accuracy. (Fisher Controls)

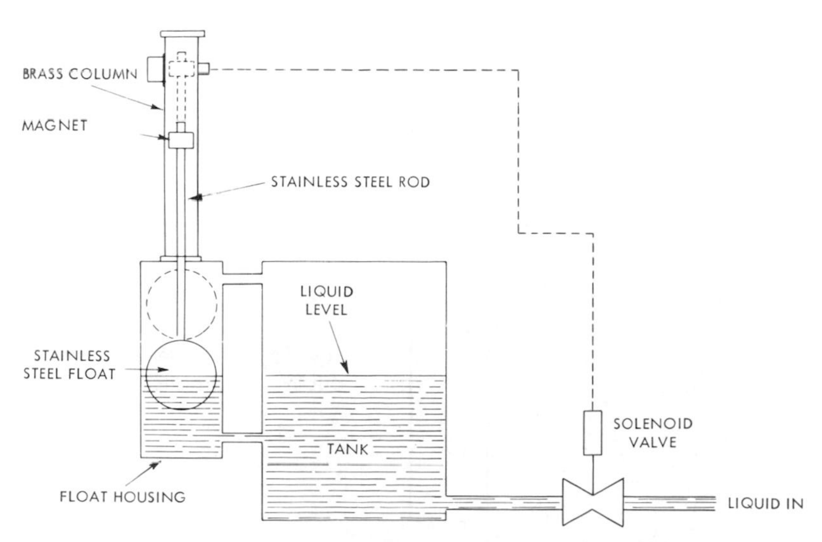

Fig. 15-15. The movement of the float in this liquid level controller operates the electric switch magnetically.

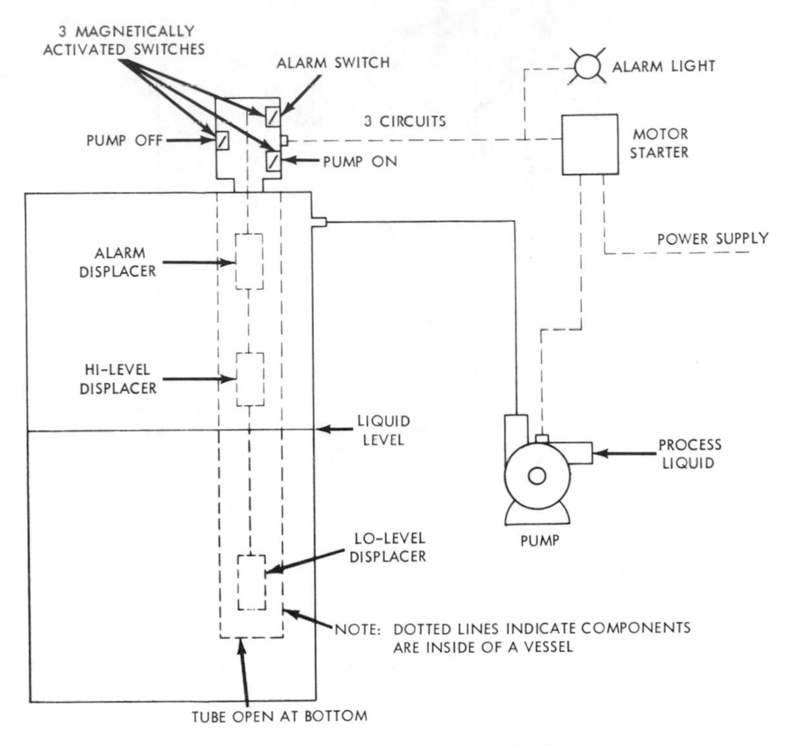

Fig. 15-16A. A level controller using displacer elements.

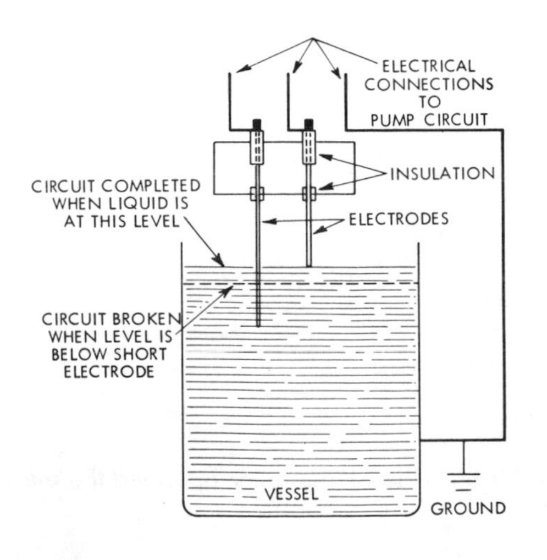

Fig. 15-16B. A level controller using conductive electrodes.

can be either pneumatic or electric. These have been described in the section on density measurement. See Fig. 15-16A.

Electrodes. Electrode liquid level controllers are used to control the level of conductive fluids such as water. The electrodes are electrically insulated from the storage vessel. The electrodes are available in different lengths. When the conductive fluid is in contact with two electrodes, an electric circuit is completed. See Fig. 15-16B. The circuit can be used to actuate a pump or operate a solenoid valve. This opens the flow of fluid into the storage vessel.

Capacitance Probes. Capacitance probes are used to control the level of various granular materials as well as many liquids. Liquid latex is a typical example. See Fig. 15-17. Capacitance probes consist of an outer shell with a center rod of metal which together serve as the plates of a capacitor. The liquid in the storage vessel serves as the insulating (dielectric) material. When the liquid falls below the level of the probe, the capacitance changes. The change in capacitance triggers an electronic circuit which operates an electric relay inside the control unit. The relay actuates the solenoid valve. The capacitance probe position establishes the point of level

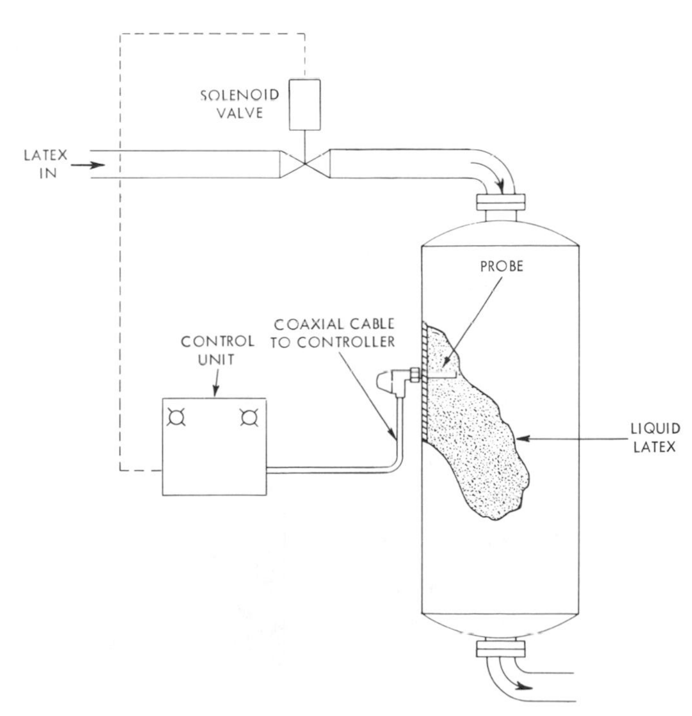

Fig. 15-17. A level controller using capacitance probe in the measurement of liquid latex. (Fisher Controls)

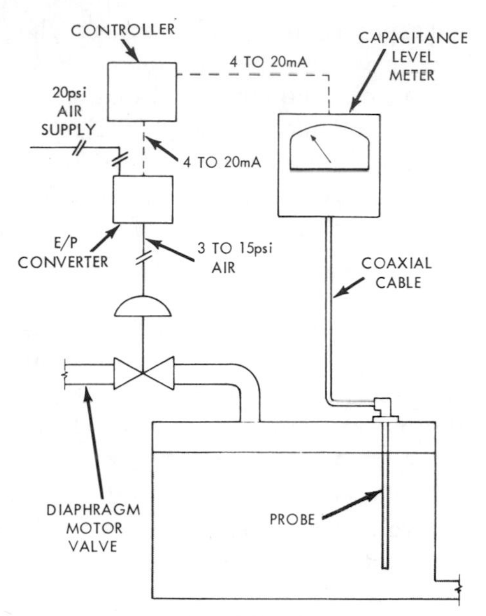

Fig. 15-18. A liquid level control system using a long capacitance probe.

control. See Fig. 15-17. Long capacitance probes are suspended in the storage vessel. The probes are used in some applications to provide proportional control of the level. They provide a gradual change in capacitance as the liquid level changes. See Fig. 15-18. This control system can be either pneumatic or electric.

Indirect-Operated Devices

Differential Pressure Controllers. Of the indirect-operated level measuring devices, the differential pressure types are most often selected for control. Their principal use is for controlling the level of volatile liquids in pressure vessels. The petroleum and chemical industries have many applications of this type. The level of liquid oxygen,

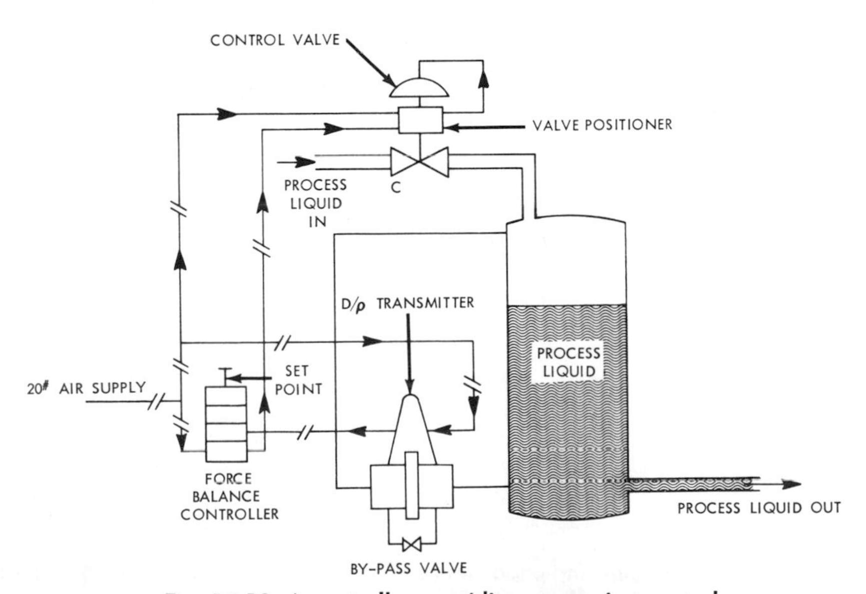

Fig. 15-19. A controller providing averaging control.

nitrogen, and hydrogen in pressure vessels can also be controlled using differential pressure instruments.

Fig. 15-19 illustrates a control system in which the level of the process liquid must be maintained at a height sufficient to enable the process itself to continue. The process liquid is stored in a pressure vessel. A differential pressure transmitter is used as the level-sensing instrument. The output of the transmitter becomes the measured variable input of a force balance pneumatic controller. The controller produces and sends a signal to the valve supplying liquid to the pressure vessel. The controller is equipped with proportional action. The proportional band uses a wide setting. This type of band setting allows the level to rise and fall with process demand. A valve positioner insures precise control. Such a system provides what is called *averaging control.*

Flow Control

Almost all industries employ flow controllers. Electric power generating stations require them for controlling fuel flow, steam flow, and water flow. Gas distribution companies need flow controllers to regulate the amount of gas being supplied to their customers. The food and beverage industries use them for cooking, heating, and cooling. Flow controllers are necessary in the steel industry to conserve water, fuel, and power. In refineries and chemical plants, flow controllers are extremely important to insure continuous processing of uniform products.

Rate Control

Differential Pressure Controllers. Various devices for measuring flow rate have been described in previous sections. The differential pressure device, which employs an orifice plate, is the type most frequently used. Fig. 15-20 shows such a device. The differential pressure produced by the orifice plate serves as the input to the measuring element. The measuring element is the bellows type. The motion of the bellows is used to actuate the control mechanism. In a pneumatic controller, the flapper is moved. This produces an output air signal used to position a diaphragm-operated control valve. The control mechanism includes a means for establishing the set point. If the measured flow rate exceeds the set point, the control valve closes to reduce the amount of flow. If the measured flow rate is below the set point, the control valve opens to increase the rate. The valve position must be corrected for any deviation in the fluid pressure because the amount of fluid that passes through a valve is affected by the pressure of the fluid. The addition of automatic reset action provides this correction. The control signal is proportional to differential pressure rather than flow rate.

Piping for differential pressure can

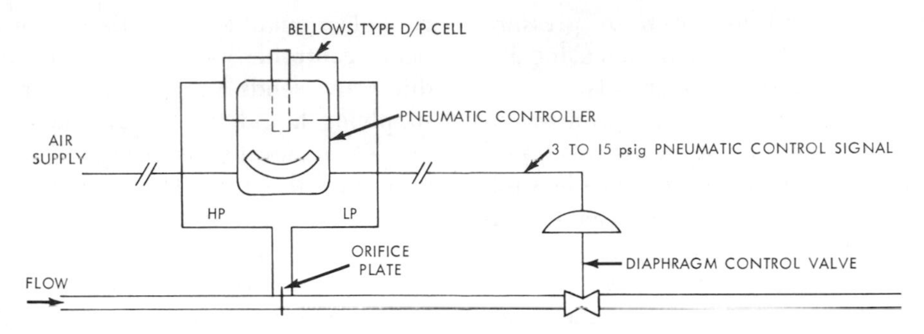

Fig. 15-20. Differential pressure controller used for rate control of flow.

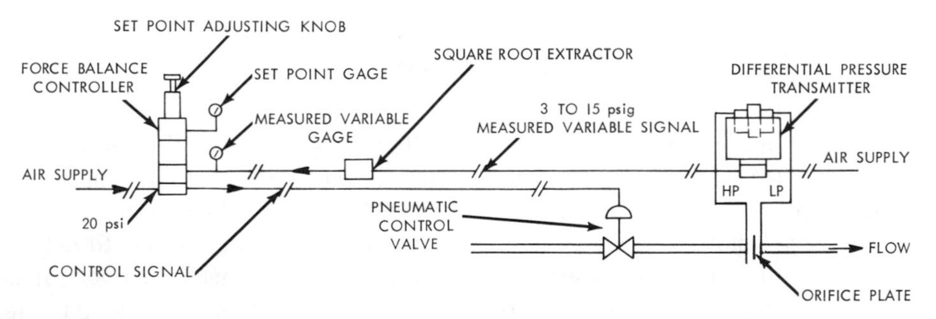

Fig. 15-21. Pneumatic system with separate transmitter and controller used for long distance rate flow control.

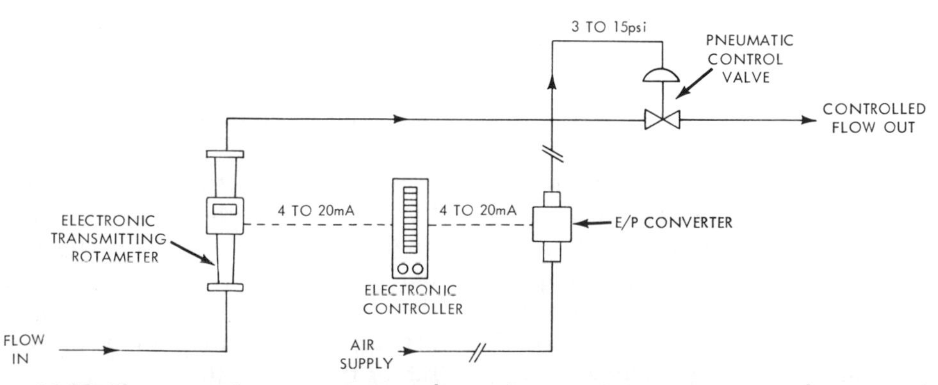

Fig. 15-22. Flow control system using an electronic transmitting rotameter with pneumatic control valve.

only be laid over a limited distance to avoid excessive measuring lag. Transmitters are often used in applications where distance is a factor. A typical installation uses a pneumatic pressure signal in the transmitter. See Fig. 15-21. The pneumatic pressure becomes the input to the force balance controller. The set point is manually applied to the controller as a pneumatic pressure. The air output of the controller goes to the control valve.

In the control system just described, the control signal varies with the differential pressure and not the flow. Since flow rate varies (approximately) as the square root of the pressure, a square root extractor is included in the system to remedy this condition.

Electric and Hydraulic Controllers. Electric or hydraulic controllers are also used for flow control. Electric systems are particularly preferred when the distance between the point of measurement and the final element is large. Although electric operators are available for use as the final element, pneumatic or hydraulic operators are commonly used because they are more powerful. Using a pneumatic operator necessitates converting the electric signal from the controller to a pneumatic signal.

Rotameters. Other flow rate measuring devices, such as rotameters, electromagnetic flowmeters, turbine-type flowmeters are also adaptable to flow control. The rotameter used is the transmitting type. See Fig. 15-22. The output of the transmitter (4 to 20 mA) becomes the input to an electronic controller. The output of the electronic controller in turn becomes the input for an electronic to pneumatic converter. The pneumatic signal (3 to 15 psig) operate a pneumatic control valve.

Turbine Flowmeter. The turbine flowmeter, shown in Fig. 15-23, acts as a transmitter. Its pulse-type output enters a digital to analog (d/a) converter which provides a 4 to 20 mA output. The output is proportional to the flow rate. An electronic controller accepts the output of the d/a converter.

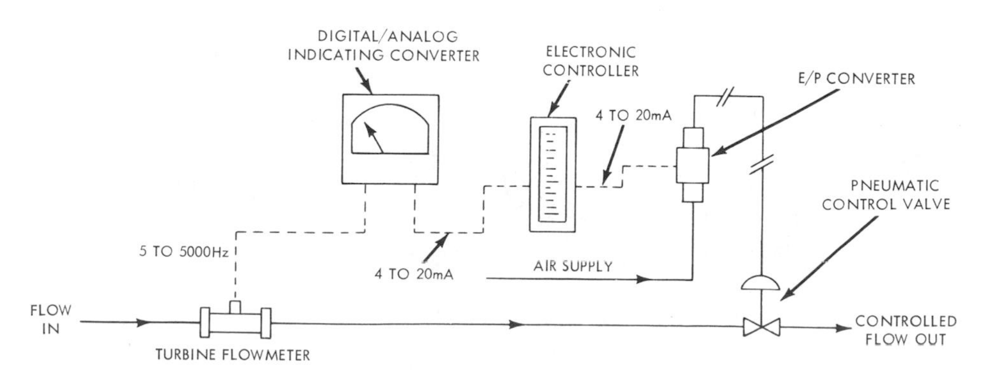

Fig. 15-23. Flow control system using a turbine flow meter and pneumatic control valve.

Ratio Control

Frequently, an industrial process requires the blending of two fluids. Flow ratio control is used to accomplish this. Uncontrolled fluid flow is called *wild flow.* Wild flow is measured by using any of the conventional methods discussed earlier. In the control system shown in Fig. 15-24, the measurement of flow is converted into a signal by a transmitter. This signal can be used to establish the set point of the controller which regulates the flow of the controlled fluid. The flow rate of the controlled fluid varies with the measured rate of the wild flow.

Another flow ratio control system is shown in Fig. 15-25. In this system, the flow that is not controlled is called the *independent flow,* and the controlled fluid is termed the *dependent flow.* The measured independent flow becomes the input to a pneumatic flow

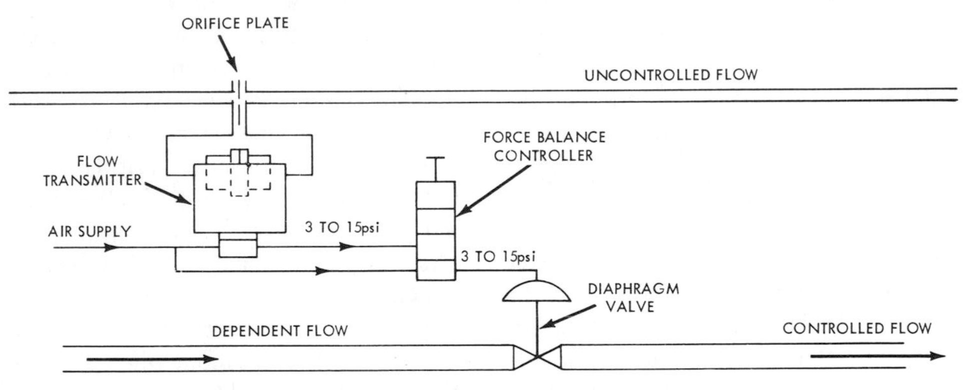

Fig. 15-24. Controlled flow depends on controlled flow measurement.

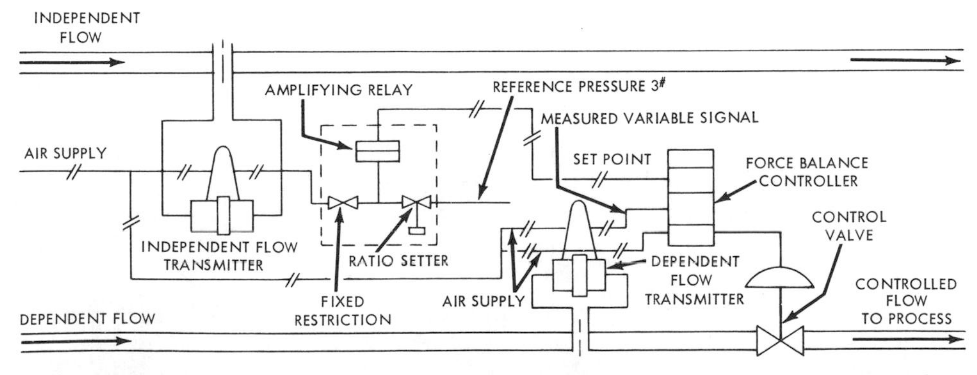

Fig. 15-25. Ratio setter determines relationship between independent and dependent flow.

transmitter which converts the measurement to an air pressure. The air pressure passes through a fixed restriction and enters the pneumatic ratio unit. This pneumatic unit is an adjustable pressure divider. A second input to the pressure divider is an adjustable reference pressure. The output of the pressure divider becomes the input to an amplifying relay. When the ratio set needle valve is fully open, the input pressure to the amplifying relay equals the pressure from the transmitter. When the needle valve is fully closed, the input pressure to the amplifying relay equals 1/25th of the pressure from the transmitter. This means that the relationship between the independent flow measurement and the output of the ratio unit can be effected by adjusting the needle valve. For example, the output of the ratio unit can vary its full range of 3 psi to 15 psi for a variation of only 25% of the measured independent flow. It can also require a change of 200% in the measured flow to produce the full range ratio unit output.

The output of the ratio unit is used as the set point input of a force balance controller. The dependent flow is also measured. This measurement is converted by the pneumatic transmitter into an air pressure. The pneumatic output of the flow transmitter becomes the measured variable pressure of the force balance controller. The controller output then regulates a diaphragm-operated valve in the dependent flow line so that the dependent flow is held at the set point. The set point represents a particular percentage of the independent flow. The dependent flow is held to this percentage of the independent flow.

Ratio control is not limited to flow applications although this is the most frequent use. The ratio of two temperatures or two pressures can be measured with similar systems.

Analysis Control

Controlling processes by using analyzers is perhaps one of the most widespread and important applications of instrumentation. This importance stems from the continued development of equipment and techniques.

Column Chromatography

Chromatographic analyzers are used in the control of petroleum refinery processes, among other industries. One of the critical phases of this technique is the creation of sampling systems. Metering pumps feed the fluids into the measuring apparatus. After the sampling and analyzing systems have been installed, the components which require control are selected by using a device called, interestingly enough, a *peak picker*. (It should be recalled that the chromatographic record is a continuous curve with a series of peaks.) A

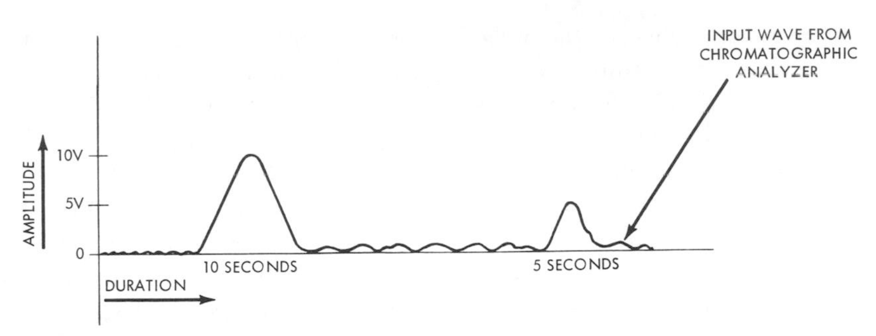

Fig. 15-26. Typical chromatographic signal pattern showing peak from one component.

millivolt signal, the output signal used for each component, varies with the height of the peak for each particular component. Fig. 15-26 shows an example of a typical chromatographic signal pattern. The output signal becomes the input to a controller. The controller receives the signal and positions a final element accordingly. The controller is maintained until a new signal is received indicating the need for a new positioning of the final element. The interval between control impulses is longer when more than one component is to be controlled. The controller used in this installation is the direct type which requires no feedback signal from the final element.

Viscosity

Another example of analysis control is the continuous control of viscosity. Coating or dipping processes, using such materials as plastics or adhesives, are improved by the application of viscosity control.

An example of viscosity control might be a process in which the viscosity of a fluid is controlled by regulating the flow rate of the solids and liquids which make up the process mixture. In this example, it is assumed that only one solid and one liquid are combined to form the process mixture. Fig. 15-27 shows a schematic of such a process, with the required instrumentation indicated. Continuous viscosity measurement is made with a rotating spindle apparatus. The electrical signal, which is proportional to the viscosity, is in the form of a capacitance change. The signal enters a capacitance bridge recorder/controller. The controller provides an output signal which operates both the feeder (dry material) and the liquid flow valve. When the viscosity becomes too low, the controller increases the flow of dry material. When the viscosity is too high, the controller increases the liquid flow. If the viscosity is within the desired range, the dry material and the liquid move into the process at a predetermined rate (ratio). The ratio can be established by employing timers to regulate the time-On and time-Off cycles of the feeder and the liquid flow valve. If the materials come into a mixing tank before entering the process, a level control device is needed. This controller cuts off the flow of materials when level reaches a predetermined point.

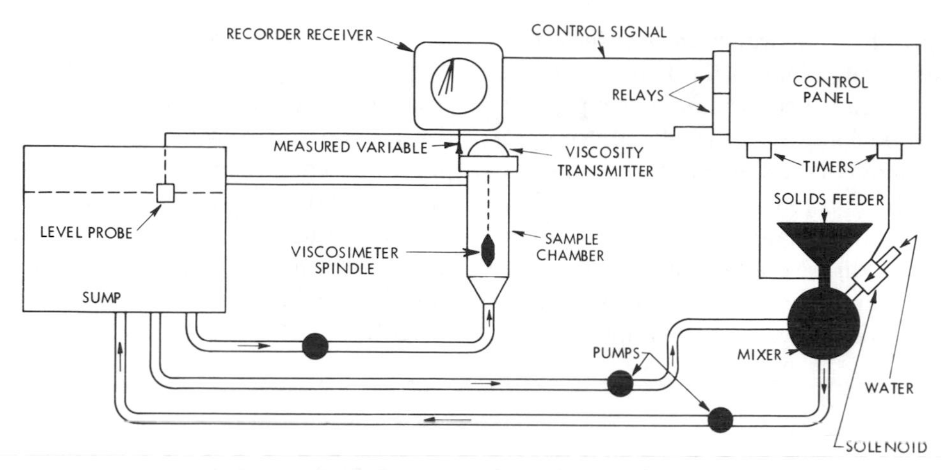

Fig. 15-27. Viscosity control system.

Electrolytic Conductivity

Another analysis instrument suitable for control applications is the electrolytic conductivity measurement device. An example of this instrument is shown in Fig. 15-28. A common application is controlling the purity of boiler feed water.

The electrolytic conductivity element is inserted into the feed water line. This enables the element to detect changes in the electrolytic conductivity of the water resulting from impurities. When the element senses that the impurity level is too high, the feed water is diverted to a treatment section. At the same time, a new sup-

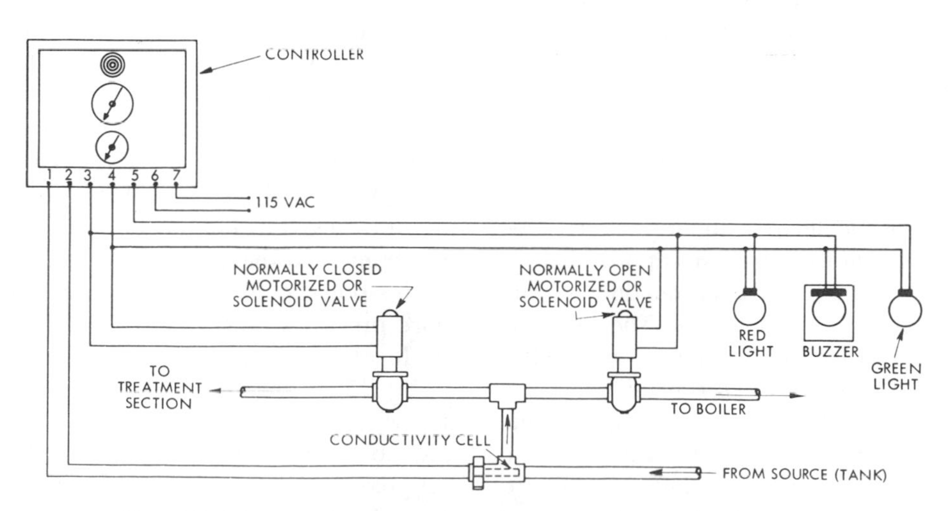

Fig. 15-28. System for monitoring the purity of boiler feed water.

ply of feed water moves into the feed water line to the boiler. The instrument for controlling such a process can be comprised entirely of electrical components, as shown in Fig. 15-28.

The controller is an On-Off mechanism. It actuates the solenoid valves which direct the flow of feed water to the boiler or to the treatment section. The solenoid valves also supply treated water to the system as needed.

There are many other analysis instruments which can be adapted for control. The instrument selected must provide a continuous measurement of the variable, using a transducer. The output of the transducer must be suitable for use as the input for a conventional controller. Electronic controllers are best suited to provide such service.

Cascade Control

The applications described above have involved measuring and controlling one variable. There are applications which achieve greater control of the process by measuring one variable and using this measurement as the set point to regulate a second variable. This type of control is called *cascade control*. Cascade control is used when the disturbances within the process make simpler methods of control ineffective.

Cascade control can be used in various applications. For example, it can be used to provide precise control of the liquid level inside a pressure vessel. Liquid level, in this application, is essential for the continuation of the process. See Fig. 15-29. The level of the process liquid is measured with an electronic differential pressure transmitter. The current output of the transmitter provides the input for an electronic level controller. The output of this controller, in turn, is used as the set point input to an electronic flow controller. The flow transmitter measures inlet flow to the pressure vessel. The current output of this transmitter, which is linear with the flow rate, serves as the variable input to the electronic flow controller. An electric to pneumatic converter converts the output of the flow controller to a pneumatic signal. The pneumatic signal positions the valve which controls the flow of liquid into the pressure vessel. The valve positioner provides better control. The flow controller regulates the quantity of liquid entering the pressure vessel as the level in the vessel changes to meet the requirements of the process. This system provides more precise level control than a simple level control system.

Fig. 15-30 illustrates another example of cascade control. Temperature and flow are combined to control the temperature of the process liquid by regulating the supply of steam heat.

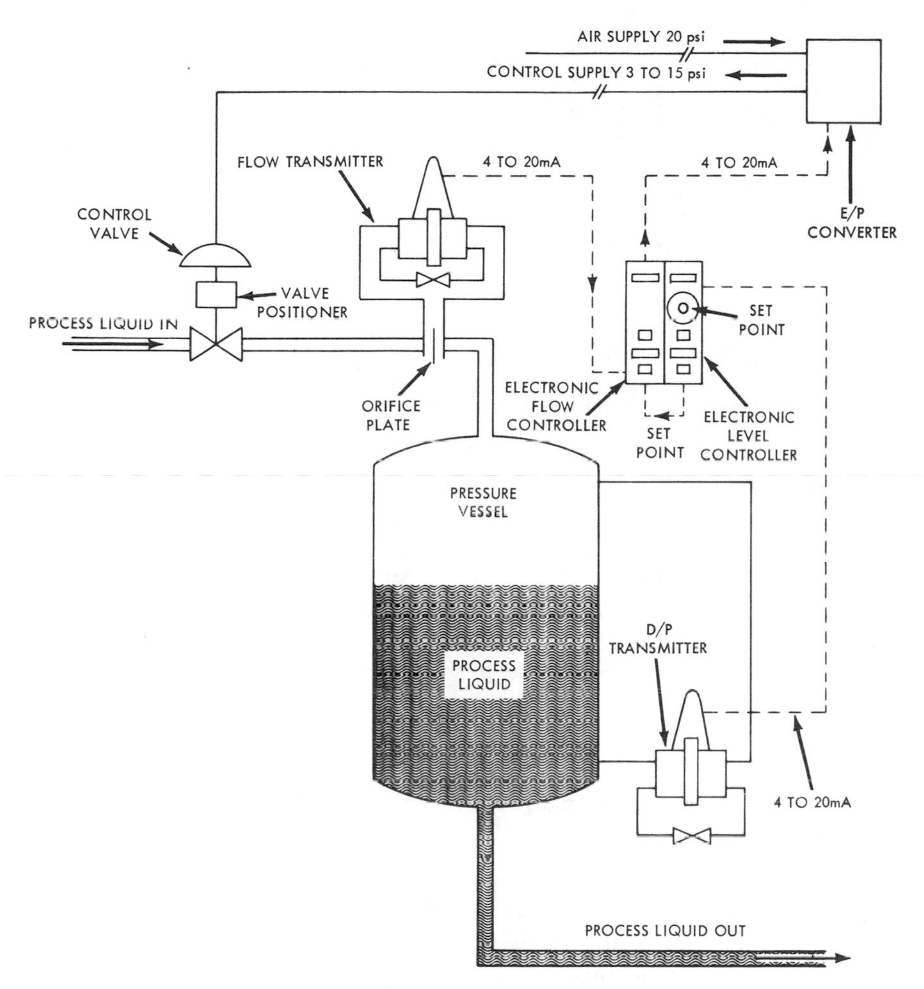

Fig. 15-29. Level-flow cascade control system.

Various process conditions dictate the need for cascade control. For example, the temperature of the process liquid can respond very slowly to disturbances. Or, a change in the process might cause a major upset in the temperature. Other variables might be disturbed by a change in the temperature of the process liquid.

In the cascade control system shown in Fig. 15-30, a filled-system temperature transmitter provides a pneumatic signal. This signal is proportional to the temperature of the process liquid. The signal serves as the input to a pneumatic force balance temperature controller. The output of the controller is used as the set point of the pneumatic force balance flow controller. An air regulator, set manually, provides

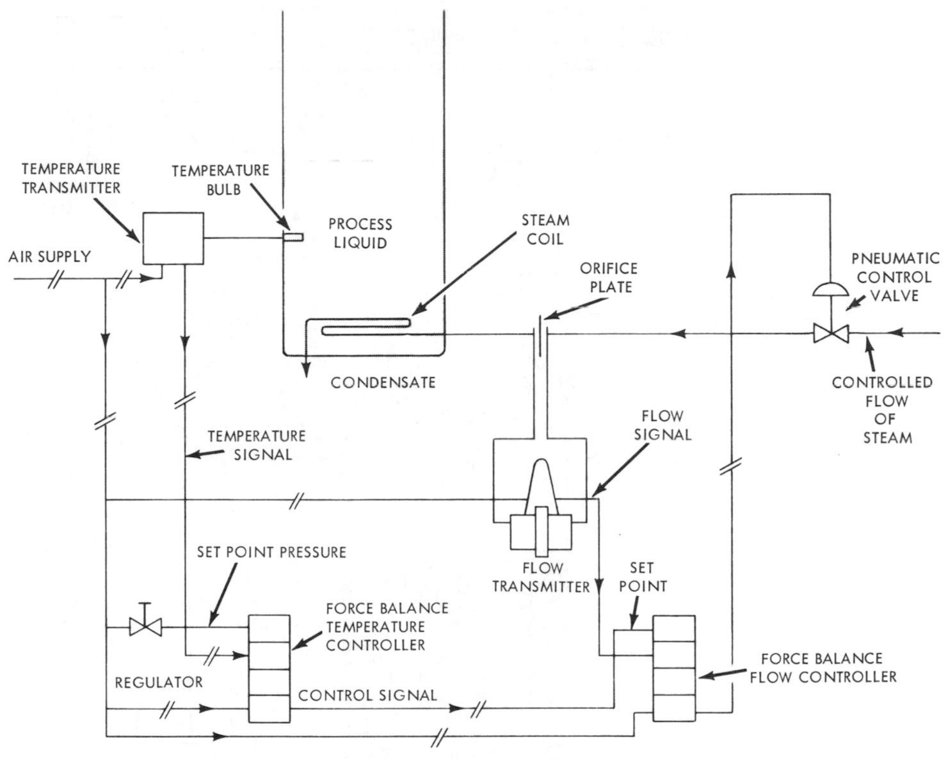

Fig. 15-30. Temperature-flow cascade control system.

the set point of the temperature controller. The differential pressure flow transmitter provides a pneumatic flow signal. The air output of the flow controller positions the pneumatic control valve. The valve, in turn, regulates the flow of steam to the coil, which produces a change in the temperature of the process liquid. The temperature controller used should be equipped with automatic reset and flow controller proportional response.

Words to Know

batch process
continuous process
overshoot
bumpless transfer
saturable reactor
pilot
relief valve
averaging control
rotameter
wild flow
independent flow
dependent flow
column chromatography
peak picker
cascade control

Review Questions

1. Give two examples of products which might be manufactured in a *batch process* and two products which might be manufactured in a *continuous process.*
2. In temperature control applications, what are some of the possible hazards which might be avoided with the proper selection of controls?
3. What is the function of the pilot on a pilot-operated pressure regulator?
4. Which level detecting devices are applicable to both liquids and solids?
5. What is meant by the term *averaging liquid* as used in liquid level control?
6. Using instrumentation catalogs, select the equipment necessary for a pneumatic ratio flow control system.
7. Select a process which is usually done in the home and show how it could be regulated, using industrial instruments. Provide a sketch of what specific equipment you would use.
8. Make a sketch of a typical cascade control system in which a temperature controller and a level controller are interconnected. Use electrical components, where possible.

Appendix

A

Instrumentation Symbols

Every scientific and technical field develops a specialized language to convey information and ideas. The specialized language not only includes particular meanings for words, it also includes abbreviations and symbols peculiar to a given field. Anyone who seeks training in such a technical or scientific field must become familiar with the specialized language of that field.

The Instrument Society of America has prepared a standard, entitled *Instrumentation, Symbols and Identification, ISA - S5.1**, for individuals in the field of instrumentation. The purpose of this standard is to provide a satisfactory system of symbols and identification for industrial process instrumentation equipment. The standard designates and identifies equipment in various applications and sets up rules for identifying new equipment. This is designed to promote a uniformity of

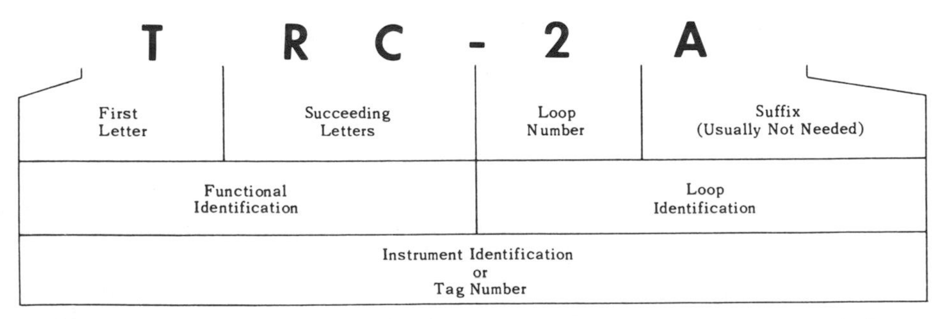

This chart shows system for identifying or constructing instrumentation symbols.

**Instrumentation, Symbols and Identification, ISA-S5.1,* (1973). Instrument Society of America, 400 Stanwix Avenue, Pittsburgh, Penna. 15222.

practice and expedite communication in the field of instrumentation. The symbols, tables, and text in this appendix are adapted from *ISA - S5.1*.

Outline of the Identification System

General. Each instrument shall be identified first by a system of letters used to classify it functionally. To establish a loop identity for the instrument, a number shall be appended to the letters. This number will, in general, be common to other instruments of the loop of which this instrument is a part. A suffix is sometimes added to complete the loop identification. A typical tag number for a temperature recording controller is shown on P. 310.

The instrument tag number may include coded information such as plant area designation. Each instrument may be represented on diagrams by a symbol. The symbol may be accompanied by an identification.

Functional Identification. The functional identification of an instrument shall consist of letters from Table 1, p. 315, and shall include one first-letter, covering the measured or initiating variable, and one or more succeeding-letters covering the functions of the individual instrument. An exception to this rule is the use of the single letter *L* to denote a pilot light that is not part of an instrument loop.

The functional identification of an instrument shall be made according to the *function* and not according to the construction. Thus, a differential-pressure recorder used for flow measurement shall be identified as an *FR*, a pressure indicator and a pressure switch connected to the output of a pneumatic level transmitter shall be identified as *LI* and *LS*, respectively.

In an instrument loop, the first letter of the functional identification shall be selected according to the *measured or initiating variable* and not according to the manipulated variable. Thus, a control valve varying flow according to the dictates of a level controller is an *LV*, not an *FV*.

The succeeding-letters of the functional identification designate one or more readout or passive functions, or output functions, or both. A modifying-letter may be used, if required, in addition to one or more other succeeding-letters. Modifying letters may modify either a first-letter or other succeeding-letters, as applicable.

The sequence of identification letters shall begin with a first-letter. Readout or passive functional letters shall follow in any sequence, and output functional letters shall follow these in any sequence, except that output letter *C* (control) shall precede output letter *V* (valve); e.g., *HCV*, a hand-actuated control valve. However, modifying letters, if used, shall be interposed so that they are placed immediately following the letters they modify.

An instrument tagging designation on a flow diagram may be drawn with as many tagging balloons as there are measured variables or outputs. Thus, a flow-ratio recording transmitter with a flow-ratio switch may be identified on a flow diagram by two tangent circles, one inscribed *FFRT-3* and the other *FFS-3*. The instrument would be designated *FFRT/FFS-3* for all uses in writing and reference. If desired, however, the abbreviated *FFRT-3* may serve for general identification or purchasing while *FFS-3* may be used for electric circuit diagrams.

The number of functional letters grouped for one instrument should be kept to a minimum according to the judgment of the user. The total number of letters within one group should not exceed *four.* The number within a group may be kept to a minimum by these means:

1. The functional letters are arranged into subgroups. This practice is used for instruments having more than one measured variable or output, but it may also be done for other instruments.
2. If an instrument indicates and records the same measured variable, then the *I* (indicate) may be omitted.

All letters of the functional identification shall be uppercase.

Loop Identification. The loop identification of an instrument shall generally use a number assigned to the loop of which the instrument is a part. Each instrument loop shall have a unique number. An instrument common to two or more loops may have a separate loop number, if desired.

A single sequence of loop numbers shall be used for all instrument loops of a project or sections of a project regardless of the first letter of the functional identification of the loops. A loop numbering sequence may begin with the number *1* or with any other convenient number, such as *301* or *1201,* that may incorporate coded information such as plant area designation.

If a given loop has more than one instrument with the same functional identification, then, preferably, a suffix shall be appended to the loop number, e.g., *FV-2A, FV-2B, FV-2C,* etc., or *TE-25-1, TE-25-2, TE-25-3,* etc. However, it may be more convenient or logical in a given instance to designate a pair of flow transmitters, for example, as *FT-2* and *FT-3* instead *of FT-2A* and *FT-2B.* The suffixes may be applied according to the following guidelines:

1. Suffix letters, which shall be uppercase should be used, i.e., *A, B, C.* etc.
2. For an instrument such as a multi-point temperature recorder that prints numbers for point identification, the primary elements may be numbered *TE-25-1, TE-25-2, TE-25-3,* etc. The primary element suffix numbers should correspond to the point numbers of the recorder. Optionally, they may not correspond.
3. Further subdivisions of a loop may be designated by alternating suffix letters and numbers.

An instrument that performs two or more functions may be designated by all of its functions. For example, a flow recorder *FR-2* with pressure pen *PR-4* is preferably designated *FR-2/PR-4.* Alternatively, it may be designated *UR-7,* a two-pen pressure recorder may be *PR-7/8,* and a common annunciator window for high- and low-temperature alarm may be *TAH/L-9.*

Instrument accessories, such as purge rotameters, air sets, and seal pots that are not explicitly shown on a flow diagram but that need a tagging designation for other purposes should be tagged individually according to their function and shall use the same loop number as that of the instrument they directly serve. Application of such a designation does not imply that the accessory must be shown on the flow diagram. Alternatively, the accessories may use the identical tag number as that of their associated instrument, but with clarifying words added, if required. Thus, an orifice flange union associated with orifice plate *FE-7* should be tagged *FX-7,* but may be tagged *FE-7 flanges.* A purge rotameter-regulator associated with pressure gage *PI-8* should be tagged *FICV-8* but may be tagged *PI-8 purge.* A thermowell used with thermometer *TI-9* should be tagged *TW-9,* but may be tagged *TI-9 thermowell.*

Symbols. The drawings below illustrate the symbols that are intended to depict instrumentation on flow diagrams and other drawings, and cover their application to the variety of processes. The applications shown were chosen to illustrate principles of the methods of symbolization and identification. Additional applications that adhere to these principles may be devised as required. The examples show numbering that is typical for the pictured instrument interrelationships, but the numbering may be varied to suit the situation. The symbols indicating the various locations of instruments have been applied in typical ways in the illustrations. This does not imply, however, that the applications or the designations of the instruments are therefore restricted in any way. No inference should be drawn that the choice of any of the schematics constitutes a recommendation for the illustrated methods of measurement or control. Where alternative symbols are shown, without a statement of preference, the relative sequence of the symbols does not imply a preference.

The circular balloon may be used to tag distinctive symbols, such as that for a control valve, when such tagging is desired. (In such instances, the line connecting the baloon to the instrument symbol shall be drawn close to but not touching the symbol.) In other instances, the balloon serves to represent the instrument proper.

A distinctive symbol, whose relationship to the remainder of the loop is easily apparent from a diagram, need not be individually tagged on the diagram. For example, it is expected that an orifice plate or a control valve that is part of a larger system will not usually be shown with a tag number on a diagram. Also, where there is an electrical primary element connected to another instrument on a diagram, use of a symbol to represent the primary element on the diagram is optional.

Where the identity is ambiguous or not conveniently determined, then it is expected that the identity will be clarified by the addition of the tag number, with or without a balloon, adjacent to the symbol. In any event, the instrument shall bear a distinctive tag number in other documents and references. A brief explanatory notation may be added adjacent to a symbol in order to clarify the function of an item.

The sizes of the tagging balloons and the miscellaneous symbols shown below are the sizes generally recommended; however, the optimum sizes may vary, depending on whether the finished diagram is to be photographically reduced in size and on the number of characters that are expected in the instrument tagging designation. The sizes of the other symbols may be selected as appropriate to accompany the symbols of other equipment on a diagram.

Aside from the general drafting requirement for neatness and legibility, all symbols may be drawn with any orientation. Likewise, signal lines may be drawn on a diagram entering or leaving the appropriate part of a symbol at any angle. Directional arrowheads shall be added to signal lines when needed to clarify the direction of flow of intelligence. The electric, pneumatic, or other power supply to an instrument is not expected to be shown unless it is essential to an understanding of the operation of the instrument or the loop.

In general, one signal line will suffice to represent the interconnections between two instruments on flow diagrams even though they may be connected physically by more than one line.

The sequence in which the instruments of a loop are connected on a flow diagram shall reflect the functional logic. This arrangement will not necessarily correspond to the signal connection sequence. Thus, a loop using analog voltage signals requires parallel wiring, while a loop using analog current signals requires series wiring. But the diagram in both instances shall be drawn as though all the wiring were parallel. This will clearly show the functional interrelationships while keeping their aspect of the flow diagram independent of the type of instrument system installed. The literal and correct wiring interconnections are expected to be shown on a suitable electric wiring diagram.

For process flow diagrams or other applications, where it may be desired to depict only those instrumentation end-functions that are needed for the operation of the process proper, the intermediate instrumentation and other details may be omitted, provided that this is done consistently for a given type of drawing throughout a project. Minor instruments and loop components, e.g., pressure gages, thermometers, transmitters, converters, may thus be eliminated from the diagrams.

It is common practice for mechanical flow diagrams to omit the symbols of interlock-hardware components that are actually necessary for a working system, particularly when symbolizing electric interlock systems. For example, a level switch may be shown as tripping a pump. Or, separate flow and pressure switches may be shown as actuating a solenoid valve or other interlock device. In both instances, auxiliary electrical relays and other components may also be required, but these additional components may be considered details to be shown elsewhere. By the same token, the current transformer shown below will sometimes be omitted and its receiver shown connected directly to the process, in this case the electric motor.

TABLE 1
MEANINGS OF IDENTIFICATION LETTERS

This table applies only to the functional identification of instruments.

	FIRST LETTER		SUCCEEDING LETTERS		
	MEASURED OR INITIATING VARIABLE	MODIFIER	READOUT OR PASSIVE FUNCTION	OUTPUT FUNCTION	MODIFIER
A	Analysis		Alarm		
B	Burner Flame		User's Choice	User's Choice	User's Choice
C	Conductivity (Electrical)			Control	
D	Density (Mass) or Specific Gravity	Differential			
E	Voltage (EMF)		Primary Element		
F	Flow Rate	Ratio (Fraction)			
G	Gaging (Dimensional)		Glass		
H	Hand (Manually Initiated)				High
I	Current (Electrical)		Indicate		
J	Power	Scan			
K	Time or Time Schedule			Control Station	
L	Level		Light (Pilot)		Low
M	Moisture or Humidity				Middle or Inter- mediate
N	User's Choice		User's Choice	User's Choice	User's Choice
O	User's Choice		Orifice (Restriction)		
P	Pressure or Vacuum		Point (Test Connection)		
Q	Quantity or Event	Integrate or Totalize			
R	Radioactivity		Record or Print		
S	Speed or Frequency	Safety		Switch	
T	Temperature			Transmit	
U	Multivariable		Multifunction	Multifunction	Multifunction
V	Viscosity			Valve, Damper, or Louver	
W	Weight or Force		Well		
X	Unclassified		Unclassified	Unclassified	Unclassified
Y	User's Choice			Relay or Compute	
Z	Position			Drive, Actuate or Unclassified Final Control Element	

TABLE 2
FUNCTION DESIGNATIONS FOR RELAYS

SYMBOL	FUNCTION
1. 1-0 or ON-OFF	Automatically connect, disconnect, or transfer one or more circuits provided that this is not the first such device in a loop
2. Σ or ADD	Add or totalize (add and subtract) †
3. Δ or DIFF.	Subtract †
4. ± + [−]	Bias*
5. AVG.	Average
6. % or 1:3 or 2:1 (typical)	Gain or attenuate (input:output)*
7. [x]	Multiply †
8. ÷	Divide †
9. [√] or SQ. RT.	Extract square root
10. x^n or $x^{1/n}$	Raise to power
11. f (x)	Characterize
12. 1:1	Boost
13. [>] or HIGHEST (MEASURED VARIABLE)	High-select. Select highest (higher) measured variable (not signal, unless so noted).
14. [<] or LOWEST (MEASURED VARIABLE)	Low-select. Select lowest (lower) measured variable (not signal, unless so noted).
15. REV.	Reverse
16.	Convert
a. E/P or P/I (typical)	For input/output sequences of the following: **Designation** — **Signal** E — Voltage H — Hydraulic I — Current (electrical) O — Electromagnetic or sonic P — Pneumatic R — Resistance (electrical)
b. A/D or D/A	For input/output sequences of the following: A — Analog D — Digital
17. ∫	Integrate (time integral)
18. D or d/dt	Derivative or rate
19. 1/D	Inverse derivative
20. As required	Unclassified

* Used for single-input relay.
† Used for relay with two or more inputs.

TABLE 3
INSTRUMENT LINE SYMBOLS

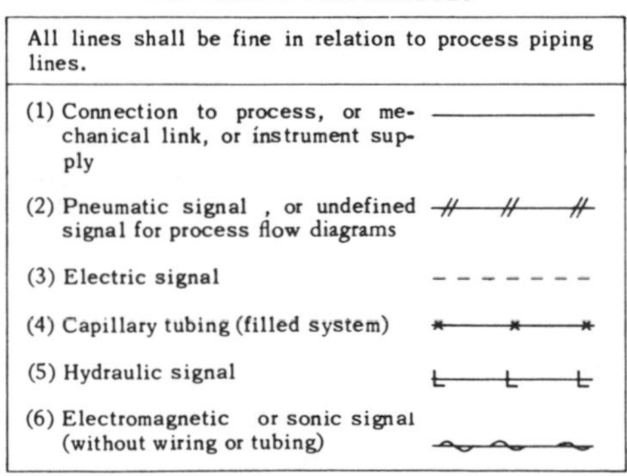

All lines shall be fine in relation to process piping lines.

(1) Connection to process, or mechanical link, or instrument supply
(2) Pneumatic signal , or undefined signal for process flow diagrams
(3) Electric signal
(4) Capillary tubing (filled system)
(5) Hydraulic signal
(6) Electromagnetic or sonic signal (without wiring or tubing)

TABLE 4
SPECIAL ABBREVIATIONS

ABBREVIATION	MEANING
A	Analog signal
ADAPT.	Adaptive control mode
AS	Air supply
AVG.	Average
C	Patchboard or matrix board connection
D	Derivative control mode Digital signal
DIFF.	Subtract
DIR.	Direct-acting
E	Voltage signal
ES	Electric supply
FC	Fail closed
FI	Fail indeterminate
FL	Fail locked
FO	Fail open
GS	Gas supply
H	Hydraulic signal
HS	Hydraulic supply
I	Current (electrical) signal Interlock
M	Motor actuator
MAX.	Maximizing control mode
MIN.	Minimizing control mode
NS	Nitrogen supply
O	Electromagnetic or sonic signal
OPT.	Optimizing control mode
P	Pneumatic signal Proportional control mode Purge or flushing device
R	Automatic-reset control mode Reset of fail-locked device Resistance (signal)
REV.	Reverse-acting
RTD	Resistance (-type) temperature detector
S	Solenoid actuator
S.P.	Set-point
SQ.RT.	Square root
SS	Steam supply
T	Trap
WS	Water supply
X	Multiply Unclassified actuator

ISA Symbols

GENERAL INSTRUMENT SYMBOLS – BALLOONS

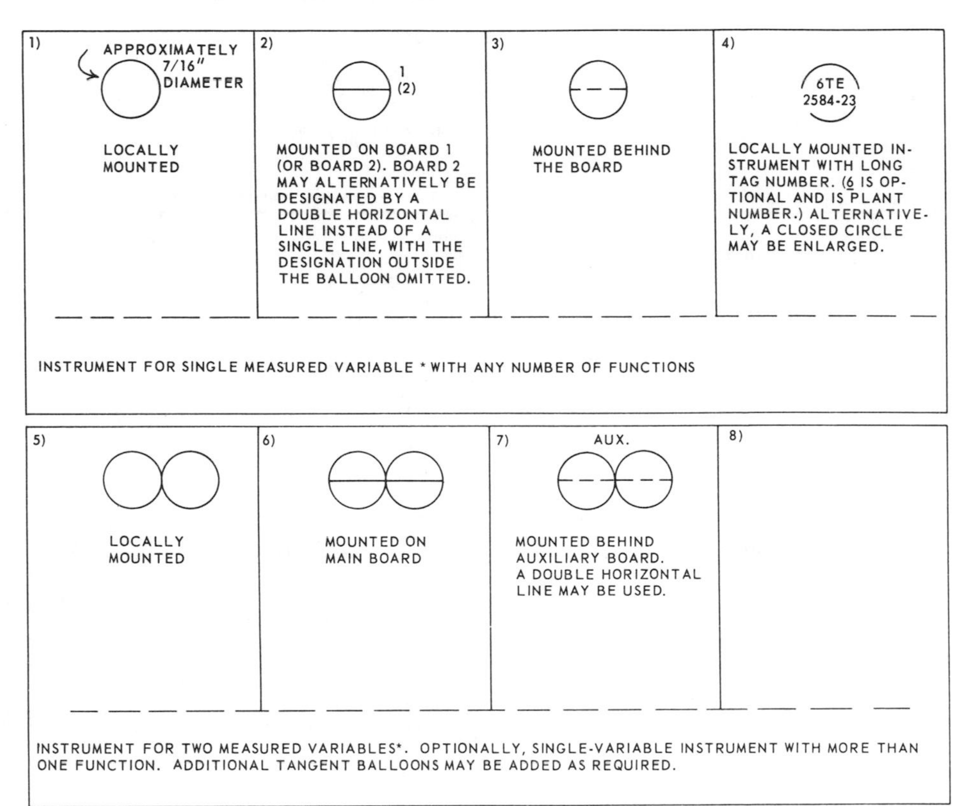

*Certain measured variables may have more than one input. An instrument that only indicates differential-pressure, for example, shall use only one balloon, tagged PDI, even though it has two inputs.

CONTROL VALVE BODY SYMBOLS

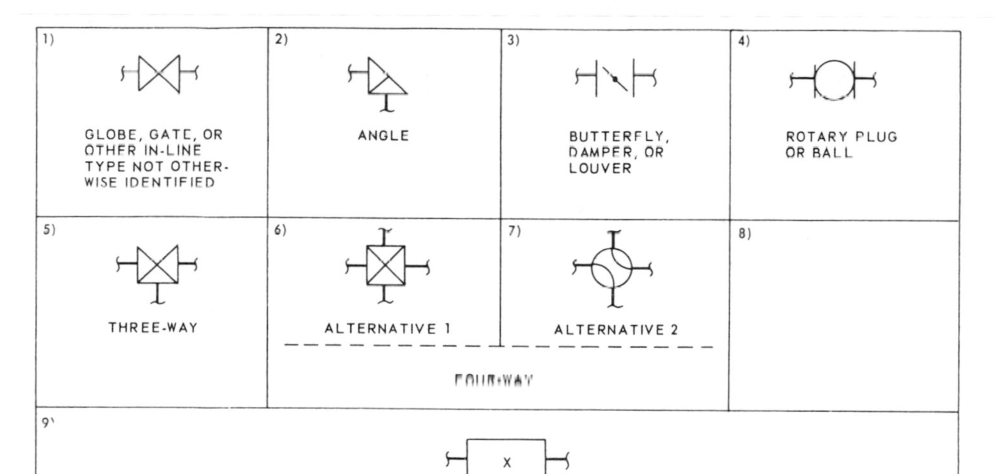

ACTUATOR SYMBOLS †

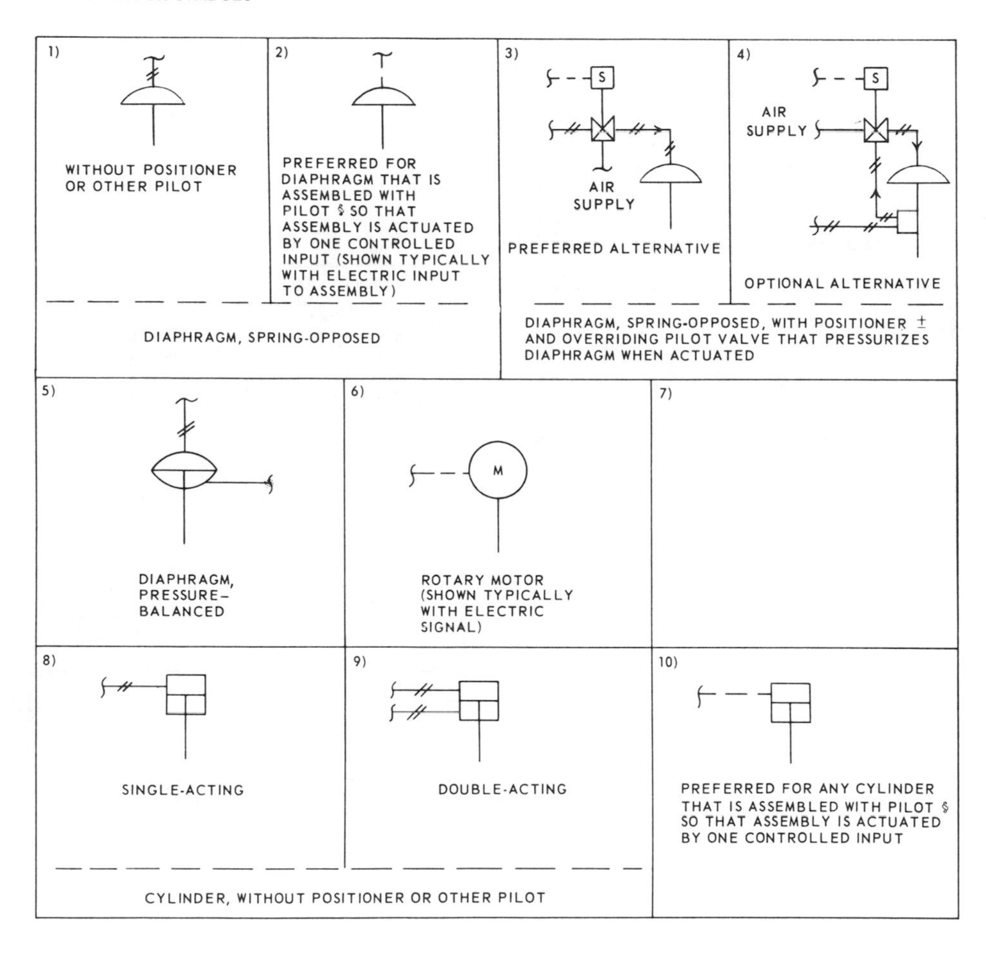

† Normally, modes of control valve action will not be designated on a flow diagram. However, an on-off valve mode may be designated, if desired, by placing the symbol 1-0 or ON-OFF near the valve symbol.

§ Pilot may be positioner, solenoid valve, signal converter, etc.

± The positioner shall preferably not be shown unless an intermediate device is on its output. The positioner tagging, ZC, shall preferably not be used even if the positioner is shown. The positioner symbol, a box drawn on the actuator shaft, is the same for all types of actuators. When the symbol is used, the type of instrument signal , i.e., pneumatic, electric, etc., shall be drawn as appropriate. If the positioner symbol is used and there is no intermediate device on its output, then the positioner output signal need not be shown.

ACTUATOR SYMBOLS (Contd.)

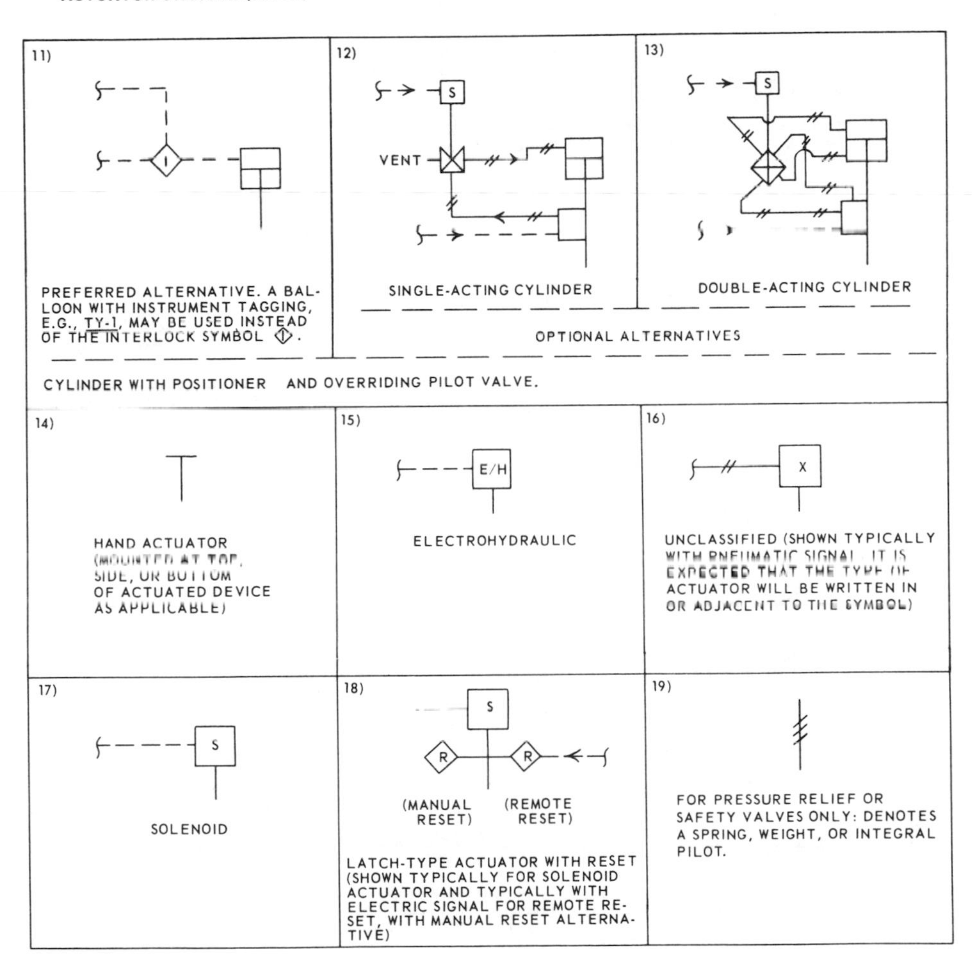

SYMBOLS FOR SELF-ACTUATED REGULATORS, VALVES, AND OTHER DEVICES

FLOW	1) 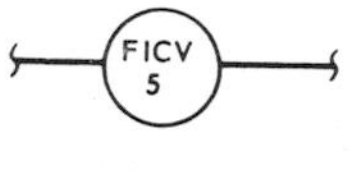 FICV 5 AUTOMATIC REGULATOR WITH INTEGRAL FLOW INDICATION. TAG REGULATOR FCV-5 IF IT DOES NOT HAVE INTEGRAL FLOW INDICATION.	2) 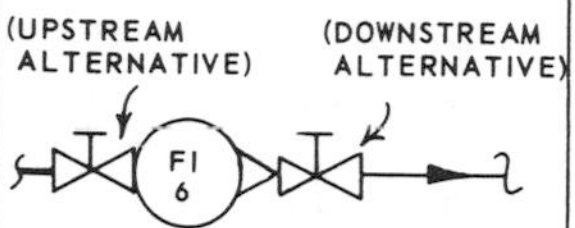 (UPSTREAM ALTERNATIVE) (DOWNSTREAM ALTERNATIVE) FI 6 INDICATING ROTAMETER WITH INTEGRAL MANUAL THROTTLE VALVE	3)
HAND	4) 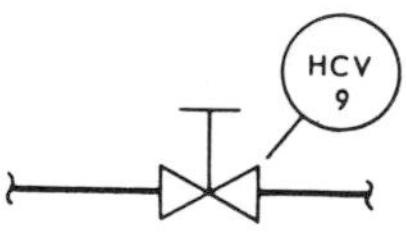 HCV 9 HAND CONTROL VALVE IN PROCESS LINE	5) 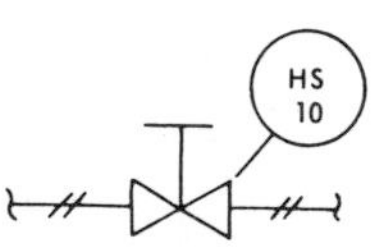 HS 10 HAND-ACTUATED ON-OFF SWITCHING VALVE IN PNEUMATIC SIGNAL LINE	6) 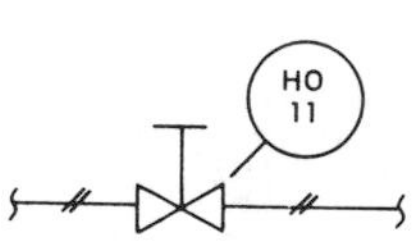 HO 11 MANUALLY ADJUSTABLE RESTRICTION ORIFICE IN SIGNAL LINE
LEVEL	7)  TANK LCV 14 LEVEL REGULATOR WITH MECHANICAL LINKAGE	8)	9)
PRESSURE	10) 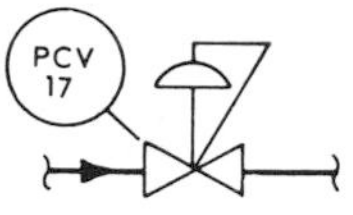 PCV 17 PRESSURE-REDUCING REGULATOR, SELF-CONTAINED	11) 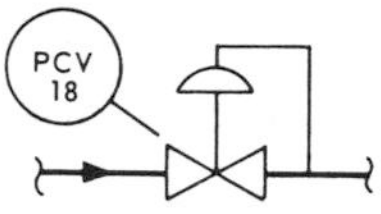 PCV 18 PRESSURE-REDUCING REGULATOR WITH EXTERNAL PRESSURE TAP	12) 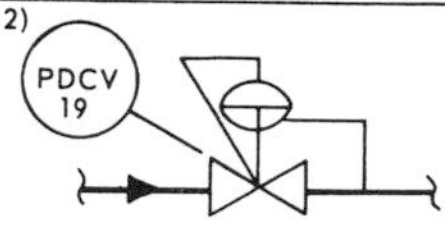 PDCV 19 DIFFERENTIAL-PRESSURE-REDUCING REGULATOR WITH INTERNAL AND EXTERNAL PRESSURE TAPS

SYMBOLS FOR SELF-ACTUATED REGULATORS, VALVES, AND OTHER DEVICES (Contd.)

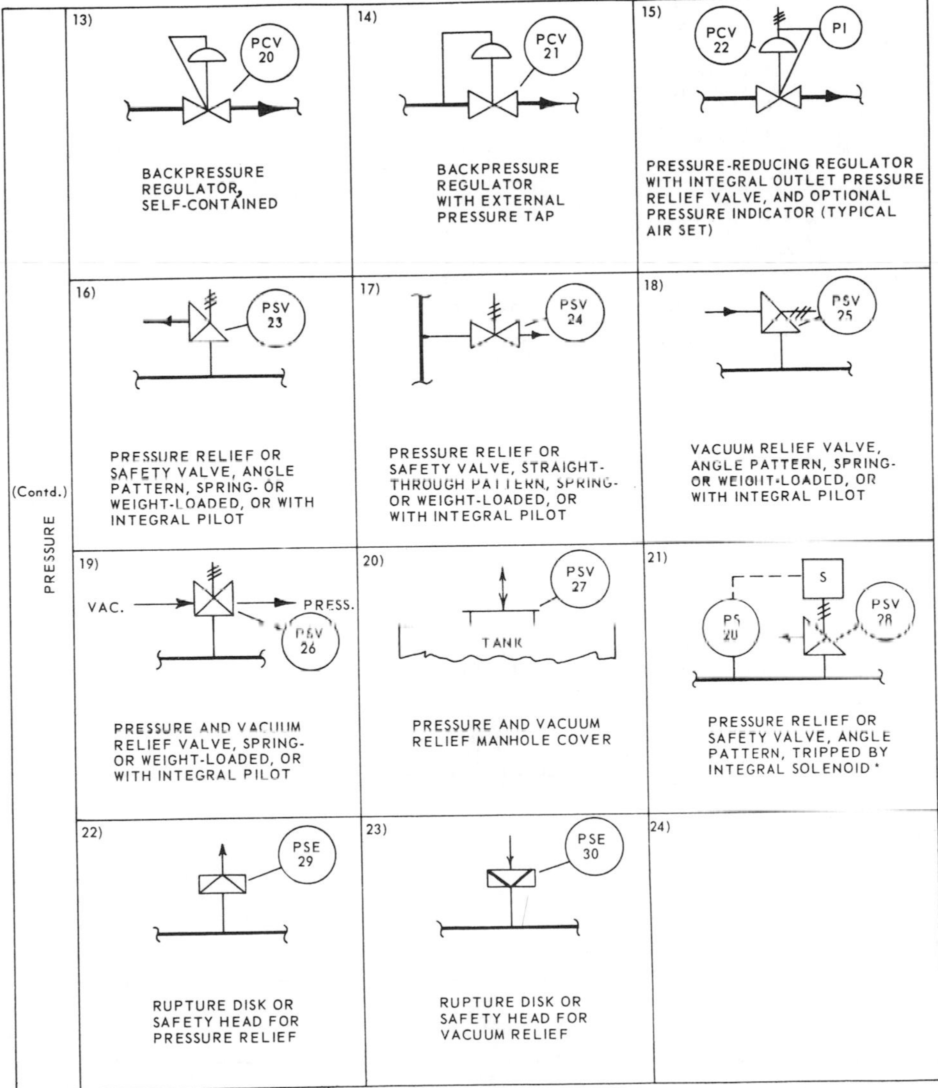

*The solenoid-tripped pressure relief valve is one of the class of power-actuated relief valves and is grouped with the other types of relief valves even though it is not entirely a self-actuated device.

SYMBOLS FOR SELF-ACTUATED REGULATORS, VALVES, AND OTHER DEVICES (Contd.)

TEMPERATURE	25) TCV 35 — TEMPERATURE REGULATOR, FILLED-SYSTEM TYPE	26) TANK, TSE 36 — FUSIBLE PLUG OR DISK	27)
TRAPS	28) T, XCV 40 — ALL TRAPS OTHER THAN BALL-FLOAT-TYPE CONTINUOUS DRAINERS	29) TANK, T, LCV 41 — CONTINUOUS DRAINER, BALL-FLOAT TYPE, WITH EQUALIZING CONNECTION	30)

SYMBOLS FOR ACTUATOR ACTION IN EVENT OF ACTUATOR POWER FAILURE.
(SHOWN TYPICALLY FOR DIAPHRAGM-ACTUATED CONTROL VALVE)

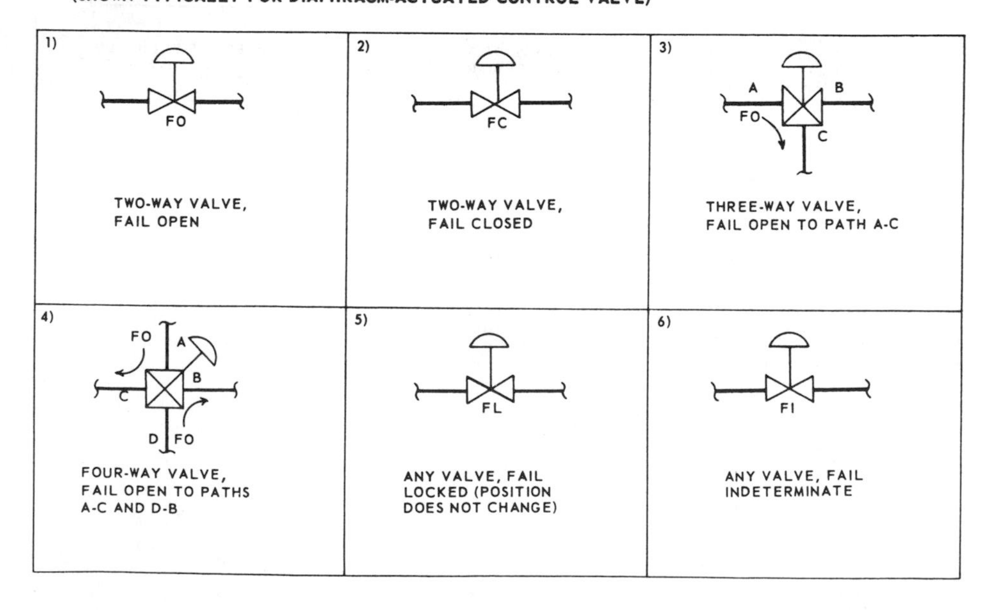

MISCELLANEOUS SYMBOLS

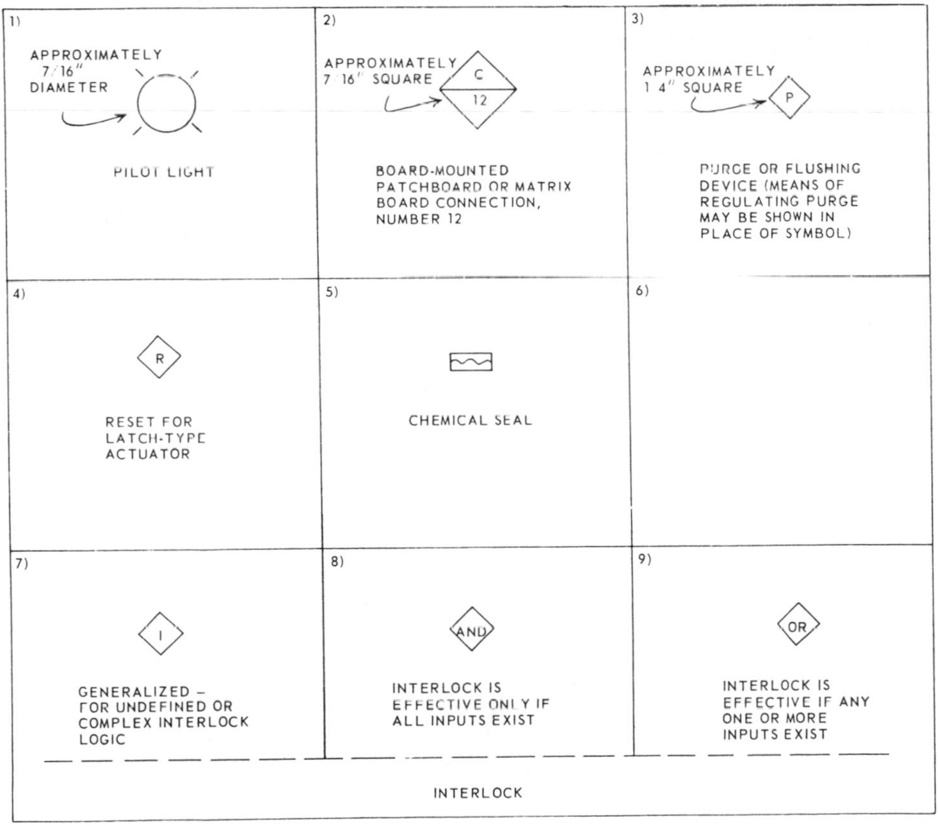

1) APPROXIMATELY 7/16" DIAMETER PILOT LIGHT	2) APPROXIMATELY 7/16" SQUARE (C / 12) BOARD-MOUNTED PATCHBOARD OR MATRIX BOARD CONNECTION, NUMBER 12	3) APPROXIMATELY 1/4" SQUARE (P) PURGE OR FLUSHING DEVICE (MEANS OF REGULATING PURGE MAY BE SHOWN IN PLACE OF SYMBOL)
4) (R) RESET FOR LATCH-TYPE ACTUATOR	5) CHEMICAL SEAL	6)
7) (I) GENERALIZED – FOR UNDEFINED OR COMPLEX INTERLOCK LOGIC	8) (AND) INTERLOCK IS EFFECTIVE ONLY IF ALL INPUTS EXIST	9) (OR) INTERLOCK IS EFFECTIVE IF ANY ONE OR MORE INPUTS EXIST
INTERLOCK		

PRIMARY ELEMENT SYMBOLS

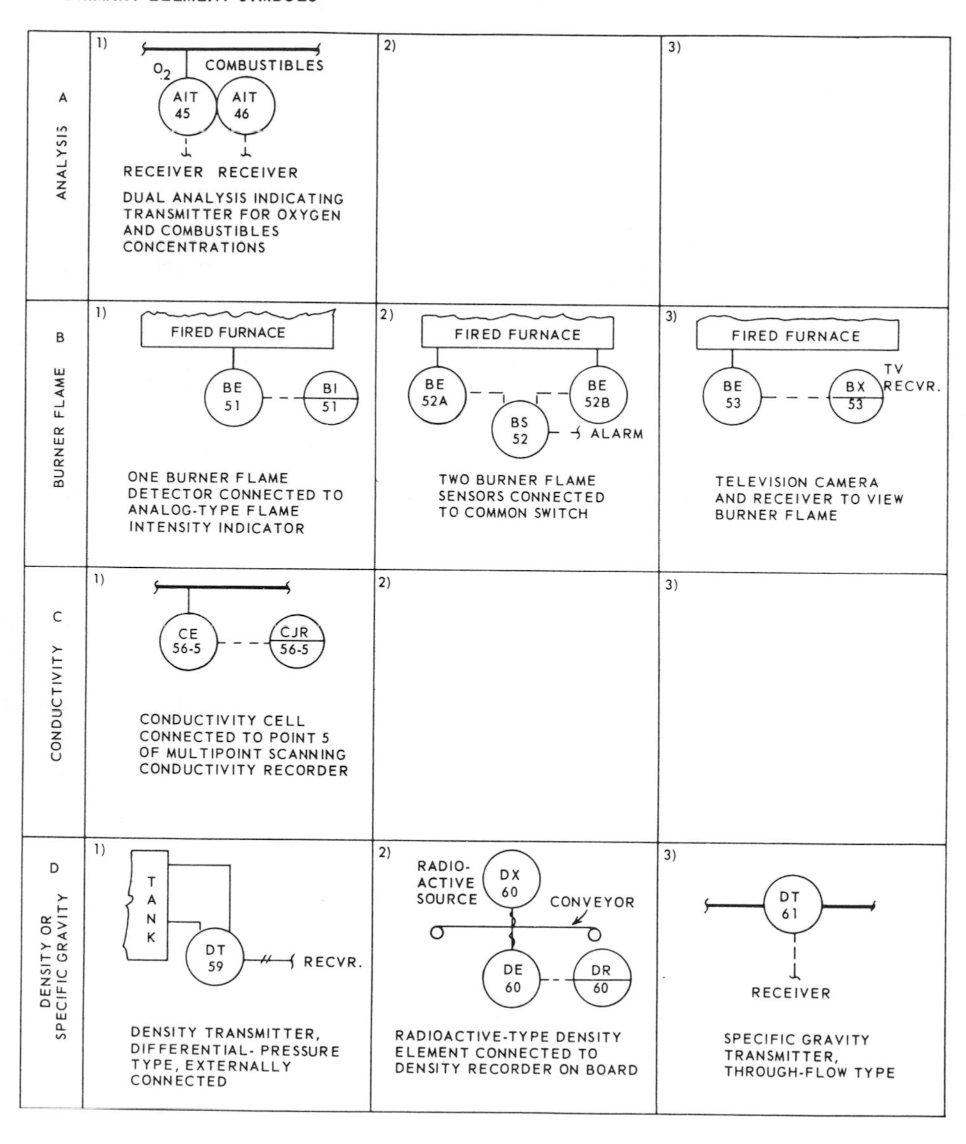

PRIMARY ELEMENT SYMBOLS (Contd.)

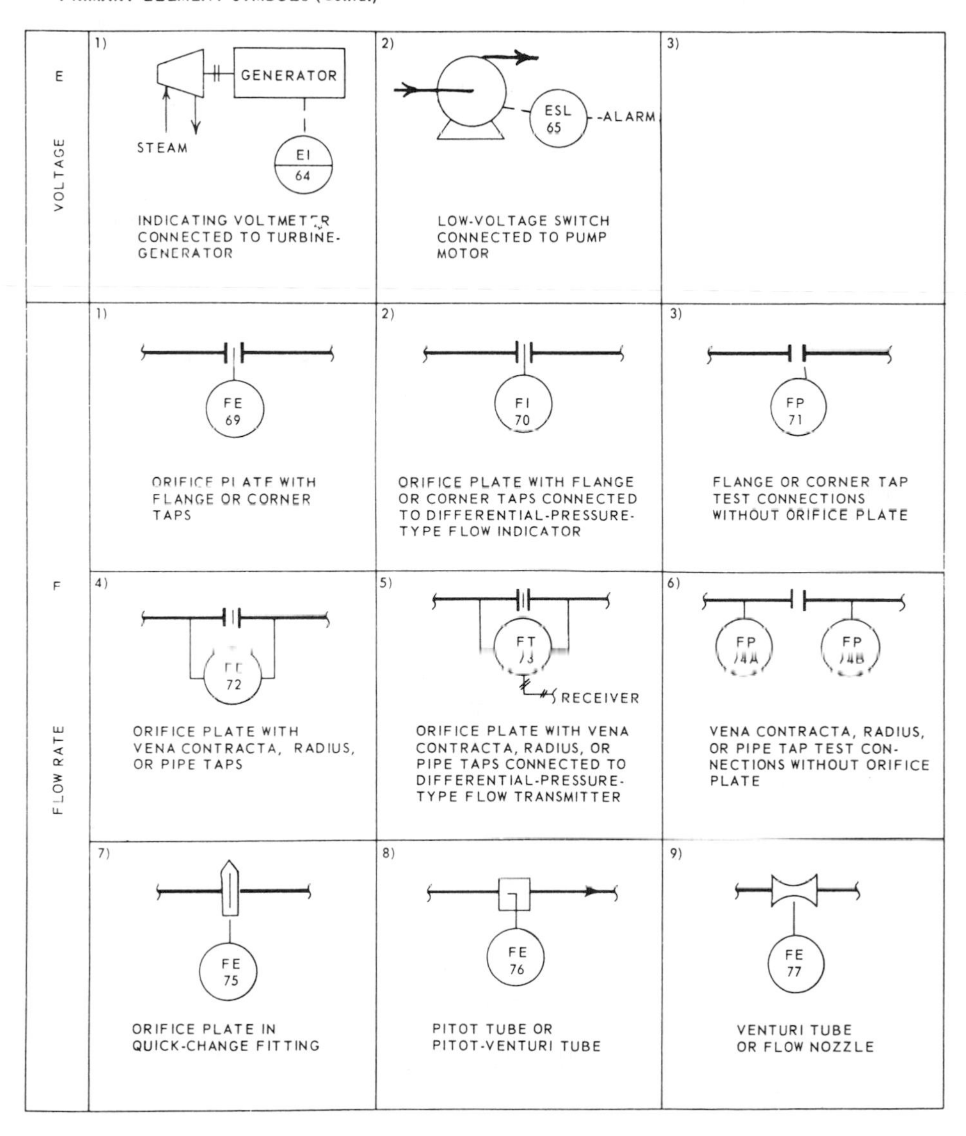

PRIMARY ELEMENT SYMBOLS (Contd.)

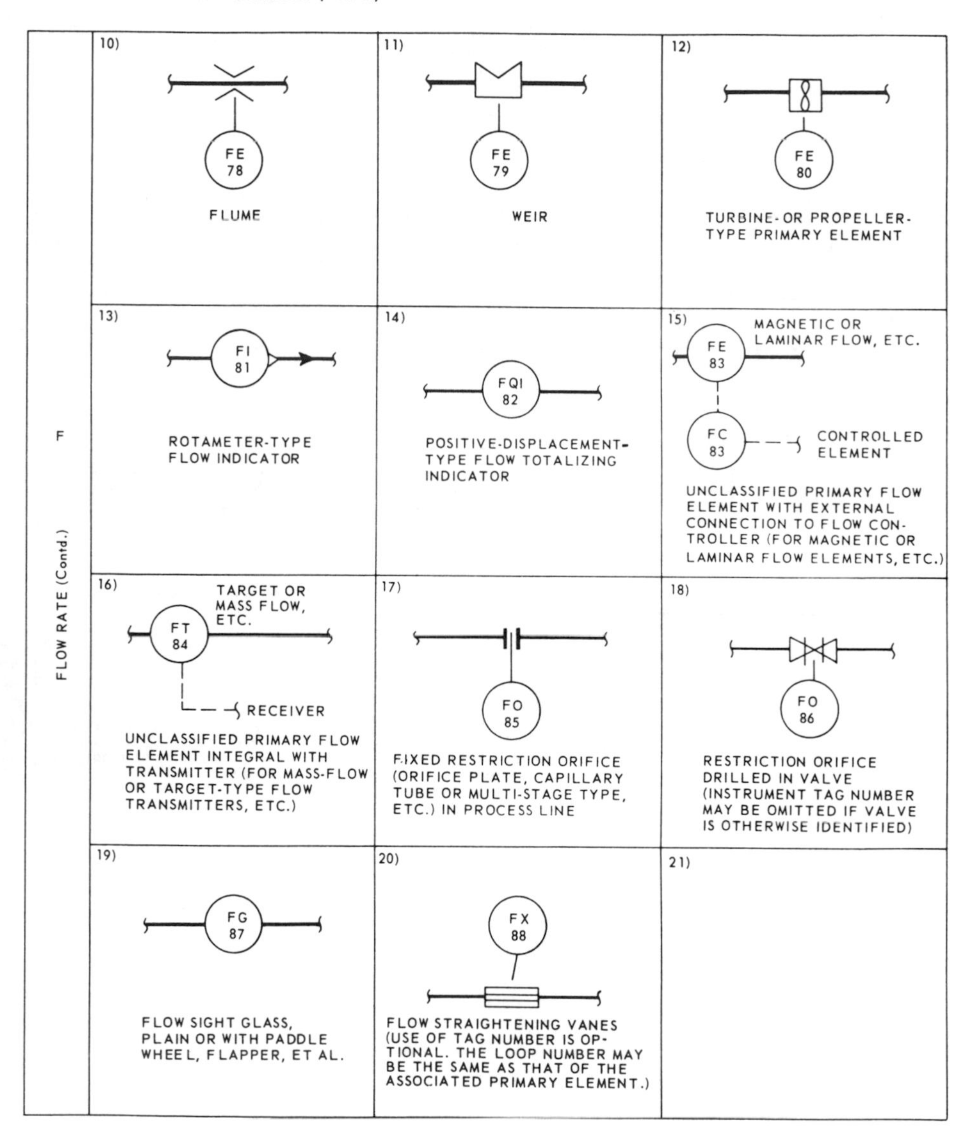

PRIMARY ELEMENT SYMBOLS (Contd.)

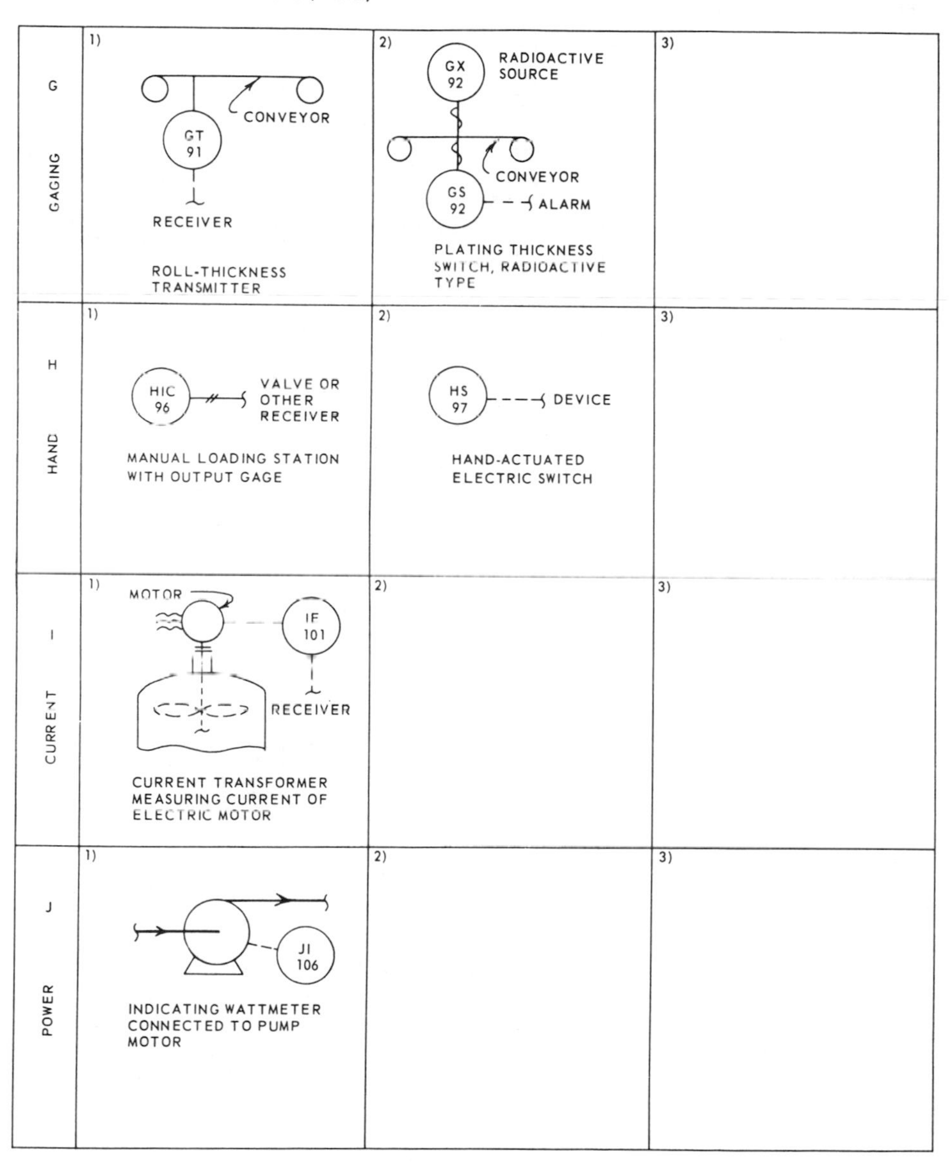

PRIMARY ELEMENT SYMBOLS (Contd.)

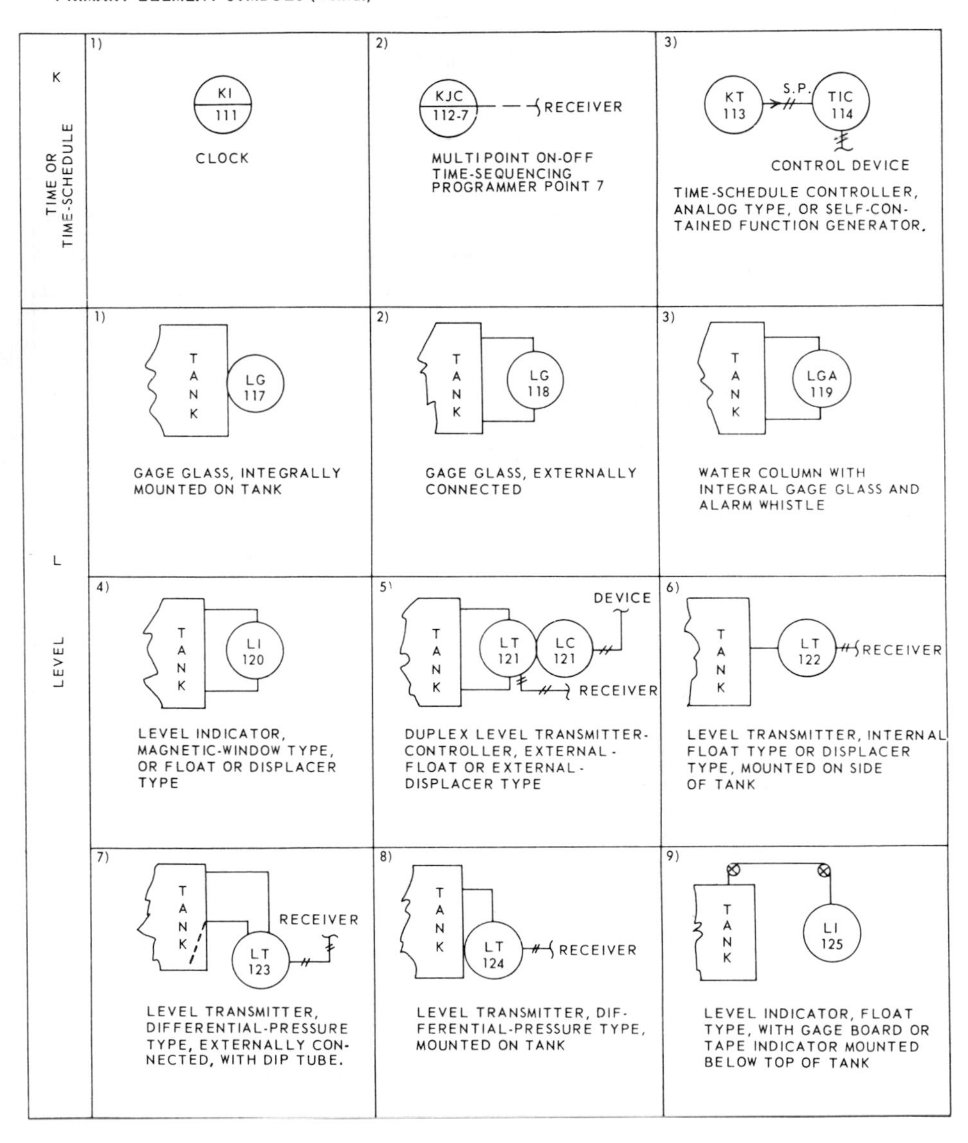

PRIMARY ELEMENT SYMBOLS (Contd.)

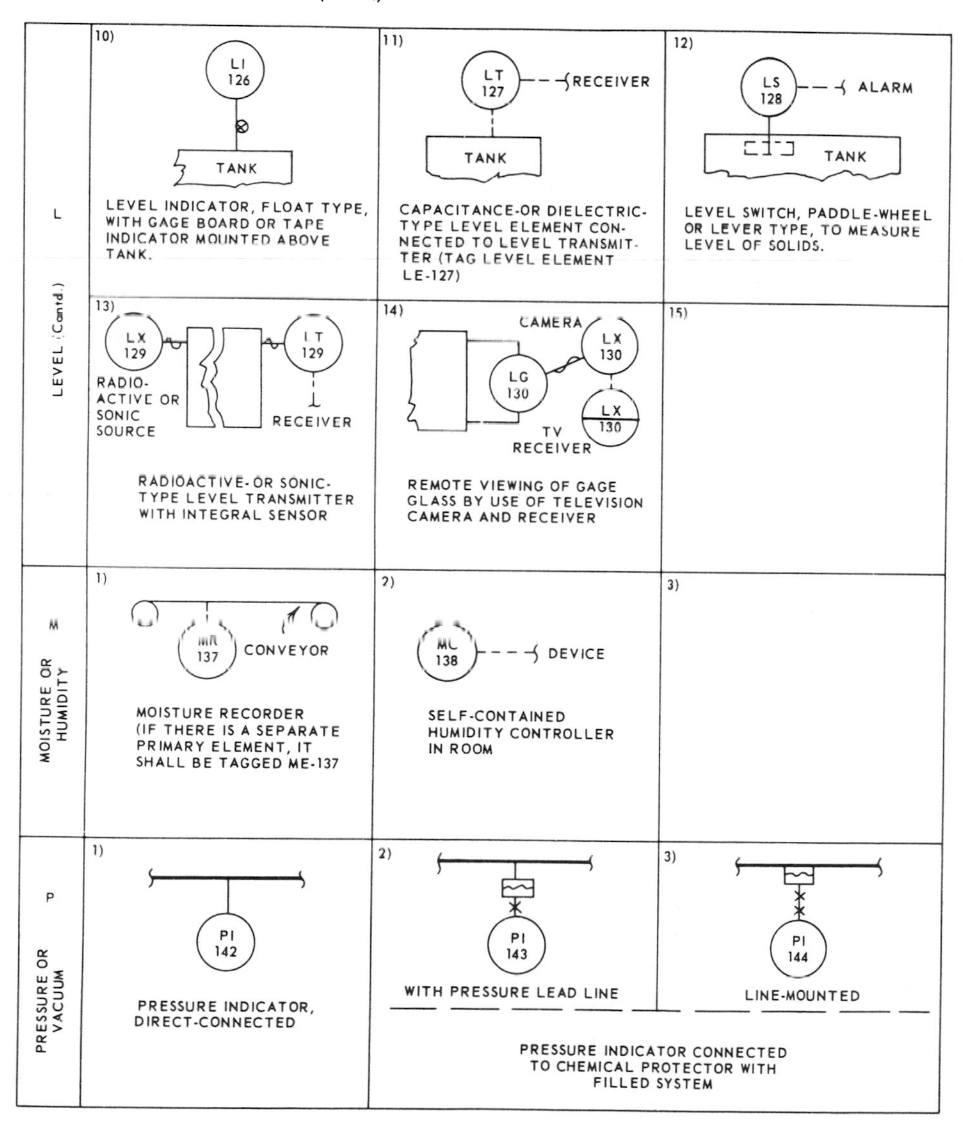

PRIMARY ELEMENT SYMBOLS (Contd.)

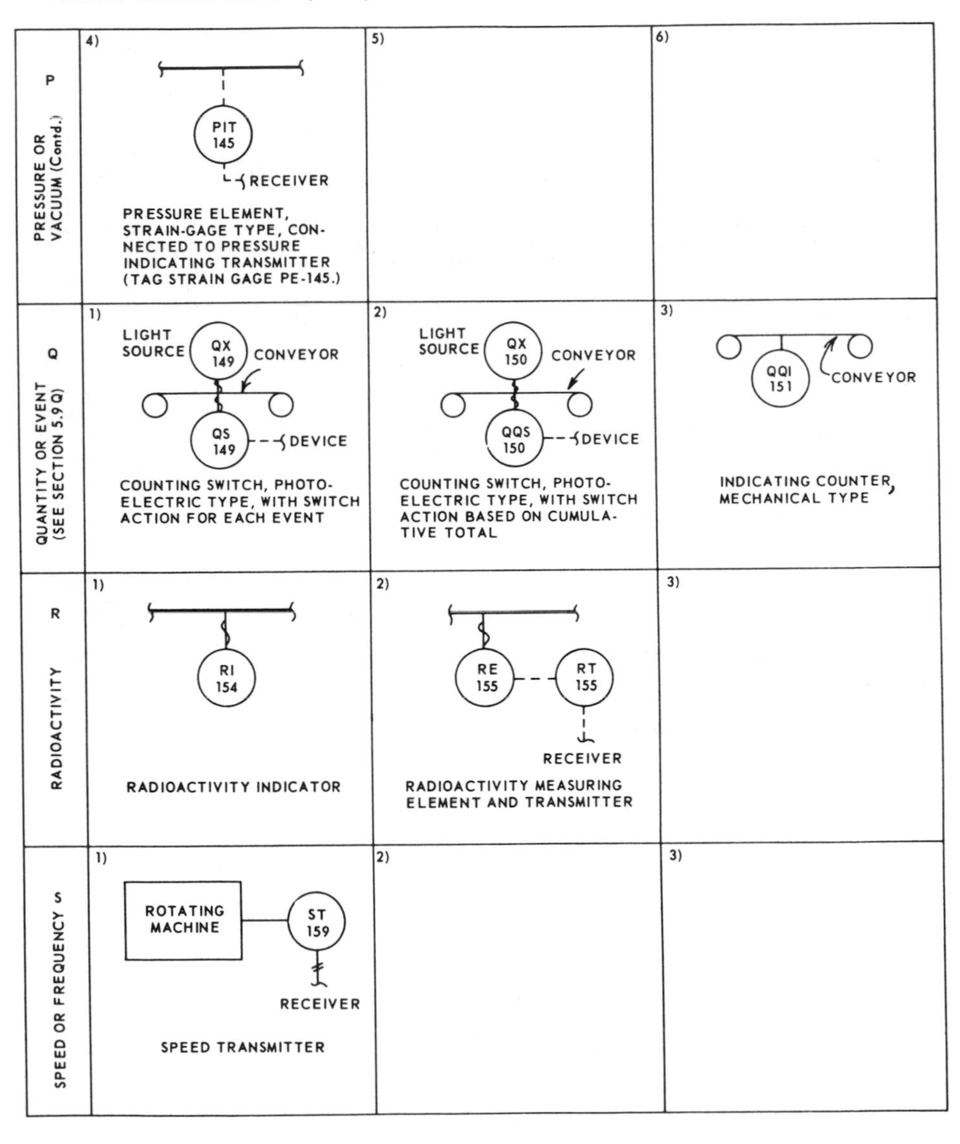

PRIMARY ELEMENT SYMBOLS (Contd.)

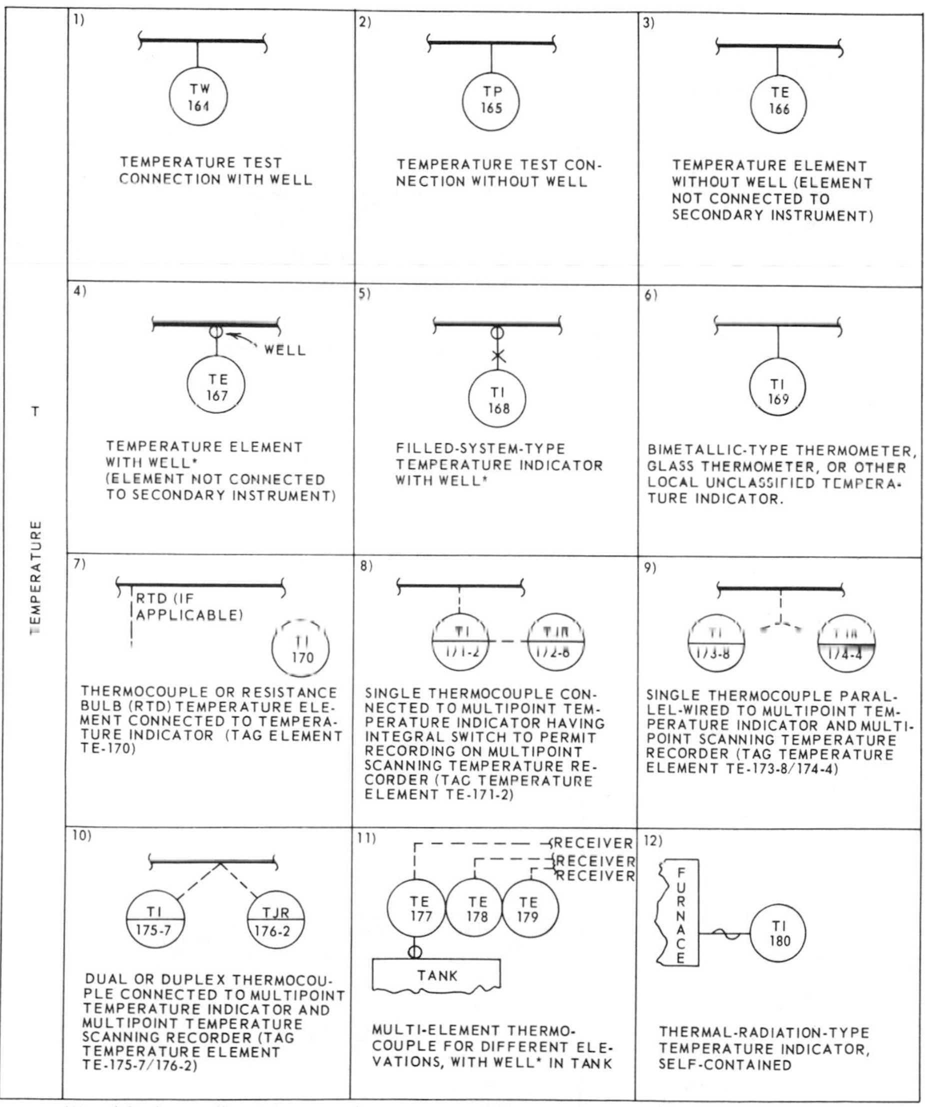

*Use of the thermowell symbol is optional. However, use or omission of the symbol shall be consistent throughout a project.

PRIMARY ELEMENT SYMBOLS (Contd.)

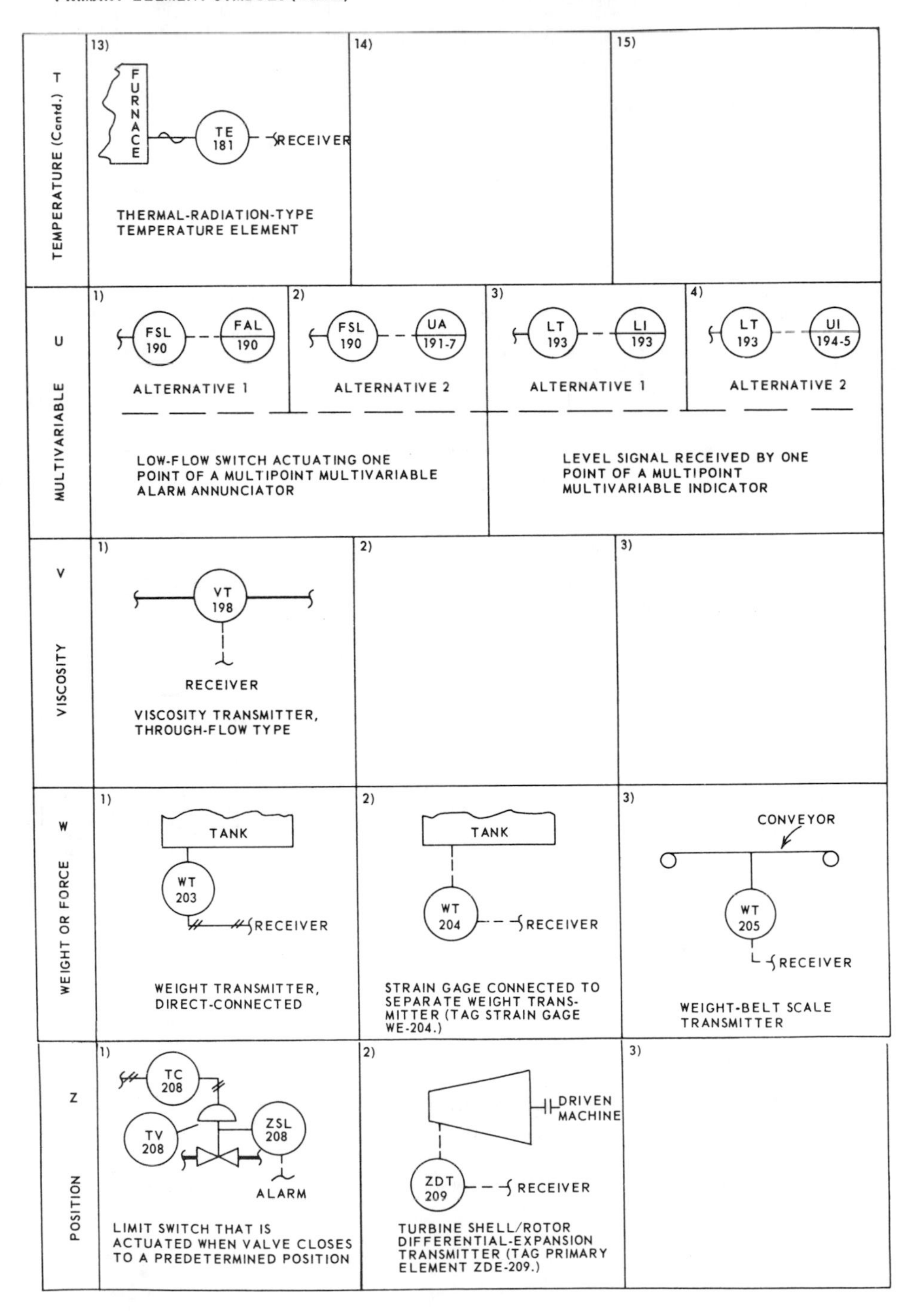

FUNCTION SYMBOLS

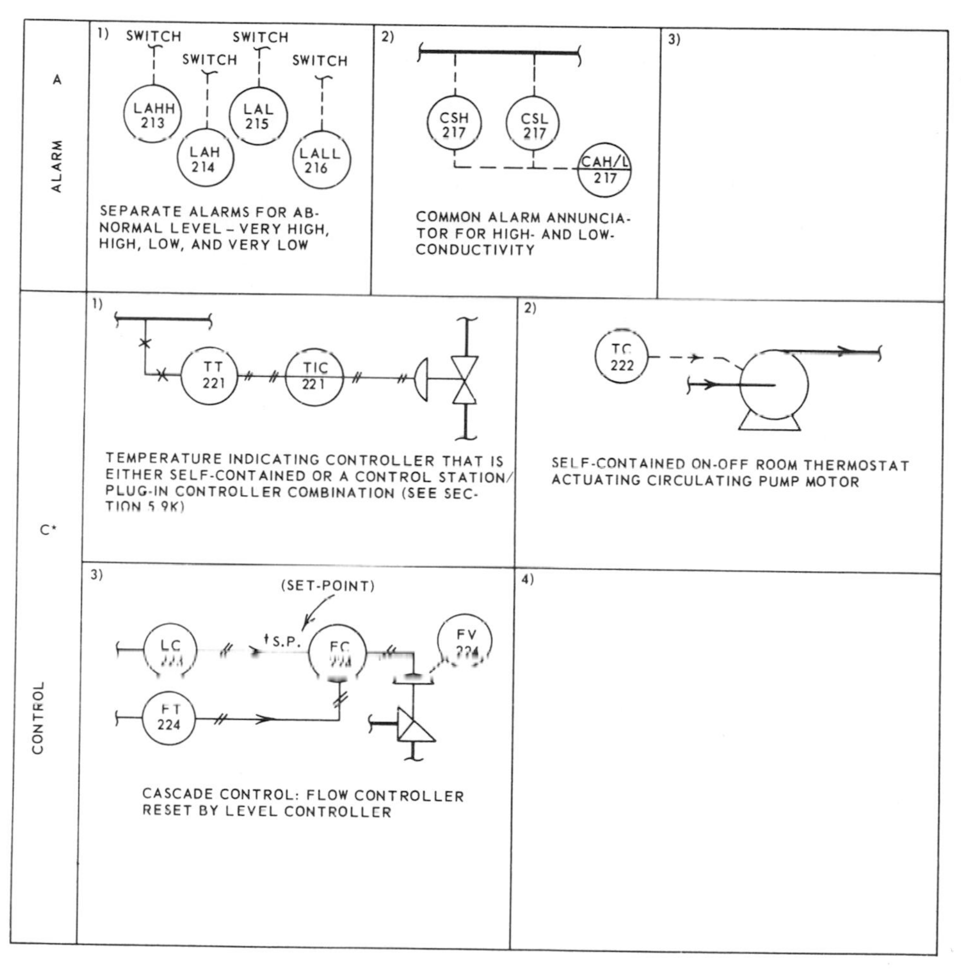

* It is expected that control modes will not be designated on a diagram. However, the following designations may be used outside the controller symbol, if desired, in combinations such as %, ∫, D.

CONTROL MODE	DESIGNATION
ON-OFF	1-0 OR ON-OFF
DIFFERENTIAL-GAP, TWO-POSITION	Δ1-0 OR ΔON-OFF
PROPORTIONAL	% OR P
AUTOMATIC RESET, FLOATING, OR INTEGRAL	∫ OR I
DERIVATIVE OR RATE	D OR d/dt
INVERSE DERIVATIVE	1/D
OPTIMIZING	OPT. OR MAX. OR MIN. (as applicable)
ADAPTIVE	ADAPT.
UNCLASSIFIED	AS REQUIRED
DIRECT ACTING	DIR.
REVERSE ACTING	REV.

† A controller is understood to have integral manual set-point adjustment unless means of remote adjustment is indicated.

FUNCTION SYMBOLS (Contd.)

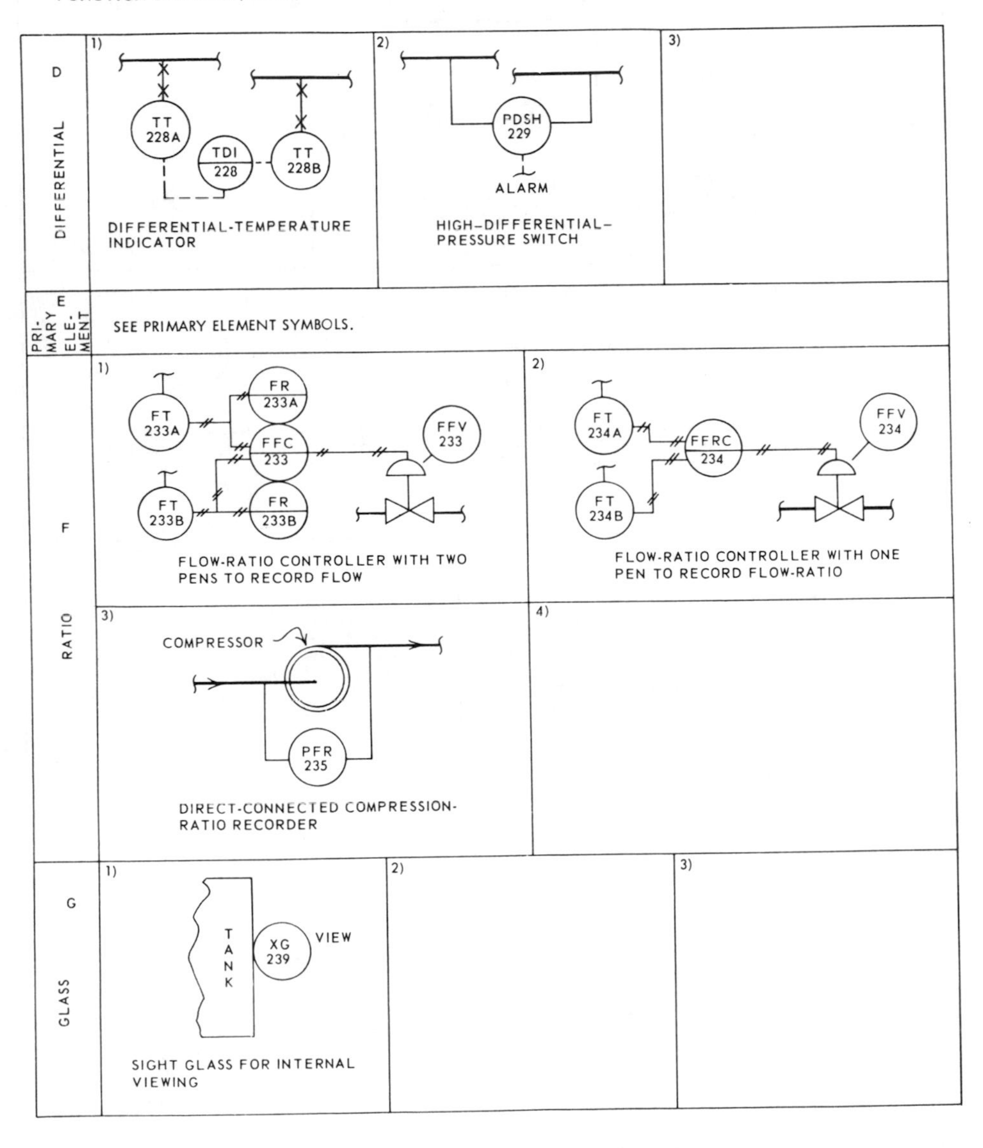

FUNCTION SYMBOLS (Contd.)

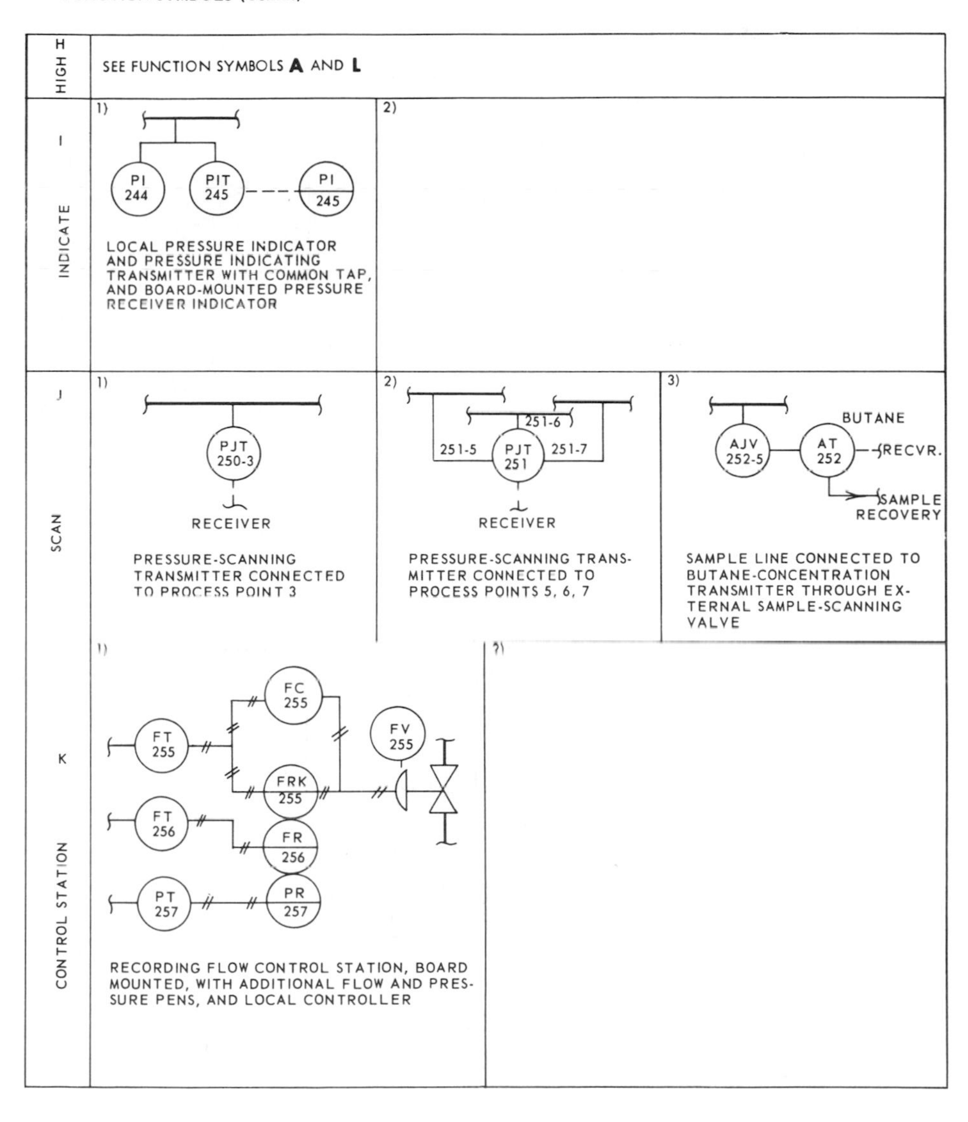

FUNCTION SYMBOLS (Contd.)

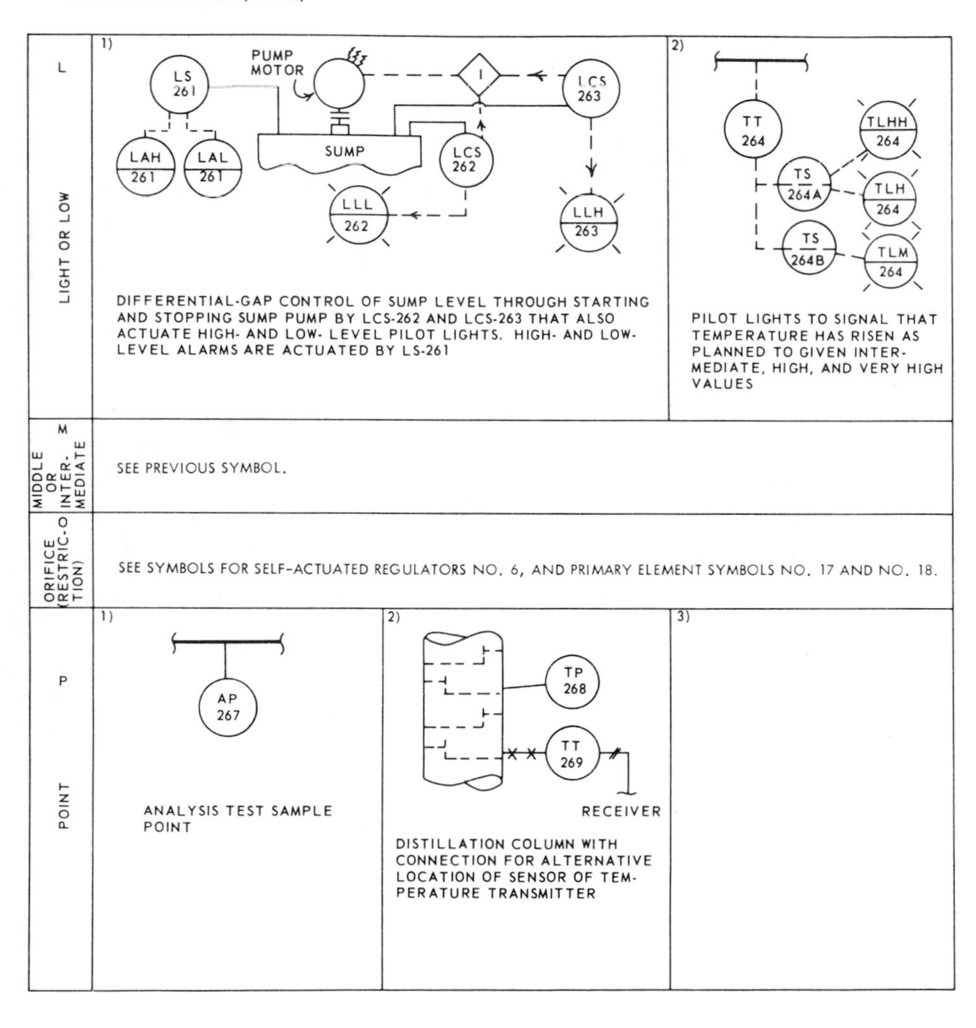

FUNCTION SYMBOLS (Contd.)

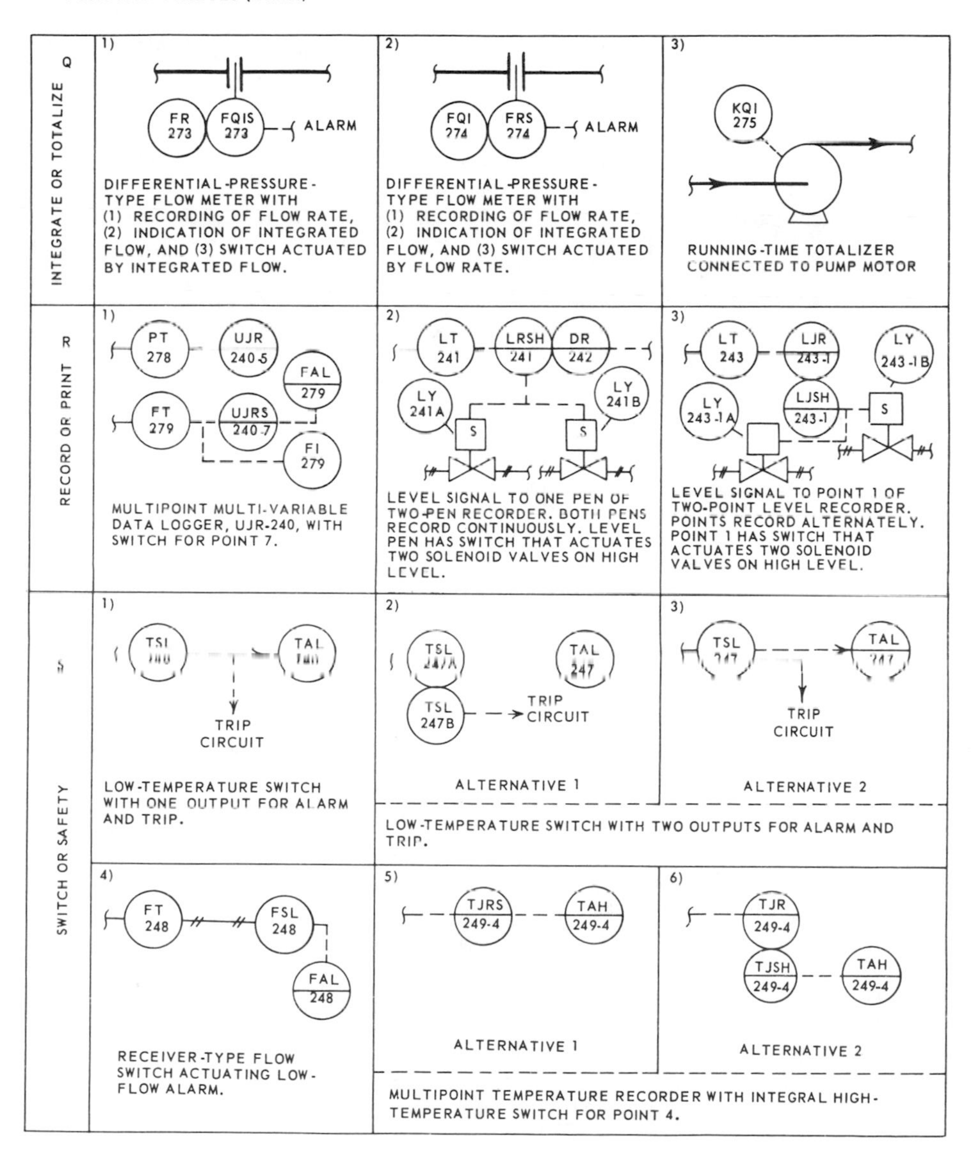

FUNCTION SYMBOLS (Contd.)

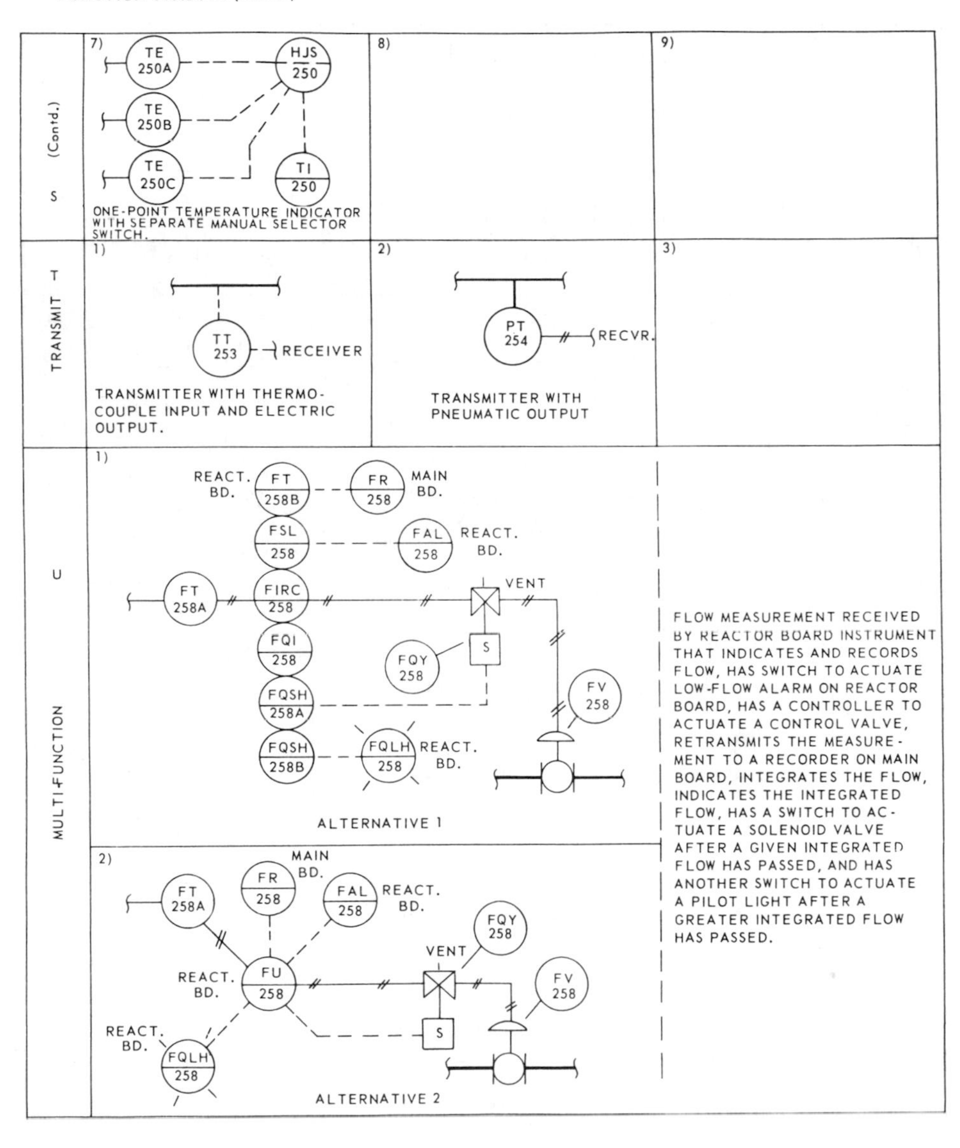

FUNCTION SYMBOLS (Contd.)

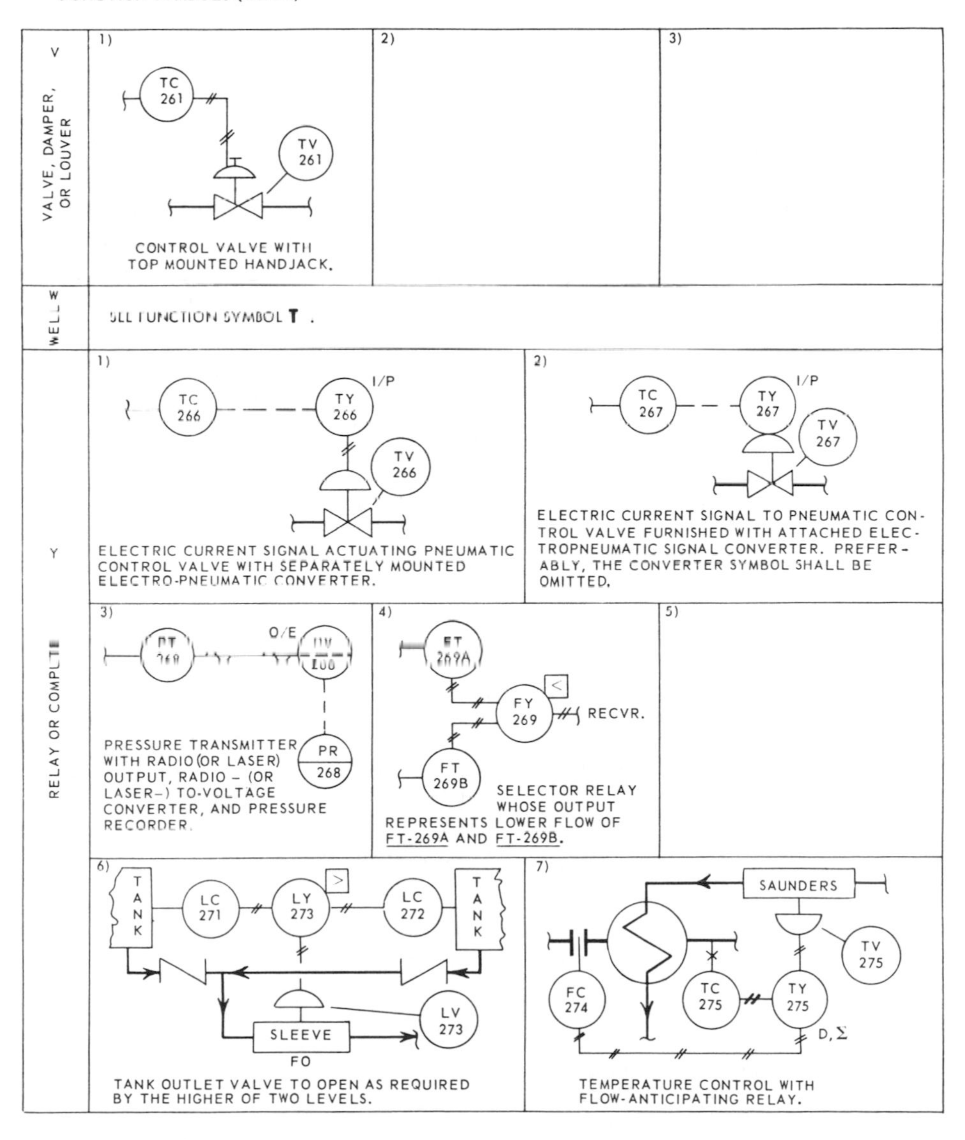

FUNCTION SYMBOLS (Contd.)

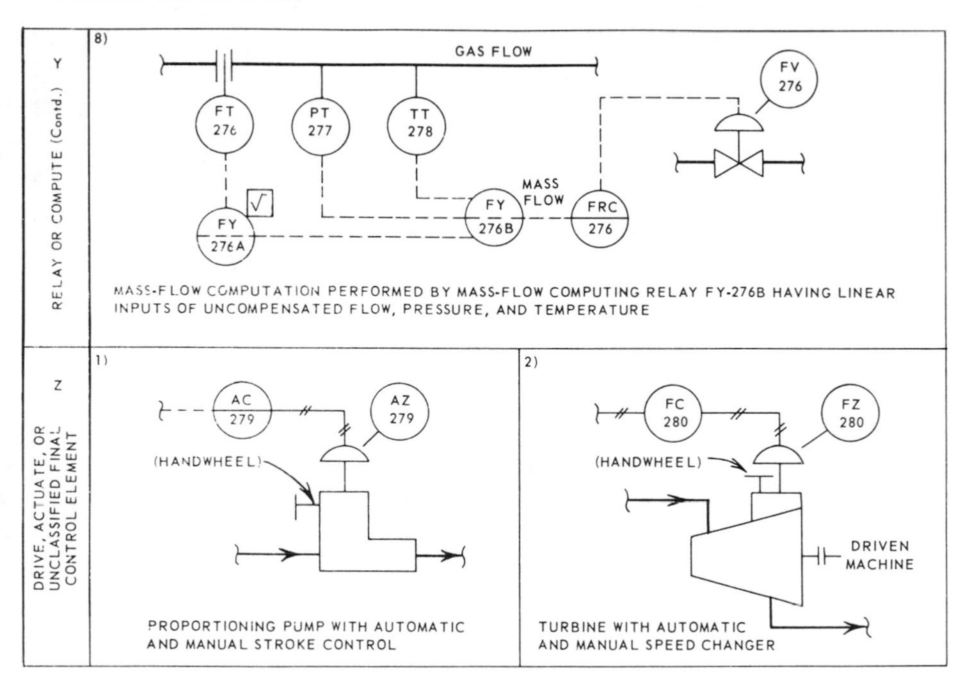

MISCELLANEOUS SYSTEMS

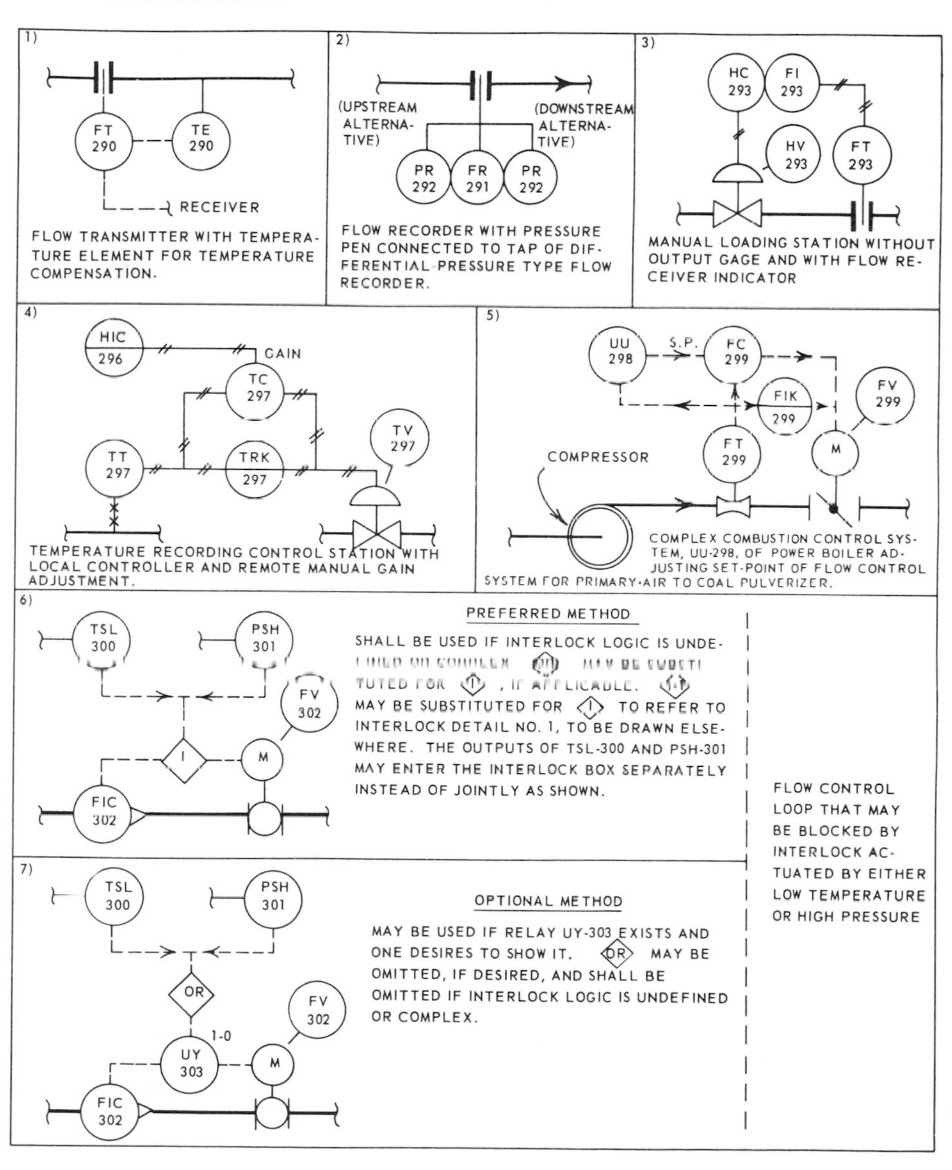

MISCELLANEOUS SYSTEMS (Contd.)

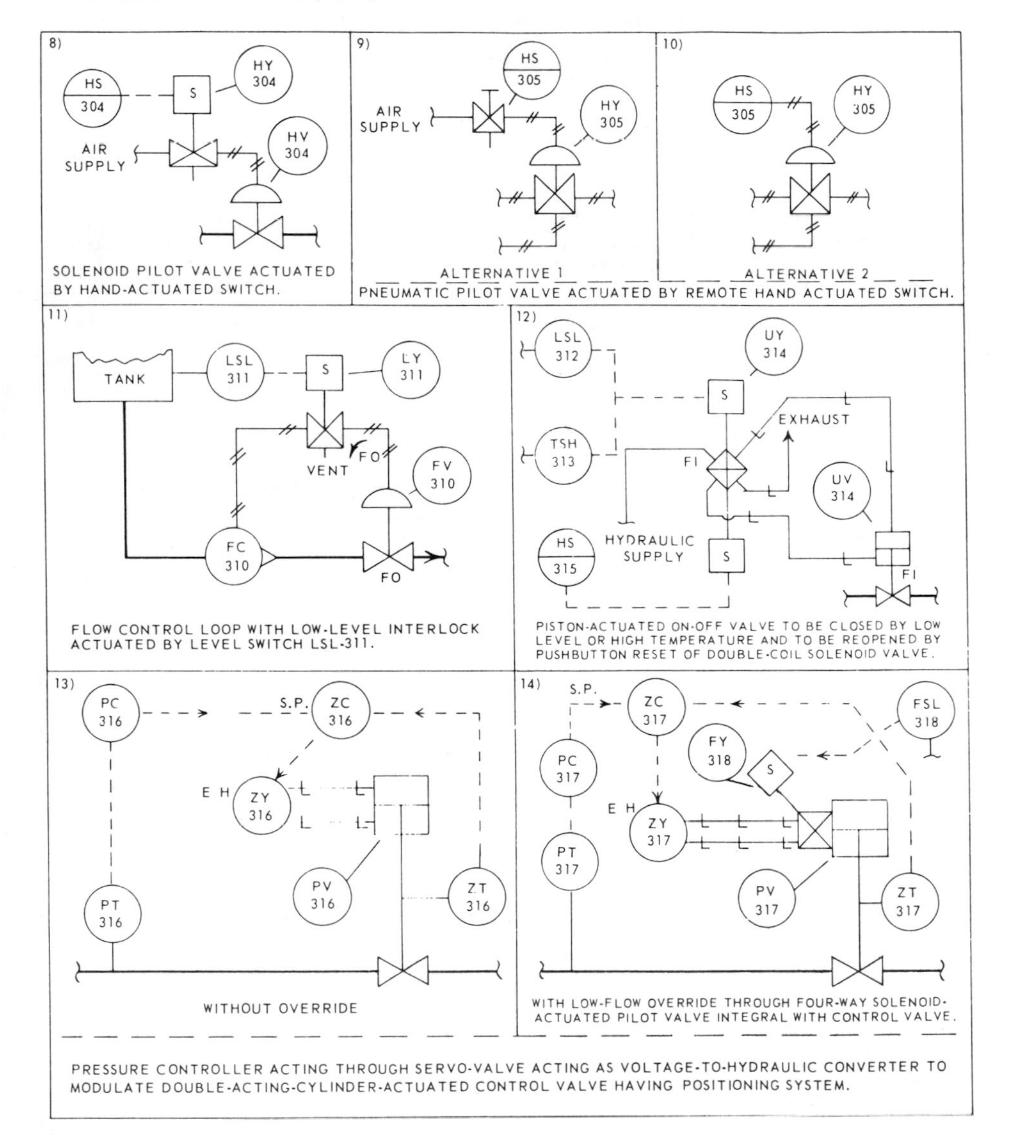

MISCELLANEOUS SYSTEMS (Contd.)

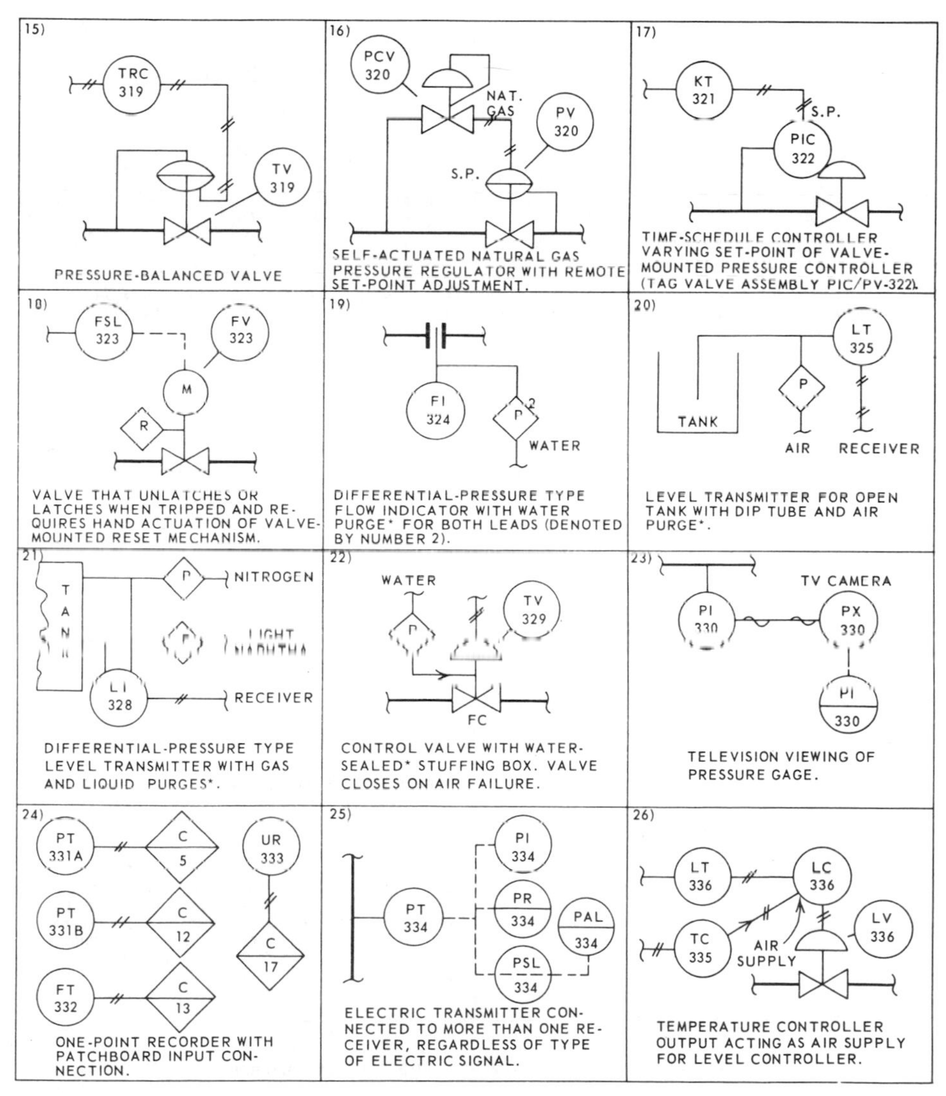

*The purge fluid supplies may optionally use the same abbreviations as for instrument power supplies.

MISCELLANEOUS SYSTEMS (Contd.)

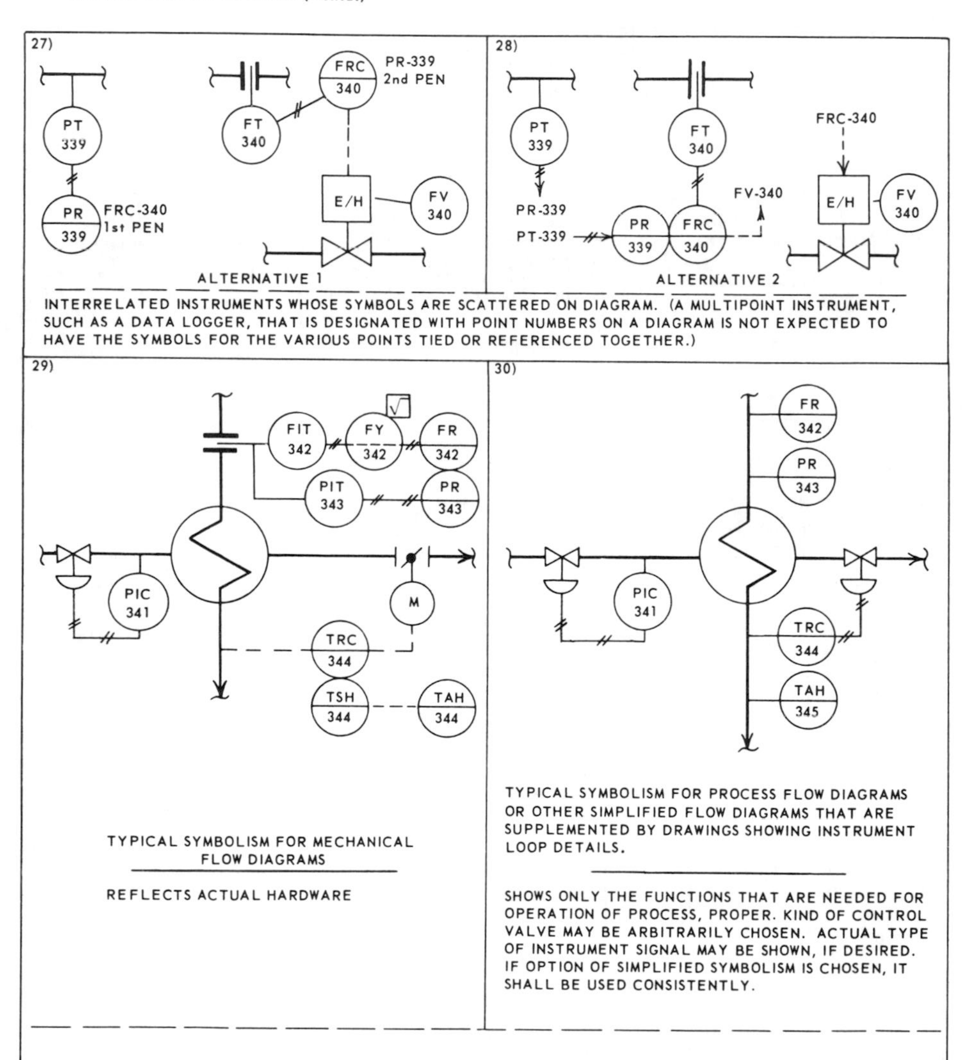

METHODS OF DEPICTING INSTRUMENT LOOPS ON FLOW DIAGRAMS

Electrical Symbols

Appendix **B**

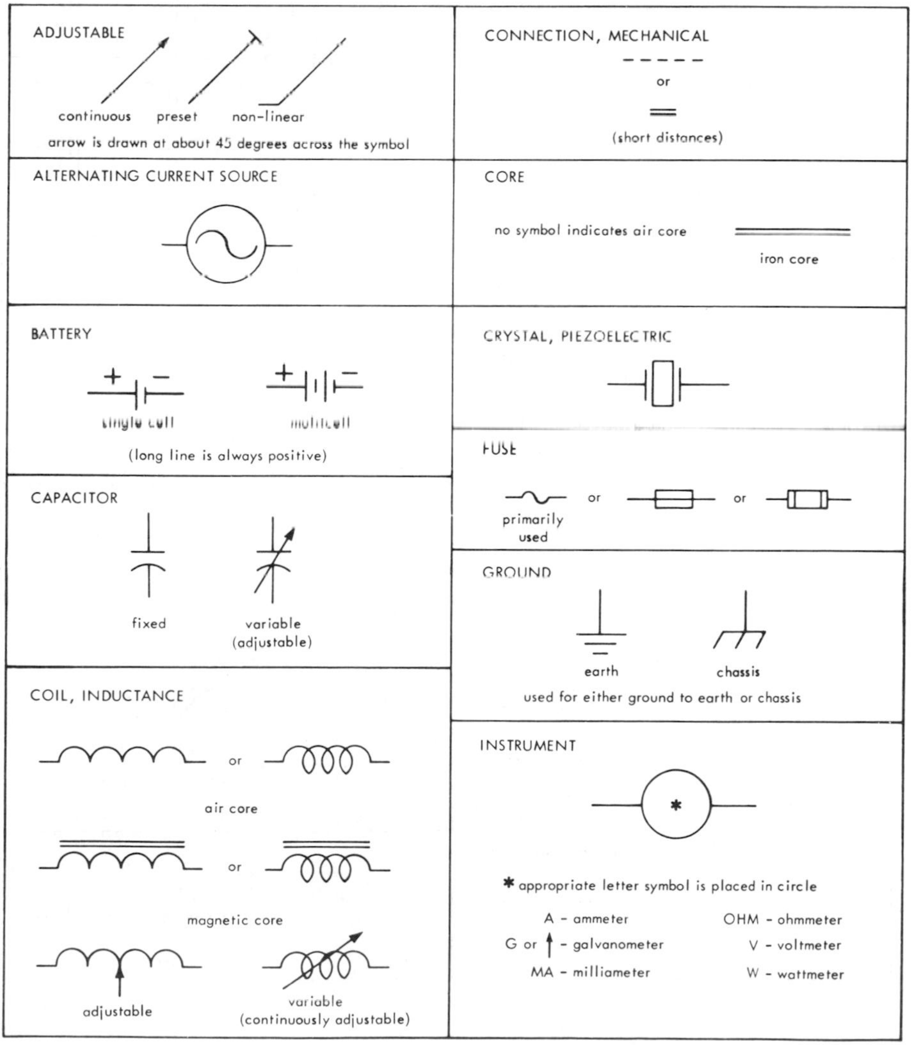

ADJUSTABLE

continuous preset non-linear

arrow is drawn at about 45 degrees across the symbol

ALTERNATING CURRENT SOURCE

BATTERY

+ − single cell

+ − multicell

(long line is always positive)

CAPACITOR

fixed

variable (adjustable)

COIL, INDUCTANCE

or

air core

or

magnetic core

adjustable

variable (continuously adjustable)

CONNECTION, MECHANICAL

or

(short distances)

CORE

no symbol indicates air core

iron core

CRYSTAL, PIEZOELECTRIC

FUSE

primarily used or or

GROUND

earth

chassis

used for either ground to earth or chassis

INSTRUMENT

*

* appropriate letter symbol is placed in circle

A - ammeter

G or ↑ - galvanometer

MA - milliameter

OHM - ohmmeter

V - voltmeter

W - wattmeter

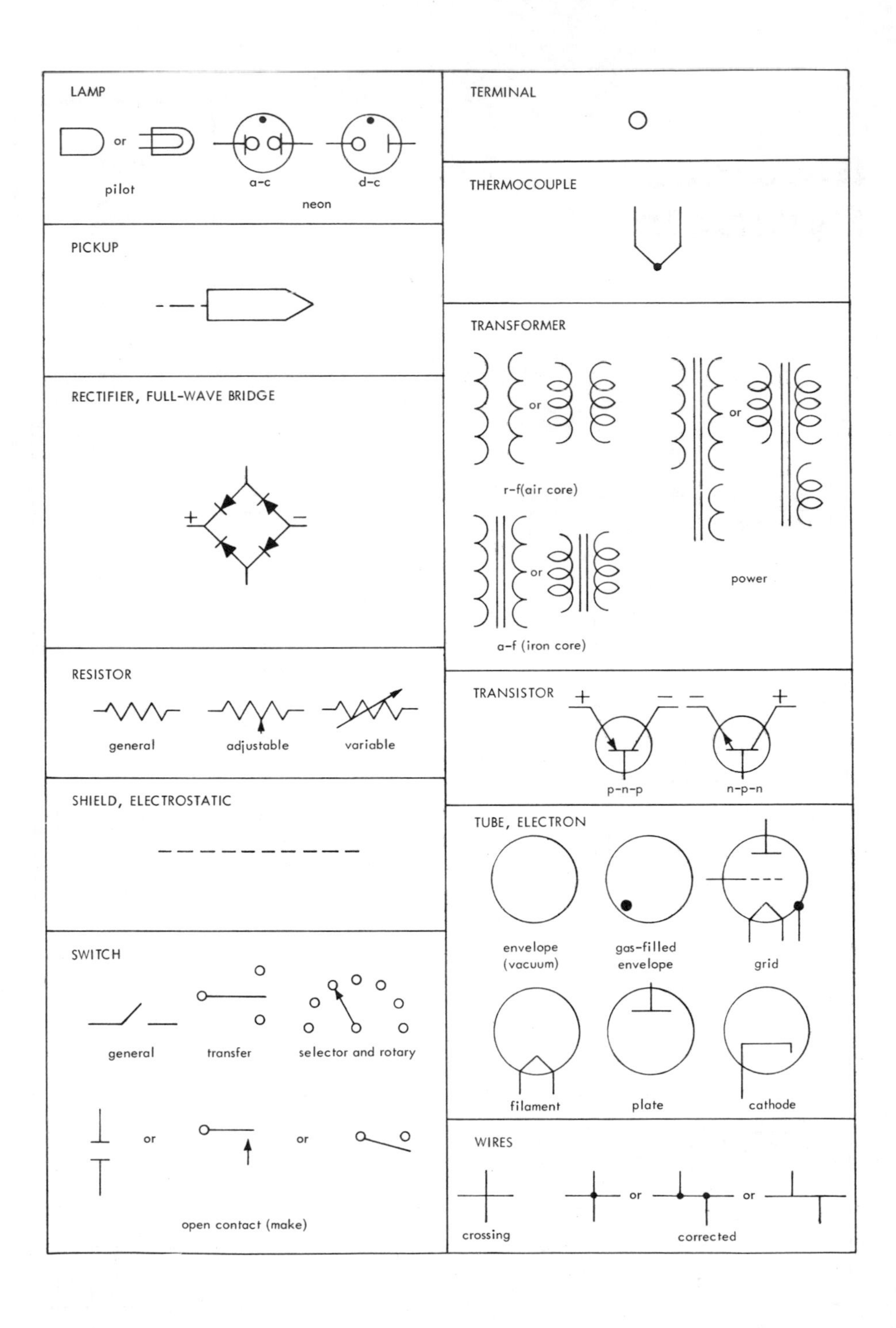
LAMP
or
pilot
a-c
d-c
neon
TERMINAL
THERMOCOUPLE
PICKUP
TRANSFORMER
or
or
r-f(air core)
or
a-f (iron core)
power
RECTIFIER, FULL-WAVE BRIDGE
+
–
RESISTOR
general
adjustable
variable
TRANSISTOR
+
–
–
+
p-n-p
n-p-n
SHIELD, ELECTROSTATIC
TUBE, ELECTRON
envelope
(vacuum)
gas-filled
envelope
grid
filament
plate
cathode
SWITCH
general
transfer
selector and rotary
or
or
open contact (make)
WIRES
crossing
or
or
corrected

PUSH BUTTON SWITCHES		TOGGLE SWITCHES	
SYMBOL	OBJECT	SYMBOL	OBJECT
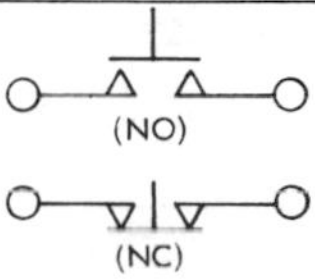 (NO) (NC) SPST (SINGLE POLE, SINGLE THROW)		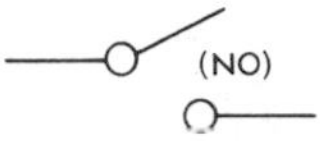 (NO) SPST (SINGLE POLE, SINGLE THROW)	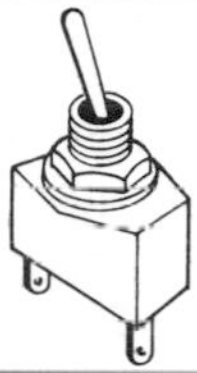
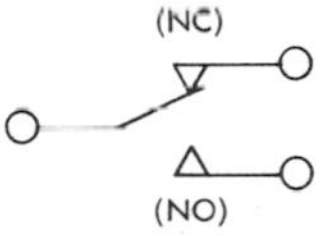 (NC) (NO) SPDT (SINGLE POLE, DOUBLE THROW)		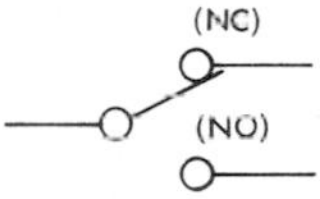 (NC) (NO) SPDT (SINGLE POLE, DOUBLE THROW)	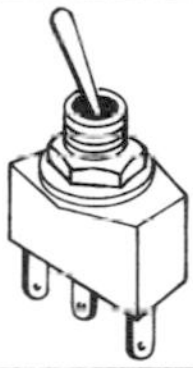
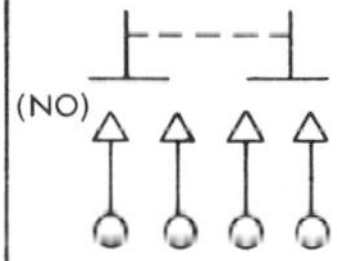 (NO) DPST (DOUBLE POLE, SINGLE THROW)		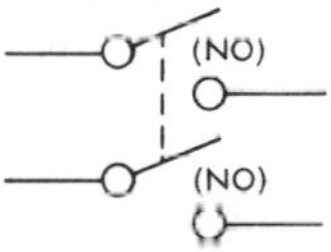 (NO) (NO) DPST (DOUBLE POLE, SINGLE THROW)	
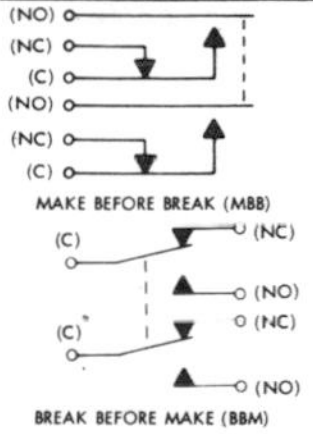 (NO) (NC) (C) (NO) (NC) (C) MAKE BEFORE BREAK (MBB) (C) (NC) (NO) (C) (NC) (NO) BREAK BEFORE MAKE (BBM) DPDT (DOUBLE POLE, DOUBLE THROW)		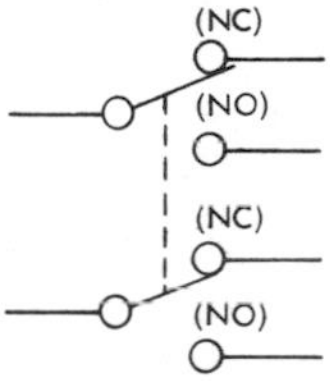 (NC) (NO) (NC) (NO) DPDT (DOUBLE POLE, DOUBLE THROW)	

NOTE: (NO) MEANS NORMALLY OPEN, (NC) MEANS NORMALLY CLOSED, (C) MEANS COMMON

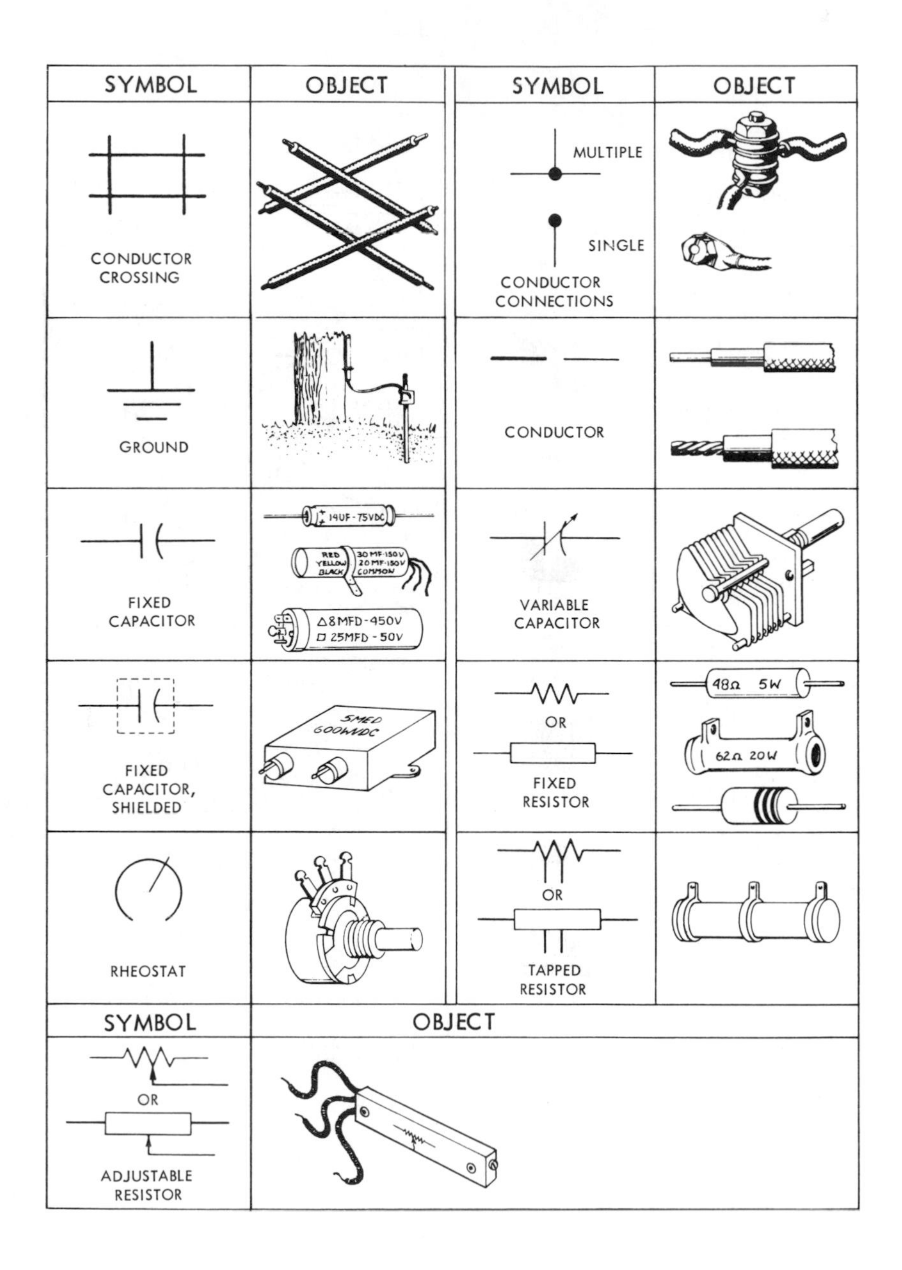
SYMBOL
OBJECT
SYMBOL
OBJECT
CONDUCTOR CROSSING
MULTIPLE
SINGLE
CONDUCTOR CONNECTIONS
GROUND
CONDUCTOR
14UF-75VDC
RED
YELLOW
BLACK
30 MF-150V
20 MF-150V
COMMON
△8MFD-450V
□ 25MFD-50V
FIXED CAPACITOR
VARIABLE CAPACITOR
FIXED CAPACITOR, SHIELDED
5MFD
600WVDC
OR
FIXED RESISTOR
48Ω 5W
62Ω 20W
RHEOSTAT
OR
TAPPED RESISTOR
SYMBOL
OBJECT
OR
ADJUSTABLE RESISTOR

General Instrumentation Data

Appendix

C

TABLE 1

DENSITY AND SPECIFIC GRAVITY OF SOME LIQUIDS AT VARIOUS TEMPERATURES

LIQUID	TEMP °C	TEMP °F	DENSITY # 1 cu ft	*SPECIFIC GRAVITY
ACETONE	15.5	60	49.4	.79
ALCOHOL (ETHYL)	20	68	49.4	.79
AMMONIA (SATURATED)	10	50	40.9	.655
BENZENE	0	32	56.1	.9
BRINE (10% Na Cl)	0	32	67.24	1.08
BUNKER (FUEL OIL)	15.5	60	63.25	1.01
FUEL OIL #2	15.5	60		
FUEL OIL #6	15.5	60	61.92	.99
GASOLINE	15.5	60	46.81	.75
GLYCERIN	0	32	78.6	1.26
KEROSENE	15.5	60	50.85	.81
MERCURY	15.5	60	846.3	13.55
PENTANE	15	59	38.9	.623
SAE 30 LUBE OIL	15.5	60	56.02	.9
WATER	4	39.2	62.43	1.0

* THE SPECIFIC GRAVITY OF A LIQUID REPRESENTS ITS METRIC DENSITY
1 cm^3 OF WATER WEIGHS 1 GRAM

TABLE 2

EQUIVALENTS USED IN INSTRUMENTATION COMPUTATION

LENGTH

1 inch = 2.54 cm = 25.4 mm
1 foot = 30.48 cm = 304.8 mm
1 millometer (mm) = .03937 in
1 micron = .001 mm
1 meter = 3.28 feet = 1.0936 yds

AREA

1 sq in = 6.4516 sq centimeters (cm)
1 sq cm = .155 sq in
1 sq ft = 929.03 sq cm

VOLUME

1 cu in = 16.387 cu cm
1 cu cm = .061 cu in
1 cu ft = 7.481 gallons
1 gallon = .1337 cu ft
3.785 liters
1 liter = .2641 gallons

WEIGHT

10 grams = .3527 ounces
1 kilogram = 2.2 pounds
1 cu ft water = 62.36 pounds
1 cu ft mercury = 848.719 pounds
1 gallon water = 8.377 pounds
1 cu in water = .073551 cu in mercury
1 cu in mercury = .4905 pounds

DENSITY

1 pound per cu in = 27.68 grams per cc
1 gram per cu cm = .03613 pounds per cu in

PRESSURE

1 inch of water = .0361 psi
= .07348 inches of mercury
1 inch of mercury = .4912 psi
= 13.62 inches of water
1 psi = 27.707 inches of water
2.309 feet of water
2.036 inches of mercury
1 torr = 1 millimeter of mercury

All mercury equivalents at 0° C or 32° F
All water equivalents at 60° F

Glossary

A

absolute humidity: An expression of water vapor content. The amount of water present in a specified volume of air or gas.

absolute pressure: The pressure of a liquid or gas measured in relation to a vacuum (zero pressure).

absolute zero: The temperature at which the molecular motion that constitutes heat ceases, and at which an ideal gas, kept at constant volume, would exert no pressure. This temperature (−273.15°C or −459.67°F) is theoretically the coldest temperature possible and is the 0° point on the absolute temperature scale.

acceleration: The time rate of change of velocity, expressed in feet (or centimeters) per second per second.

accuracy: The difference between the reading of an instrument and the true value of what is being measured, expressed as a percent of full instrument scale.

acidity: A measure of hydrogen ion content of a solution.

action: Generally refers to the action of a controller, and defines what is done to regulate the final control element to effect control. Types of action include On-Off, proportional, rate, and reset.

actuator: The final element of a control system. The actuator performs the action indicated by the controller.

alkalinity: A measure of hydroxyl ion content of a solution.

alternating current: An electric current which flows alternately in one direction then the other direction. It is commonly expressed by the abbreviation *ac*.

ambient conditions: The conditions of temperature, pressure, humidity, etc., existing in the medium that surrounds an instrument.

ambient temperature: The temperature surrounding auy particular object.

ammeter: An instrument that indicates the quantity of flow of an electric current in a circuit.

amplification: The process of obtaining an output signal that is greater than the input signal through auxiliary power controlled by the input signal.

amplification factor: The ratio of a change in plate voltage to a change in grid voltage, with the plate current held constant.

amplifier: A device for increasing the strength of a signal.

amplitude: The greatest distance through which a vibrating body moves from the mid-point.

analog computer: An apparatus that converts mathematically expressed variables into mechanical or electrical equivalents.

angstrom unit: A unit of wave length equal to one ten-billionth of a meter.

anode: The electrode that is charged positively in an electron (vacuum) tube or in an electrolytic solution, such as those used in electro-plating or in batteries. It is the electrode that receives electrons.

armature: A part of the moving member of an instrument on which the magnetic flux reacts to provide deflecting torque.

atmospheric pressure: The pressure exerted on a body by the air, equal at sea level to about 14.7 pounds per square inch.

attenuation: The amount of decrease of the amplitude of a signal as it passes through any part of a control system.

automatic control: The process of using the differences between the actual value and desired value of a variable to take corrective action without human intervention.

automatic control system: Any combination of automatic controllers connected in closed loops with one or more processes.

automatic controller: A device that measures the value of a variable and operates to correct or limit deviation from a selected reference. It includes means for both measurement and control.

automation: The employment of devices which automatically control one or more functions in an industrial process.

AWG: (American Wire Gage.) Letters used to indicate that equipment is suitable for use with air, water, or gas.

B

base: Center section of a junction, or bipolar, transistor. Small current flow through the base is used to control the output of the transistor.

batch process: Applies where a given amount of material is processed in one operation to produce what is required.

bellows: A pressure sensing element consisting of a convoluted metal cylinder closed at one end. Pressure difference between the outside and inside of the bellows causes it to expand or contract along its axis.

beta ratio: Diameter of orifice: internal diameter of pipe.

bias: The fixed voltage applied between the control grid and the cathode of electron tubes.

bimetallic element: The temperature sensitive bimetal of a bimetallic thermometer. The bimetal is composed of two or more metal alloys mechanically associated so that the bimetal bends in a particular direction when heated.

blackbody: The ideal radiator of thermal energy, which emits as much thermal energy as it absorbs but which reflects none.

bluff body: An obstruction placed in a piped fluid flow stream to produce vortices:

bourdon tube: A pressure sensing element that consists of a curved tube having a flattened elliptical cross-section which is closed at one end. A positive pressure difference between the inside and outside of the tube tends to straighten the tube.

bridge circuit: A network in which the value of an unknown component is obtained by balancing one circuit against another. It consists of a detecting device and four resistances connected in series to form a diamond.

British thermal unit: A unit of heat. The amount of heat required to raise the temperature of one pound of water one degree Fahrenheit. It is abbreviated *BTU*.

buffer: An industrial process solution having a definite resistance to change of pH when an acid or base is added to it.

bumpless transfer: Change from manual to automatic control, or vice versa, without change in control signal.

C

calibration: The procedure laid down for determining, correcting, or checking the absolute values corresponding to the graduations on a measuring instrument.

capacitance: A measure in units of quantity that is determined by the type of quantity contained and the type of reference variable.

capacitor: A device for storing electrical energy.

capacity: A measure of the maximum amount of energy or material that may be stored in a given system.

cascade control: In an automatic control system, the resetting of the control point of a secondary controller by the output of a primary controller.

cathode: An electrode charged negatively in an electron (vacuum) tube or in an electrolytic solution.

Celsius scale: The centigrade temperature scale on which the freezing point of water is 0° and the boiling point is 100°.

Charles' law: Pressure increase in a closed system is proportional to temperature increase, given a constant volume.

chopper: Device used to interrupt dc to produce a pulsating current.

chromatography: The method of analyzing a fluid stream by separating its constituents by passing it through an adsorbent column.

circuit: The complete path through which an electric current can flow.

closed loop: A combination of control units in which the process variable is measured and compared with the desired value (or set point). If the measured value differs from the desired value, a corrective signal is sent to the final control element to bring the controlled variable to the proper value.

coefficient of linear expansion: Increase in unit length per degree of temperature rise.

coefficient of volumetric expansion: Increase in unit volume per degree of temperature rise.

cold junction: The point at which a pair of thermocouple is held at a fixed temperature. It is also called the *reference junction.*

collector: Term used to indicate an end terminal of a junction transistor. Analagous to the anode of a vacuum tube.

combustibility: A material's ability to burn.

common-base amplifier: A transistor amplifier circuit arrangement in which the base is common to both input and output.

common-emitter amplifier: A transistor amplifier circuit arrangement in which the emitter is common to both input and output.

compensated thermometer system: Thermometer system which has been compensated to nullify temperature variations at the recording point which tend to distort accurate process temperature readings.

compensation: Provision of a supplemental device to counteract known sources of error.

computer: A machine for carrying out mathematical calculations automatically.

conduction of heat: The transfer of heat from one part of a body to another part in direct contact with it, the heat energy being passed from molecule to molecule.

continuous process: Applies to a process where raw material is treated by flowing continuously through a series of operations.

control agent: The process energy or material which is manipulated to hold the controlled medium at its desired value. In heating water with steam, the steam is the control agent.

control elements: The portion of the feedback control system which is required to produce the manipulated variable from the actuating signal.

control point: The value of controlled variable which, under any fixed set of conditions, an automatic controller operates to maintain.

control spring: A spring with characteristics predetermined so that its torque is equal and opposite to that of the instrument for any deflection of the pointer within the scale range.

control system: An assemblage of control apparatus coordinated to execute a planned set of control functions.

controlled medium: The process, energy or material which is to be adjusted to a definite value.

controlled variable: A quantity or condition of the controlled system that is directly measured and controlled.

controller lag: The delay in the response of a controller to a change in its input signal.

convection: The transfer of heat by the movement of heated particles.

coulomb: The quantity of electricity conveyed by a 1 ampere current in 1 second.

current: Electrical flow expressed in amperes. The symbol for current is *I*.

current amplification: The ratio between the current produced in the output circuit of an amplifier and the current supplied to the input circuit.

cycle: One complete oscillation. In the measurement of electrical frequency, one cycle per second is called a *hertz*.

D

damping: The decrease in the amplitude of an oscillation or wave motion with time.

dead time: A specific delay, measured in units of time, between two related actions of an instrument. Also referred to as distance velocity lag.

density: Mass per unit volume of a substance.

dependent flow: The flow to be controlled in a ratio system.

derivative action: An action in which there is a predetermined relationship between the time derivative of the controlled variable and the position of the final control element. It is also called *rate action*.

desired value: The value of the controlled variable which is to be maintained.

deviation: The difference between the instantaneous value of the controlled variable and the set point.

dew point: The temperature at which the amount of water present saturates the air.

diaphragm: A thin, flexible partitioning used to transmit pressure from one substance to another while keeping them from direct contact.

diaphragm motor: A pneumatic diaphragm mechanism used to position a valve or other final control element in response to the action of a pneumatic controller or a pneumatic positioning relay.

diaphragm motor valve: A pneumatic-powered valve which regulates fluid flow in response to a pneumatic signal.

dielectric: A non-conductor capable of sustaining an electrical field.

dielectric constant: A measure of the effectiveness of a dielectric material.

differential gap: Applies to two position (On-Off) controller action. It is the smallest range of values through which the controlled variable must pass in order to move the controller output from its On to its Off position (or vice versa).

differential pressure: The difference in pressure between two pressure sources, measured relative to one another.

differential pressure cell: (d/p cell) A device for measuring the difference between two pressures, using diaphragms or bellows. d/p cells are used to measure flow and liquid level.

digital computer: A type of computer that makes calculations using digits rather than continuous signals.

digital counter: A counter in which numerical information is represented directly in digits.

digital meter: A meter with a display on which the numbers are expressed directly as digits.

diode: A two-electrode device containing an anode and a cathode.

direct current: An electric current which only flows in one direction. It is commonly expressed by the abbreviation *dc*.

displacer: A level sensor operating on buoyant force.

draft gage: A low pressure sensor used in combustion systems.

drain: Term used to describe a main terminal of a field effect, or unipolar transistor.

drift: Gradual departure of instrument output from the correct value.

dry bulb: A temperature sensor used in humidity measurement.

dynamic characteristic: A quality of an instrument affected by time.

dynamic error: The difference between the true value of a quantity or condition changing with time and the instrument reading or controller action.

dynamic response: The behavior of an output in response to a changing input.

E

elastic chamber: The container or chamber in which the pressure medium is confined. It can be a single flat or corrugated diaphragm, a Bourdon tube, a bellows, a piston moving against a spring, or a combination of any of these.

elastic deformation element: A pressure sensor which changes shape with pressure.

electric controller: A device or a group of devices which serves to govern the electric power delivered to the apparatus to which it is connected.

electric thermometer: An instrument which uses electrical means to measure temperature.

electric transducer: A transducer in which all the signals concerned are electric.

electrical conductivity: The ability of a substance to carry electrical current-expressed in mhos.

electrode: A conducting element that performs one or more of the functions of emitting, collecting, or controlling by an electric field the movements of electrons or ions.

electromotive force: Electrical potential measured in volts.

electron tube: An electron device in which conduction by electrons takes place through a vacuum or gaseous medium within a gas-tight envelope. It is also referred to as a *vacuum tube.*

emissitivity: The rate at which a substance will radiate thermal energy.

emitter: Term used to describe one of the end terminals of a junction, or bipolar, transistor.

end device: The final system element that performs the final conversion of measurement to an indication, record, or the initiation of control.

end-point control: Quality control through continuous or periodic analysis of the final product of a process (sometimes referred to as stream analysis). In highly automatic operations, the final product is analyzed and corrected continuously and automatically.

error: In automatic control terminology, the difference between the actual controlled variable and the set point. The margin by which an automatic controller misses its target value.

error signal: A measurement of the error by an automatic controller.

F

Fahrenheit scale: A temperature scale on which the freezing point of water is 32° and the boiling point is 212°F.

farad: A unit of capacitance. In practical applications, the microfarad is used as the unit of

feedback: Part of a closed loop system which provides information about a given condition for comparison with the desired condition.

feedback controller: A mechanism which measures the value of the controlled variable, accepts the value of the command, and, as a result of a comparison, manipulates the controlled system in order to maintain an established relationship between the controlled variable and the command.

feedback control system: A control system which tends to maintain a prescribed relationship of one system variable to another by comparing functions of these variables and using the difference as a means of control.

feedback elements: The portion of the feedback control system which establishes the relationship between the primary feedback and the controlled variable.

feedback signal: A signal that is returned to the input of the system and compared with the reference signal to obtain an actuating signal which returns the controlled variable to the desired value.

feed-forward control: Open-loop control.

fidelity: The ability of an instrument to follow changes in the input value.

field effect transistor: A transistor made from a single crystal of silicon (P-type), with two small regions on opposite sides of the crystal composed of N-type material. These two small regions are electrically connected and are called the gate. One end of the main crystal is called the source, the other the drain. It is also called a *unipolar transistor*.

filament: A cathode of a thermionic tube, usually in the form of a wire or ribbon, to which heat may be supplied by passing current through it.

filter: An electronic circuit composed of one or more resistors, inductors, or capacitors, or any combination of these. The filter accepts or rejects the frequencies for which it is designed.

final control element: Unit of a control loop (such as a valve) which manipulates the control agent.

flapper: The device which acts upon the nozzle in a pneumatic controller.

flapper-nozzle: A combination of elements in a pneumatic controller.

flexivity: Term used to indicate the movement of a bimetal strip when heated. The flexivity of a bimetal strip is proportional to temperature, length, and thickness.

float: A level sensor for level measurement.

floating action: Occurs where there is a predetermined relation between the deviation and the rate of motion of a final control element.

floating control: A term that describes a control action in which the rate of motion of the final control element is determined by the deviation of controlled variable from set point.

flow diagram: A graphical representation of a sequence of operations.

flow, laminar: (or streamline) Any part of a fluid in which the velocity is smooth and constant.

flow nozzle: A primary element for flow measurement.

flow rate: The weight or volume of flow per unit of time.

flow, turbulent: Any part of a fluid in which the velocity at a given point varies more or less rapidly in magnitude and direction with time.

fluid: A liquid or a gas.

flume: A primary flow element for open channel measurement.

forward bias: State where diode or transistor is so connected in a circuit that it will conduct a current.

frequency: A measure of the number of times a cycling quantity repeats itself. Expressed in hertz.

frequency response: The response of a component, instrument, or control system to input signals at varying frequencies.

frequency response analysis: A method of systematically analyzing process control problems, based on introducing cyclic inputs to a device or system and measuring the resulting output signals at various frequencies.

full scale: The maximum value of the rate or range of an instrument.

full scale cycle: A complete transversal of range of an instrument from minimum reading to a full scale and back to minimum reading.

full scale deflection: The maximum reading on a measuring indicator or recorder. Percentage accuracy generally refers to FSD.

full scale value: The largest value of the actuating quantity indicated on the scale.

full wave rectifier: A combination of rectifier elements so arranged that the output is unidirectional.

G

gage: A device or instrument, containing the primary measuring elements, applied to the point of measurement.

gage pressure: The pressure of a liquid or gas measured relative to the ambient atmospheric pressure.

gain: Amount of increase in a signal as it passes through any part of a control system. If a signal gets smaller, it is said to be *attenuated*. If it gets larger, it is said to be *amplified*.

galvanometer: An instrument for indicating or measuring a small electric current or a function of the current by means of a mechanical motion derived from electromagnetic or electrodynamic forces which are set up as a result of the current.

gate: Term for one of the main terminals of a field effect transistor. See *field effect transistor*.

graduation: A division or mark on an instrument to indicate degree or quantity.

granular solid: Particulated matter like sand, grain, pellets, etc.

graphic panel: A control panel which pictorially displays and traces the relative position and function of measuring and control equipment to process equipment. Graphic panels can represent a total plant operation.

grid: An electrode having one or more openings for the passage of electrons or ions.

grid circuit: A circuit which includes the grid-cathode path of an electron tube in a series connection with other elements.

ground: An electrically conducting connection, accidental or intentional, to the earth or to some other conducting body at zero potential with respect to the earth.

H

half-wave rectifier: A rectifier which delivers unidirectional output current only during the half cycle when the applied alternating current voltage is of the polarity at which the rectifier has low resistance. During the opposite half cycle, the rectifier passes no current. Hence a half-wave rectifier rectifies only one half of the alternating current wave.

head: Pressure resulting from gravitational forces on liquids. Measured in terms of the depth below a free surface of the liquid which is the reference zero head.

helix: A spiral formed from wire and used in instruments such as a bimetallic thermometer.

Henry: A unit of electrical inductance. The abbreviation for a henry is *H*.

Hertz: A unit of electrical frequency. The abbreviation for a hertz is *Hz*.

holes: Positive charges, the movement of which in the P-type material of transistors constitutes the main current-carrying activity within the transistor.

hot junction: The joined ends of a thermocouple. It is also called the *measuring junction*.

hydrometer: An instrument for measuring the specific gravity of a liquid.

hydrostatic pressure: The pressure at the bottom of a column caused by the weight of the material in the column.

hygrometer: An instrument for measuring humidity.

hygroscopic: Descriptive of material which readily absorbs and retains moisture.

hysteresis: The total difference between the response of a unit or system to an increasing signal and the response to a decreasing signal.

I

impedance: The complex ratio of a force-like quantity (force, pressure, voltage, temperature, or electric field strength) to a related velocity-like quantity (velocity, volume velocity, current, heat flow, or magnetic field strength).

inclined tube manometer: A manometer with one arm at an angle, permitting the scale on that arm to be expanded for more precise readings of low pressure.

independent flow: Measured but uncontrolled flow in a ratio system. It is also called *wild flow*.

inductance: The property of an electrical circuit which tends to oppose change of current in the circuit.

inductor: A component, usually a coil, that possesses the property of inductance.

industrial control: The methods and means of governing the performance of a device, apparatus, equipment, or system used in industry.

input: Incoming signal to a control unit or system.

instrument: Used broadly to connote a device incorporating measuring, indicating, recording, controlling, and/or operating abilities.

instrumentation: The instruments that are used in a process system, usually including the control valves. Also refers to the science of applying instruments to manufacturing processes.

integrator: A device which continually totalizes or adds up the value of a quantity for a given time.

inverse derivative action: Control which produces a corrective operation inversely proportional to the rate at which the process variable deviates from the set point. For instance, if there is a sudden process change, this action causes the final control element to lag behind the process in producing any corrective action.

ion: Any electrically charged particle of molecular, atomic, or nuclear size.

ionization: The process in which a neutral atom or molecule is split into positive or negative ions.

isothermal: Without change in temperature.

J

junction transistor: A transistor consisting of three sections, the base, emitter, and collector, joined end to end. The sections can be arranged in an NPN or PNP configuration. The points where the sections join are called *junctions.* Also called *bipolar transistor.*

K

Kelvin temperature scale: A thermodynamic absolute temperature scale, having as its zero the absolute zero of temperature (−273.15°C).

Kirchoff's law: A statement in physics: in an electric network, the algebraic sum of the currents in all the branches that meet at any point is zero.

L

lag: Refers to delay, and is expressed in seconds or minutes. Lag is caused by conditions such as capacitance, inertia, resistance and dead time, either separately or in combination.

laminar flow: Streamline flow in a viscous fluid.

law of intermediate metals: A statement in thermoelectricity: the use of a third metal in a thermocouple circuit has no effect as long as the temperature of the junction of the three metals is constant.

law of intermediate temperatures: A statement in thermoelectricity concerning the independence of reference temperature.

linear: The relationship between two quantities in which a change in one is proportional to the change in the other quantity.

linearity: The degree to which the calibration curve of a device matches a straight line. The linearity error is generally the greatest departure from the best straight line that can be drawn through the measured calibration points.

load: The amount of energy or material that a device or machine must deliver or handle.

load cell: A device for measuring pressure.

load change: A change in process demand.

load control: A method of regulating the energy to a process.

logger: An instrument which automatically scans conditions (temperature, pressure, humidity) and records or logs findings on a chart, usually with respect to time. A digital logger records numerical values in tubular form by such means as an automatic typewriter.

M

magnetic amplifier: A device similar in construction and appearance to a transformer which is used to perform the amplifying functions of electron tubes or transistors in some applications.

magnetic field: A state of the medium in which moving electrified bodies are subject to forces by virtue of both their electrifications and motion.

manipulated variable: That quantity of condition of the control agent which is varied by the automatic controller so as to affect the value of the measured (controlled) variable. In heating water with steam, the flow of steam is the manipulated variable.

manometer: A gage for measuring pressure of gases and vapors.

manual controller: A controller having all

its basic functions performed by devices that are operated by hand.

mass: The amount of matter contained in an object and measured by the inertia of that object.

mass flow measurement: The measurement of flow in weight units rather than conventional volumetric units.

measured variable: Analogous to controlled variable when used in connection with control applications.

measuring element: The primary device for measuring the process variable.

measuring junction: The junction of the two dissimilar wires of a thermocouple which is exposed to the temperature to be measured.

measuring lag: The delay of a measuring instrument to a change in process variable.

measuring means: Those elements of an automatic controller which are involved in ascertaining and communicating to the controlling means either the value of the controlled variable, the error, or the deviation.

medium: A solid or a fluid through which a force or effect is conveyed.

megohm: 1,000,000 ohms. The symbol for a megohm is $M\Omega$.

mho: A unit of electric conductivity. The reciprocal of an ohm (resistance).

microfarad: One millionth of a farad. The symbol for a microfarad is μF.

microsecond: One millionth of a second. The symbol for a microsecond is μs.

microvolt: One millionth of a volt. The symbol for a microvolt is μV.

milliammeter: A meter for measuring electric current. The meter is calibrated in milliamperes.

milliampere: One thousandth of an ampere. The symbol for a milliampere is mV.

millimeter: One thousandth of a meter, or one-tenth of a centimeter. The symbol for a millimeter is *mm*.

millisecond: One thousandth of a second. The symbol for a millisecond is *ms*.

millivoltmeter: A meter for measuring small amounts of electrical voltage.

miniaturization: Method of reducing the size of instruments to minimize panel space requirements. Permits more instruments to be mounted in smaller spaces.

molecule: The smallest particle into which a substance can be divided and still retain its chemical properties and identity.

motor operator: A portion of the controlling means which applies power for operating the final control element.

multiple action: Motion in which two or more controller actions are combined.

multiple helix bimetal thermometer: Thermometer in which the bimetal measuring element is arranged in the form of a helix coil within a helix coil, permitting the use of long bimetal strips within a compact space for greater sensitivity.

multiposition action: Movement in which a final control element is moved to one of three or more predetermined positions, each corresponding to a definite range of values of the controlled variable.

N

needle valve: Small valve inserted in the process pipeline which is subject to pulsating pressures. The valve has the effect of dampening or flattening the impulses.

negative feedback: Feedback which results in decreasing the amplification effect.

neutral zone: A range of measured values in which no control action occurs. Same as dead band.

newtonian material: A substance which exhibits a linear relation between an applied sheer stress and the rate of deformation.

noise: Meaningless stray signals in a control system, similar to radio static. Some types of noise interfere with the correctness of an output signal.

non-newtonian material: A substance in which the relationship between the applied sheer stress and the rate of deformation is not constant.

nozzle: A duct of changing cross section in which fluid velocity is increased.

npn transistor: Junction (bipolar) transistor end sections of which are composed of N-type material and center section of which is P-type material.

n-type material: Semiconductor material to which a donnor impurity is added, resulting in a quantity of loosely bonded or free electrons.

null balance: The condition of zero difference between opposing forces.

null detector: A device which detects the point of minimum signal or the point of no signal.

O

offset: A sustained deviation of the controlled variable from set point. (This characteristic is inherent in proportional controllers that do not incorporate reset action.) Offset is caused by load changes.

ohm: The unit of electrical resistance. The symbol for an ohm is Ω.

ohmmeter: A direct-reading instrument for measuring electric resistance. It is provided with a scale, usually graduated in either ohms or megohms.

Ohm's law: A statement referring to the relationship between voltage current and resistance. It can be expressed as E (voltage) = I (current) × R (resistance).

on-off control action: (Same as two position action.) Occurs when a final control element is moved from one of two fixed positions to the other with a very small change of controlled variable.

open loop: A system in which no comparison is made between the actual value and the desired value of a process variable.

open nozzle: A primary flow element for open channel measurement of flow rate.

operating pressure range: Stated high and low values of pneumatic pressure required to produce full-range operation when applied to a pneumatic intelligence-transmission system, a pneumatic motor operator, or a pneumatic positioning relay.

optimalization: Theoretical analysis of a system, including all of the characteristics of the process, such as thermal lags, capacity of tanks or towers, length and size of pipes, etc. This analysis is made, usually with the aid of frequency response curves, to obtain the most desirable instrumentation and control.

organic liquid: A liquid containing carbon.

orifice: A symmetrical aperture, having circular transverse cross sections, the diameter of the smallest of which is large in comparison with the thickness of the plate in which it is cut and which has such sudden approach curvature that contraction is fully developed or only partially suppressed.

orifice plate: A thin, circular metal plate with an opening in it. It is used for measuring flow rate.

oscillation: A change from one extreme to another.

oscillator: Circuit used to generate a constantly varying ac voltage or signal.

output: Outgoing signal of a transmitter or control unit.

overshoot: The amount by which a changing process variable exceeds the desired value as changes occur in a system.

P

peak picker: An electric circuit which remembers a maximum value.

Peltier effect: Depending on the direction of current flow, heat is either absorbed or liberated at the junction of two dissimilar metal wires through which a current is passing.

pentode: A five-electrode electron tube containing an anode, a cathode, a control electrode, and two additional electrodes that are ordinarily grids.

period: Length of time required to complete one cycle of operation.

pH: The measure of effective acidity or alkalinity of solutions based on a measurement of the concentration of hydrogen ions. pH values less than 7 are considered acidic, pH values greater than 7 are considered basic.

phase: Any stage in the cycle of an ac voltage or alternating current.

phase shift: A time difference between the

input and output signal of a control unit or system.

photoelectric cell: A device whose electrical properties undergo a change when it is exposed to light.

photon: A unit of electromagnetic radiation.

pilot: An auxiliary mechanism that actuates or regulates another mechanism.

pitot tube: A cylindrical tube with an open end pointed upstream, used in measuring impact pressure.

pnp transistor: Junction (bipolar) transistor whose end sections are composed of P-type material and whose center section is composed of N-type material. See *junction transistor.*

positioning action: Movement in which there is a predetermined relation between value of the controlled variable and position of a final control element.

positive displacement flowmeter: A flowmeter which measures total flow by counting known volumes.

potential: Usually synonymous with voltage. The electric charge of one body as compared with that of another.

potentiometer: Measures by comparing the difference between known and unknown electrical potentials. In order to measure process control variables by means of a potentiometer, these variables, such as temperature, pressure, flow and liquid level, must first be translated into electrical signals that vary proportionally with changes in the variable.

power supply: A circuit that supplies various ac and dc voltages and alternating and direct currents for a specific purpose.

power unit: A portion of the controlling means which applies power for operating the final control element.

pressure: Force per unit area. Measured in pounds per square inch (psi), or by the height of a column of water or mercury which it will support (in feet, inches, or centimeters).

pressure capsule: Two diaphragms, metallic or non-metallic, welded or otherwise joined together to form a sealed capsule which will deflect when subjected to pressure.

pressure gage: An indicating gage having a scale graduated to show pressure.

pressure potentiometer: A pressure transducer in which the electrical output is derived by varying the position of a contact arm along a resistance element.

pressure sensing element: The part of a pressure transducer which converts the measured pressure into a mechanical motion.

pressure spring thermometer: A type of thermometer which employs a bulb connected by tubing to a hollow spring. As the temperature of the liquid within the system rises the spring tends to unwind, moving an indicator. Can be mercury, liquid, gas, or vapor-filled.

pressure transducer: An instrument which converts a static or dynamic pressure input into a proportional electrical output.

primary element: The portion of the measuring means which first either utilizes or transforms energy from the controlled medium to produce an effect in response to change in the value of the controlled variable. The effect produced by the primary element may be a change of pressure, force, position, electrical potential, or resistance.

primary measuring element: A device or instrument which measures a variable. Primary measuring elements are used to convert a measurement to a signal for transmission to a controller, a recorder, or an indicator. Also known as detector, sensor or sensing element.

printout: Recording by a typewriter or printer.

process: Comprises the collective functions performed in and by the equipment in which a variable is to be controlled.

process instrumentation: The technology of measuring and controlling industrial processes.

process reaction curve: A record of the reaction of a control system to a step change.

process reaction rate: The rate at which a process reacts to a step change.

program control: A control system in which the set point is automatically varied during definite time intervals in order to make the process variable vary according to some prescribed manner.

programmed system: A control system which regulates the occasion and duration of a process change.

proportional action: Produces an output signal proportional to the magnitude of the input signal. In a control system proportional action produces a value correction proportional to the deviation of the controlled variable from set point.

proportional band: The amount of deviation of the controlled variable from set point required to move the final control element through the full range (expressed in % of span). An expression of gain of an instrument (the wider the band, the lower the gain).

proportional plus derivative action: Proportional-position action and derivative action are combined.

proportion plus reset action: Proportional-position action and proportional-speed floating action are combined.

proportional plus reset plus derivative action: Proportional-position action, proportional-speed floating action, and rate action are combined.

psia: Pounds per square inch *absolute.*

psid: Pounds per square inch *differential.*

psig: Pounds per square inch *gage.*

psychrometer: An instrument for measuring humidity.

psychrometric chart: A chart showing relationship between temperature, humidity, and dew point.

p-type material: Semiconductor material to which an acceptor impurity is added, resulting in a quantity of holes or positive charges.

pulse: A variation of a quantity whose value is normally constant. Pulse is characterized by a rise and fall and of finite deviation.

purge pressure: A large, constant pressure introduced into the process pressure line to increase the pressure level and accentuate low process pressure changes.

purging: Elimination of an undesirable gas or material from an enclosure by means of displacing the undesirable material with an acceptable gas or material.

pyrometer: An instrument for measuring temperature. Usually refers to temperature measuring instruments used to measure flame temperature, temperatures above 1000°F.

R

radiation: Emission and propagation of energy from a source.

radiation pyrometer: A pyrometer in which the radiant power from the object or source to be measured is used in the measurement of its temperature. The radiant power within wide or narrow wavelength bands filling a definite solid angle impinges upon a suitable detector. The detector is usually a thermocouple or thermopile, a bolometer responsive to the heating effect of the radiant power, or a photo-sensitive device connected to a sensitive electrical instrument.

range: The difference between the maximum and minimum values of physical output over which an instrument is designed to operate normally.

rangeability: Describes the relationship between the range and the minimum quantity that can be measured.

range resistor: An electrical resistance used for establishing the measuring range of an instrument.

Rankine scale: Absolute Fahrenheit temperature scale.

rate action: A control action which produces a corrective signal proportional to the rate at which the controlled variable is changing. Rate action produces a faster corrective action than proportional action alone. Also called *derivative action.*

rate time: Amount of time (expressed in minutes) by which proportional action is advanced by the addition of rate action.

ratio control: Maintains the magnitude of a controlled variable at a fixed ratio to another variable.

reactance: The component of the impedance of an electrical circuit, not due to resistance, which opposes the flow of alternating current. The reactance is the algebraic sum of that due to inductance in the circuit with value in ohms equal to the product 2π, the frequency in hertz, and the inductance in henrys and that due to capacitance in the circuit with a value in ohms equal to the reciprocal of the product 2π, the frequency in hertz, and the capacitance in farads.

readout: Visual or printed display of measured values.

record: Printed or written information.

recorded value: The value recorded by the marking device on the chart, with reference to the division lines marked on the chart.

rectification: Conversion from ac to dc current.

rectifier: A device which is used for converting an alternating current into a continuous or direct current by permitting the passage of current in only one direction.

reference junction: The junction of a thermocouple which is at a known or reference temperature.

reflectance: Ratio of reflected energy to total energy received.

regulator: A device which varies or prevents variation in a desired characteristic.

relative humidity: Ratio between the amount of water actually present and the maximum amount air could hold at the same temperature.

relay: A device which enables the energy in one circuit (generally of high power) to be controlled by the energy in another.

relay-operated controller: One in which the energy transmitted through the primary element is either supplemented or amplified to operate the final control element by employing energy from another source.

relief valve: A valve which opens on overpressure.

reproducibility: The ability of an instrument to duplicate, with exactness, measurements of a given value. Usually expressed as a percent of span of the instrument. Also referred to as *repeatability*.

reset action: A control action which produces a corrective signal proportional to the length of time the controlled variable has been away from the set point. Takes care of load changes. Also called *integral action*.

reset rate: Repeats per minute. Expresses the number of times proportional response is repeated or duplicated in one minute.

reset response: A corrective control action in which there is a relationship between offset and the rate of motion of the final element. It is also called *integral action*.

reset time: Time required for reset action to match proportional position action.

resistance: Property that impedes flow or motion of a quantity. As applied to electrical circuits, resistance is a property that impedes the flow of electricity. The symbol for resistance is R.

resistance bridge: A bridge circuit employing electrical resistors.

resistance thermometer: An instrument which measures temperature by measuring the varying resistance of a sensing element whose resistance varies with temperature.

resistance thermometer element: The temperature-sensitive unit of a resistance thermometer bulb comprising a material whose electrical resistance changes with temperature, its supporting structure, and means for attaching conductors.

resistivity: The resistance of a material expressed in ohms per unit length and unit cross section.

resistor: A device which conducts electricity but converts part of the electrical energy into heat.

responsiveness: Ability of an instrument to follow changes.

Reynolds number: The product of the density of the fluid, the flow velocity, and the internal diameter of the pipe, divided by the viscosity of the fluid.

rheostat: A resistor which is provided with means for readily adjusting its resistance.

reverse bias: State where a diode or transistor is so connected in a circuit that no current can flow through the compound.

rotameter: A device for measuring fluid flow using a tapered tube and a float.

rotary switch. A switch that operates when the control knob is rotated.

S

sampling action: Occurs when the difference between set point and the value of the controlled variable is measured and correction made at intermittent intervals.

saturable reactor: A device for regulating ac power by using dc current variation.

scanner: An instrument which automatically checks a number of measuring points for the purpose of collecting information which automatically *scans* a number of measuring points and indicates which have deviated too far from the desired values.

screen grid: A grid placed between a control grid and an anode, usually maintained at a fixed positive potential, for the purpose of reducing the electrostatic influence of the anode in the space between the screen grid and the cathode.

secondary chamber: A seat or socket for an industrial thermometer which permits safe, convenient replacement or removal without disturbing the process pipeline.

secondary winding: A winding on the output side of a transformer.

Seebeck effect: Current will flow through a thermocouple loop formed of two dissimilar wires, provided the two junctions are at different temperatures.

self-operated controller: One in which all the energy necessary to operate the final control element is derived from the controlled medium through the primary element. This type of automatic controller must have both self-operated measuring means and self-operated controlling means.

self-operated measuring means: All the energy necessary to actuate the controlling means of an automatic controller is derived from the controlled medium through the primary element.

self-regulation: A property of a process or instrument by which, in the absence of control, equilibrium is reached after a disturbance.

semiconductor: Material, such as silicon or germanium which has a greater resistance to current flow than a conductor, but not as great a resistance as an insulator.

sensing element: The part of a transducer mechanism which is in contact with the medium being measured and which responds to changes in the medium.

sensitivity: (1) Ratio of change of output to change of input. (2) The least signal input capable of causing an output signal having desired characteristics.

servo-mechanism: A closed-loop system in which the error or deviation from a desired or preset norm is automatically corrected to zero, and in which mechanical position is usually the controlled variable.

set point: The position at which the control point setting mechanism is set. This is the same as the desired value of the controlled variable.

shunt: A device or component that is connected in parallel with another device or component.

signal: Information conveyed from one point in a transmission or control system to another. Signal changes usually call for action or movement.

silicon controlled: A solid state controlled rectifier consisting of an anode, a cathode, and the gate. The abbreviation for a silicon controlled rectifier is *SCD*.

sine wave: A wave made up of instantaneous values which are the product of a constant and the sine of an angle having values varying linearly.

sinusoidal vibration: A cyclical motion in which the object moves linearly. The instantaneous position is a sinusoidal function of time.

slidewire: An electrical resistor used with a contacting slider which permits resistance adjustment.

slurry: A liquid containing suspended particulate matter.

solenoid magnet: An electromagnet having an energizing coil approximately cylindrical in form, and an armature whose motion is reciprocal within and along the axis of the coil.

Sorteberg bridge: A pneumatic multiplier-divider.

source: Main terminal of a field effect (unipolar) transistor.

span: The difference between the top and bottom scale values of an instrument. On instruments starting at zero, the span is equal to the range.

specific gravity: Ratio between the density of liquid to the density of water, or the density of gas to the density of air.

specific heat: The ratio of the thermal capacity of any substance to the thermal capacity of water is called the *specific heat* of that substance.

specific humidity: The amount of water vapor per unit mass of moist air.

square root extraction: The electrical, mechanical, or pneumatic process whereby the square root of a measurement is derived. In flow measurement, for example, the square root of differential pressure equals flow.

stability: Freedom from undesirable deviation, a measure of the controllability of a process.

standard cell: A cell which serves as a standard of electromotive force.

standardization, automatic: A method of comparing potentiometer circuit voltage with standard cell voltage.

static characteristic: A quality of a measuring instrument independent of time.

stator: The stationary plates of a variable capacitor, or the stationary field of an electric motor.

Stefan-Boltzmann law: States that the thermal energy radiated per second per unit area from a blackbody (ideal radiator) is proportional to its absolute temperature raised to the fourth power.

strain gage: An element (wire) which measures a force by using the principle that electrical resistance varies in proportion to tension or compression applied to the element.

static error: The difference between the true value of a controlled variable and the instrument reading.

strip chart: A recording instrument chart made in the form of a long strip of paper.

suppressed range: A suppressed range is an instrument range which does not include zero. The degree of suppression is expressed by the ratio of the value at the lower end of the scale to the span.

synchro: A rotating electrical device for transmitting angular position. A minimum synchro system employs two units: a transmitter and a receiver.

systems engineering: Control engineering in which a process and all the elements affecting a process and all the possibilities for introducing automatic controls are considered during the design and installation of processing equipment.

T

telemetering: Transmission of measurements over very long distances, usually by electrical means.

temperature The relative hotness or coldness of a body as determined by its ability to transfer heat to its surroundings. There is a temperature difference between two bodies if, when they are placed in the thermal contact, heat is transferred from one body to the other. The body which loses heat is said to be at the higher temperature.

thermal capacity: The amount of heat required to raise the temperature of one pound of any substance one degree Fahrenheit.

thermal conductivity: Transfer of heat from one part of a solid to the remainder of the solid.

thermionic tube: An electron tube in which one of the electrodes is heated for the purpose of causing electron or ion emission from that electrode.

thermistor: A resistor whose resistance varies with temperature in a definite desired manner. Used in circuits to compensate for temperature variation, to measure temperature, or as a nonlinear circuit element.

thermocouple: A pair of dissimilar conductors so joined that an electromotive force is developed by the thermoelectric effects when the two junctions are at different temperatures.

thermocouple well: Device used for protecting thermocouples by eliminating direct contact of the thermocouple with possibly corrosive substances being measured.

thermometer: A device for measuring temperature.

thermometer time constant: The time required for a thermometer to reach 63.2% of its final reading.

thermopile: A group of thermocouples connected in series. This term is usually applied to a device measuring radiant power or used as a source of electric energy.

Thomson effect: If there is a temperature gradient along a current carrying conductor, heat will be liberated or absorbed at any point where current and heat flow in the same direction, depending on the type of metal used as a conductor.

transducer: An electro-mechanical device which converts a physical quantity being measured (such as temperature or pressure) to a proportional electrical output.

transfer lag: A process characteristic caused by the passage of energy from one capacity, through a resistance, to another capacity.

transistor: A tiny semiconductor amplifying device which performs the same functions as an electron tube.

triode: A three-electrode electron tube, containing an anode, a cathode, and a control electrode.

turbulent flow: Rapid flow of a fluid during which the Reynolds number is higher than 4000.

two-position action: Action in which a final control element is moved from one of two fixed positions to the other. *Open and shut action,* and *On-Off action* are synonymous terms.

U

u-tube: A form of manometer used for pressure measurement.

V

variable: A process condition, such as pressure, temperature, flow, or level, which is susceptible to change and which can be measured, altered, and controlled.

velocity meter: An instrument for measuring fluid velocity, or rate of flow.

vena contracta: The smallest cross section of a fluid jet which issues from a freely discharging aperture or is formed within the body of a pipe owing to the presence of a constriction.

Venturi tube: A short tube of varying cross section. The flow through the venturi tube causes a pressure drop in the smallest section, the amount of the drop being a junction of the velocity of flow.

viscosity: The resistance of a fluid to flow.

volt: The unit of electrical potential. The symbol for a volt is *V*.

voltage divider: A network consisting of impedance elements connected in series to which a voltage is applied and from which one or more voltages can be obtained across any portion of the network.

voltmeter: An instrument having circuits so designed that the magnitude either of voltage or of current can be measured on a scale calibrated in terms of each of these quantities.

W

weir: An obstruction placed across an open liquid stream to raise the level of the liquids. It is used for flow measurement.

wet bulb: A temperature sensor covered with a wet cloth. It is used with a dry bulb for humidity measurement.

wheatstone bridge: A four arm bridge, all arms of which are predominately resistive.

wild flow: Uncontrolled fluid flow in a flow ratio system.

X

xylene: A benzene derivative used in liquid-filled thermometers.

Z

zero shift: (zero error) The output error, expressed as a percent of span, at zero input.

Index

Numerals in **bold type** refer to illustrations.

R